Springer Series in Measurement Science and Technology

Series editors

The Springer Series in Measurement Science and Technology comprehensively covers the science and technology of measurement, addressing all aspects of the subject from the fundamental physical principles through to the state-of-the-art in applied and industrial metrology. Volumes published in the series cover theoretical developments, experimental techniques and measurement best practice, devices and technology, data analysis, uncertainty, and standards, with application to physics, chemistry, materials science, engineering and the life sciences.

The series includes textbooks for advanced students and research monographs for established researchers needing to stay up to date with the latest developments in the field.

More information about this series at http://www.springer.com/series/13337

Masahiko Hirao · Hirotsugu Ogi

Electromagnetic Acoustic Transducers

Noncontacting Ultrasonic Measurements using EMATs

Second Edition

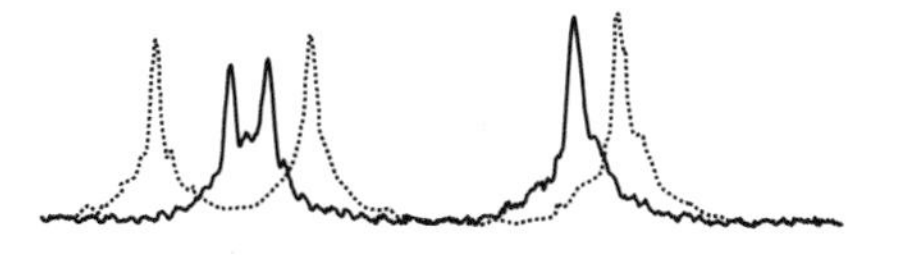

Masahiko Hirao
Osaka University
Toyonaka
Japan

Hirotsugu Ogi
Osaka University
Toyonaka
Japan

ISSN 2198-7807 ISSN 2198-7815 (electronic)
Springer Series in Measurement Science and Technology
ISBN 978-4-431-56760-8 ISBN 978-4-431-56036-4 (eBook)
DOI 10.1007/978-4-431-56036-4

Softcover reprint of the hardcover 2nd edition 2016

Printed on acid-free paper

This Springer imprint is published by Springer Nature
The registered company is Springer Japan KK

Preface to the Second Edition

The original version of this book, *EMATs for Science and Industry: Noncontacting Ultrasonic Measurements*, was published by Kluwer Academic Publishers in 2003. The main theme of the book remains unchanged, but this second edition contains new research products that have emerged since then and that we were unaware of when the original book was prepared.

The thematic chapters and sections are rearranged to accommodate new entries of guided torsional waves (Sects. 3.4 and 14.3), point-focusing EMATs (Sects. 3.10 and 14.4), antenna transmission technique (Sects. 3.13 and 8.9), point-defect dynamics in aluminum (Sect. 7.3), high-temperature measurement of elastic constants (Sects. 8.5 and 8.9), resonant ultrasound microscopy (Chap. 9), and EMAR-based nonlinear acoustic measurements (Chap. 10). We created Chap. 18 to summarize the field applications of EMATs. It is enriched by the addition of monumental works, corrosion detection of heat exchanger tube, and in-process weld inspection. Chapter 2 becomes more comprehensive for describing coupling mechanisms on EMAT phenomena. Errors in equations have been corrected: Eqs. (2.51), (2.60), (13.2), (13.5), and (13.8) in the original version.

EMATs of low transduction efficiency have been thought to be useful only when the users need their unique advantages, that is, the noncontact nature and the ability to transmit and receive elastic wave modes that are otherwise difficult. However, it is demonstrated, for instance, that the point-focusing EMAT can detect stress-corrosion cracks around the weld of stainless steel plates and showed a great latent potential for flaw detection. We hope that this publication triggers such new ideas of ultrasonic measurements and contributes to the development in materials science research and nondestructive evaluation of industrial materials.

In preparing this book, we have benefited from the advice and encouragement of many colleagues. G. Petersen (Ritec Inc.) contributed Sect. 4.3 on impedance matching. Particular gratitude is due to G. Alers (Sonic Sensors of EMAT Ultrasonics, Inc.) for his advice and support with Chap. 18. P. Nagy (University of Cincinnati) provided us with valuable discussions on the EMAT mechanisms in Chap. 2. Students fabricated the EMATs in our laboratory. Their efforts in modeling and experiments resulted in finding new phenomena and suggestions for next directions.

In March 2011, we heard with deepest regret of the passing of R. Bruce Thompson. He pioneered and guided the EMAT research over decades in developing the theoretical model and practical applications. His research work that appeared in the 1970s is the milestone through the history of EMATs development. We dedicate this book to Bruce with many thanks and as a tribute to his great accomplishment.

Toyonaka, Osaka
Spring 2016

Masahiko Hirao
Hirotsugu Ogi

Preface to the First Edition

A man with hammer sees everything as a nail.

Against this lesson, the authors of this book have kept hitting things with *electromagnetic acoustic transducers* (EMATs) for the purposes of studying the sound they made. This book comprises the physical principles of EMATs and the applications of scientific and industrial ultrasonic measurements on materials. The objects are summarized and systematized mainly from the authors' own research results in Osaka University. The choice of subjects is admittedly arbitrary and strongly influenced by the authors' own interests and views.

The text is arranged in four parts. Part I is intended to be a self-contained description of the basic elements of coupling mechanism along with practical designing of EMATs for various purposes. There are several implementations to compensate for the low transfer efficiency of the EMATs. Useful tips to make an EMAT are also presented. Part II describes the principle of EMAR, which makes the most of contactless nature of EMATs and is the most successful amplification mechanism for precise velocity and attenuation measurements. In Part III, EMAR is applied to studying the physical acoustics. New measurements emerged on three major subjects; *in situ* monitoring of dislocation behavior, determination of anisotropic elastic constants, and acoustic nonlinearity evolution. Part IV deals with a variety of individual topics encountered in industrial applications, for which the EMATs are believed to the best solutions.

Our study on EMATs started more than ten years ago. The 5th International Symposium on Nondestructive Characterization of Materials took place in May 1991, at Karuizawa, Japan. One afternoon during the Symposium, H. Fukuoka, the late C. M. Fortunko, G. L. Petersen, T. Miya, T. Yamasaki, and one of the authors (M.H.) got together and discussed the possibility of mechanizing the resonant—EMAT concept. We were trying to push forward the project of measuring the acoustoelastic response in thin metal sheets, for which extremely high-frequency resolution was required. Among several potential approaches, the pulsed CW excitation and superheterodyne detection on gated reverberation signals seemed promising. But we were not sure whether it was achievable with EMATs. In October, M.H. flew to Providence, Rhode Islands, with a bulk-wave EMAT built by the other author (H.O.), who was then a graduate student. Japan-made EMAT was connected to US-made instrument for the first time and the combination produced exactly what we wanted, the high-Q resonance spectrum with metal sheets.

This was the birth of *electromagnetic acoustic resonance* (EMAR) of the latest generation. The system was found well applicable not only to acoustoelastic stress measurements, but also to many other nondestructive evaluation issues, including the determination of attenuation in solids. Noncontact measurement with high enough signal intensity was striking. Basic preconditions of theoretical approaches were realized by eliminating artifacts caused by the contact transducers. EMAR thus illuminated antiquated theories, which were accepted to be of little use or limited to qualitative interpretation of observations. It also uncovered interesting phenomena. Continuous monitoring of attenuation and acoustic nonlinearity resulted in the detection of ongoing microstructure evolutions in deforming or fatiguing metals. Our aim of writing this is to share the knowledge of and the results obtained with EMATs with students, researchers, and practitioners who had no chance to know the value. We hope that this book provides practical answers to the needs of ultrasonic measurements and direction to open a novel methodology.

We wish to acknowledge the influence of H. Ledbetter, who over a long period taught us physical metallurgy and elastic-constant theories, which formed the background of the ultrasonic researches described herein. Furthermore, he has taken a critical reading of the whole manuscript through intensive examination and corrections. We sincerely appreciate his attention, advice, and encouragement.

Many friends and colleagues helped us with the research and this book in their ways. Of special note are the stimulating discussions with the late C. M. Fortunko, G. A. Alers, and W. Johnson on EMATs and ultrasonic measurements in general. T. Ohtani collaborated with us by designing printed coils as well as by developing measurements to sense metal's fatigue and creep damages. T. Ichitsubo extended EMAR measurements to materials science field and low-temperature solid-state physics. T. Yamasaki and K. Yaegawa deserve our many thanks. We acknowledge the invaluable support of Ritec Inc. (B. B. Chick, G. L. Petersen, M. J. McKenna) and Sonix K.K. (T. Miya). From the very beginning till now, they have provided us with the state-of-the-art instruments, which have done much to shape our ideas on ultrasonic researches. G. L. Petersen wrote Section 3.3 on impedance matching with EMATs. K. Fujisawa and R. Murayama, when they were at Sumitomo Metal Indust., introduced the magnetostrictive coupling to us during the joint work to develop online texture monitoring system on steel sheets. That was our first contact with EMATs. W. Johnson and K. Fujisawa gave criticisms by reading portions of the manuscript. Many students contributed to the EMAT studies through painstaking work in calculations and experiments; they are S. Aoki, T. Hamaguchi, T. Honda,

N. Nakamura, S. Kai, D. Koumoto, Y. Minami, K. Minoura, K. Sato, G. Shimoike, N. Suzuki, Y. Takasaki, S. Tamai, M. Tane, A. Tsujimoto, and K. Yokota.

Lastly, we wish to thank our mentor, H. Fukuoka, for his encouragement during the preparation and his foresight that the EMATs should further develop ultrasonic measurements. He suggested that we pursue acoustoelastic programs with EMATs. They were the toughest subjects, but when solved, other subjects of ultrasonic measurements became easier tasks.

Toyonaka, Osaka
Spring 2003

Masahiko Hirao
Hirotsugu Ogi

Contents

Chapter 1
Introduction

Noncontact Ultrasonic Measurements

Abstract Contactless transduction is indispensable for precise ultrasonic measurements both of velocity and attenuation in solids materials. In this chapter, the importance is demonstrated by the signal distortion induced by a piezoelectric transducer and also by comparing the received signals observed by a piezoelectric transducer and an electromagnetic acoustic transducer (EMAT) on the same metal plate. Only one weakness of EMATs is the low transduction efficiency, but incorporation with resonance operation can overcome this problem; the combination is named electromagnetic acoustic resonance (EMAR) and is the main subject of this book. Brief historical sketch is provided as for the progress of understanding the coupling mechanisms with EMATs along with the EMAR measurements of the earlier generations.

Keywords Contact/noncontact transducers · Coupling mechanism · Flaw detection · Physical acoustics

1.1 Contact and Noncontact Measurements

Propagation behavior of ultrasonic waves in solids has been extensively studied, because they can provide powerful tools for the nondestructive characterization of materials. The principal advantage is the capability of probing inside of solid materials, making a contrast with other techniques such as the optical methods, X-ray diffraction, and the magnetic testing methods, which are applicable to sensing the surface or near-surface region.

At present, there are two main areas relying on the precise ultrasonic measurements. First is the industry-oriented nondestructive evaluation for structure safety and integrity through flaw detection and measurements of stresses and microstructure changes in solid materials. The top priority goes to detecting signals reflected by cracks and obtaining quantitative information needed for the fracture mechanics calculations. Ultrasonic techniques have been already developed for this purpose. The practice is nowadays advancing from manual to automatic and from analog to digital

M. Hirao and H. Ogi, *Electromagnetic Acoustic Transducers*,
Springer Series in Measurement Science and Technology,
DOI 10.1007/978-4-431-56036-4_1

operation for the quick inspection on welded parts, for instance. Equally important is to detect the microstructure change caused by fatigue, creep, thermal aging, irradiation, corrosion, and so on, since we are facing serious problems caused by aging infrastructures. Nondestructive assessment of the deterioration degree and the remaining life is getting more and more important. The damaging process would continue evolving during service and eventually form the macroscopic cracks leading to failure. Required is the sensing of some indicative precursors well ahead of the final event and continuous monitoring of their subsequent growth for estimating the remaining life and making the replacement schedule. Since a large amount of materials is used in a structure, the nondestructive evaluation method should be simple and take minimum consumption of time and labor. Some applications such as those in nuclear power plants must be done under remote control and therefore be performed with a noncontacting way, requiring no coupling media and no surface preparations.

The second area is physical acoustics to study elastic and anelastic properties of materials. Ultrasonics can potentially provide materials microstructure information through either or a combination of phase velocity, attenuation (internal friction), and nonlinearity. The phase velocity yields the elastic constants, which represent the atomic bonding strength. Attenuation evolves with the anelastic and irreversible processes in materials, reflecting the activity of lattice defects. Acoustic nonlinearity originates from lattice anharmonicity, damping mechanism, and microcracks. There has been a fundamental problem that the existing techniques fail to measure these acoustic properties with high enough accuracy and to detect the minute indications from materials. Typical case is the velocity change associated with stress in a metal, which is known as the acoustoelastic effect described in Chap. 12. Relative velocity change is as small as 10^{-5} MPa^{-1} for common metals. Another case is the attenuation evolution caused by the anelastic dislocation vibration during metal's fatigue and creep lives as well as that caused by grain scattering in polycrystalline metals. Those are the topics in Chaps. 7 and 15–17.

All of these difficulties come from the contact transducers. The common key for solutions is a noncontacting measurement to achieve high-grade ultrasonic nondestructive characterization in both practical and scientific fields. Piezoelectric transducers need a coupling fluid to make the intimate contact between them and the specimen surfaces. The elastic waves propagate not only in the material but also in all the elements in the transducer and couplant. The transducer receives the reflection echoes from the opposite surface of the specimen together with the disturbing echoes that arise from the interfaces within the transducer. They naturally overlap each other and degrade the accuracy of time-of-flight measurement. The contact transducers pose much more serious problem in measuring the attenuation. Attenuation occurs from energy absorption and wave scattering in the material. As-measured attenuation is usually overestimated, because a contact transducer eats the wave energy on every reflection at the material surface where the transducer is mounted. The energy leakage is not negligible and often exceeds the inherent attenuation of the material. Thus, it has been unrealistic to obtain the absolute value of material's attenuation.

It is easy to illustrate how the conventional piezoelectric transducer distorts the reflection signals. The pulse echo trace in Fig. 1.1a was obtained by feeding a broadband pulse to piezoelectric transducer pressed onto a 6-mm-thick aluminum

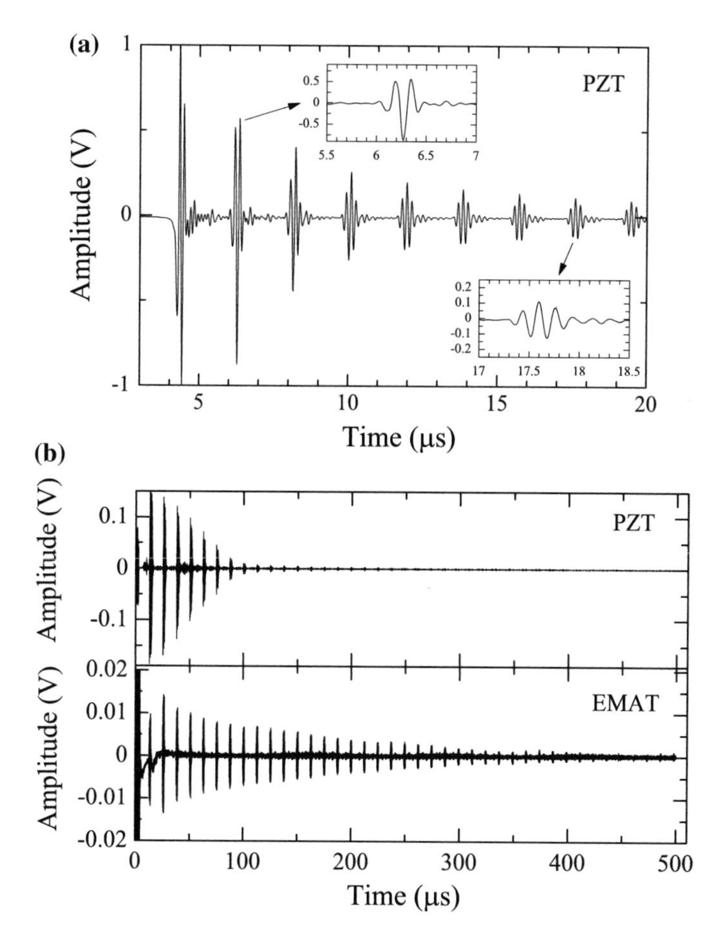

Fig. 1.1 Trains of reflection echoes in plates; **a** broadband longitudinal wave pulse excited and received by a piezoelectric transducer of 5-MHz center frequency with 6-mm-thick aluminum plate, and **b** comparison between the shear waves excited and received with a piezoelectric transducer and an EMAT, both at 5 MHz, for 19-mm-thick aluminum plate

plate. The nominal center frequency was 5 MHz, and the diameter was 8 mm. The duration of the longitudinal wave pulse, traveling in the thickness direction, kept increasing as the reflection repeated. The duration extended because of the overlapping of interface echoes with delays. Such a distortion promotes as the transfer efficiency is improved by matching the acoustic impedance to the measuring material. Figure 1.1b compares the two echo trains observed with a shear wave transducer (10-mm^2) and a bulk-wave EMAT (14 × 20 mm^2) on a 19-mm-thick aluminum plate. Burst signals of eight cycles at 5 MHz were used. It is obvious that the elastic wave lasts much longer in the plate, experiencing more number of reflections, when we use the EMAT than the piezoelectric transducer. The transducer loss makes such a big difference on multiple reflections.

It is of no doubt that contactless transducers are superior to contacting transducers for precise transit time and attenuation measurements. We have now three options: EMATs, laser ultrasonic, and air-coupled transducer. EMATs are the devices with which ultrasonic waves can be generated and detected on metals. An EMAT consists of a permanent magnet (or electromagnet) to introduce a static field and mostly a flat coil to induce a dynamic current in the surface skin. Generation and detection of ultrasound are provided by coupling between the electromagnetic fields and the elastic field in the surface skin. It transfers the electromagnetic energy to the mechanical energy and *vise versa* via an air gap of a few millimeters. The wave source is then established within the materials to be measured, making a clear contrast with the piezoelectric transducers whose vibration should be transmitted to the objects through coupling fluid. Their geometry can be designed to measure the desired mode of elastic waves on the basis of the coupling mechanisms, the Lorentz force, and the magnetostrictive force (Chap. 2). Various propagation modes can then be utilized to meet a wide range of measurement needs; some are only available with the EMAT technique. This is one of the advantages over two other noncontact ultrasonic techniques.

1.2 Brief Historical Sketch of EMAT

The origin of EMATs is vague, and it is hard to trace the literature to the right source(s). It is not important to know who discovered electromagnetic generation and detection of ultrasound in metals, who first used the effects for measurements, and who called them EMATs. The acoustic coupling with EMATs is a basic physical effect, which occurs whenever a magnetic field, dynamic current, and a metal object are located in close vicinity. People may have encountered it unintentionally in laboratories and other places. In the case of pulsed nuclear magnetic resonance (NMR), the EMAT mechanism causes a spurious ringing from the probe housing or pole faces of magnets and obscures the free induction decay signals (Buess and Petersen 1978, for instance).

The study on EMATs has a long history. It took a long time to establish the coupling mechanism, probably because it is an interdisciplinary phenomenon encompassing solid mechanics and electromagnetics. Looking back over the past, Wegel and Walther (1935) studied the low-frequency internal friction in brass by applying variable frequency current to a coil, which drove a magnetic unit bonded to one end of a cylindrical specimen. The identical magnetic unit on the other end detected the vibration. The specimen was suspended at the nodal points. They discussed the dissipation mechanism with the observed resonance spectra of longitudinal and torsional modes. Following them, Randall et al. (1939) devised a measurement setup without using pole pieces to study the grain size dependence of internal friction in brass. Figure 1.2 shows their arrangement based on the Lorentz force coupling to detect the longitudinal resonance for the Q value. This resonance EMAT approach on long specimens developed to be a commercial instrument to

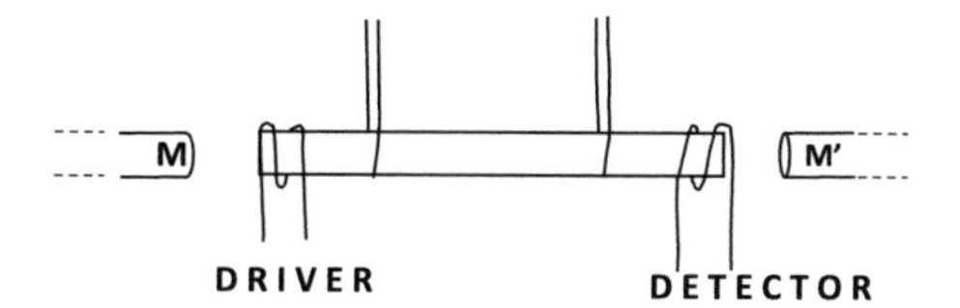

Fig. 1.2 Method of forcing and of detecting longitudinal vibrations. The specimen is supported by silk threads at nodes of vibration. The driving force is obtained by the reaction upon a permanent magnet *M* of the eddy currents induced by an alternating current in the driving coil. An alternating electromotive force is induced in a detector coil at the other end by the eddy currents arising from the motion of the specimen in the field of a second permanent magnet *M*. Reprinted with permission from Randall et al. 1939, Copyright (1939) by the American Physical Society. (http://dx.doi.org/10.1103/PhysRev.56.343)

estimate the steel sheet formability (Mould and Johnson 1973). The solenoid coil assembly shown in Fig. 1.3 generates and picks up a longitudinal resonance vibration of ribbon specimen, about 100-mm long, through the magnetostrictive coupling. The specimen is supported at the center, which is the nodal point of the fundamental mode. The square of resonance frequency is proportional to Young's modulus. Empirical formula relates Young's moduli in 0, 45, and 90° directions from the rolling direction to the so-called *r*-values, which represent the plastic anisotropy caused by texture (Chap. 11).

Apparently being independent from these early studies, the interaction between helicon waves and elastic shear waves in metals was investigated in the 1960s. Helicon is a transverse electromagnetic wave propagating in a dense electron gas along the applied magnetic field in solids. During this, Grimes and Buchsbaum (1964) and Larsen and Saermark (1967) observed resonance of acoustic shear wave in monocrystal potassium and aluminum, respectively, at liquid helium temperature. Dobbs (1970) wrote that Larsen and Saermark "were propagating helicons by placing a small coil on each side of an aluminum disk (the size of dime), exciting one coil at 1–2 MHz and observing the signal received in the second coil … As

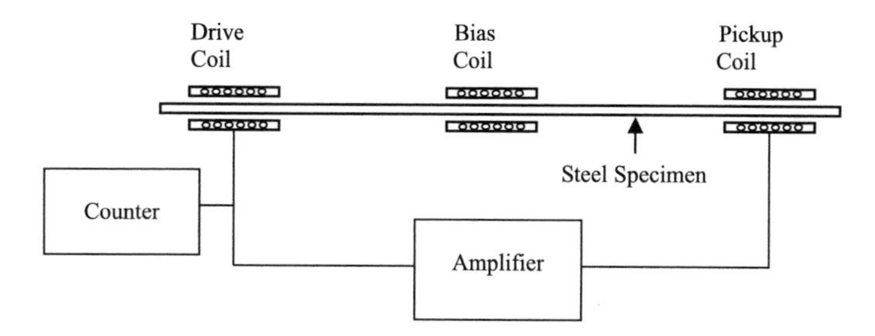

Fig. 1.3 Schematic block diagram of the magnetostrictive oscillator. (After Mould and Johnson 1973)

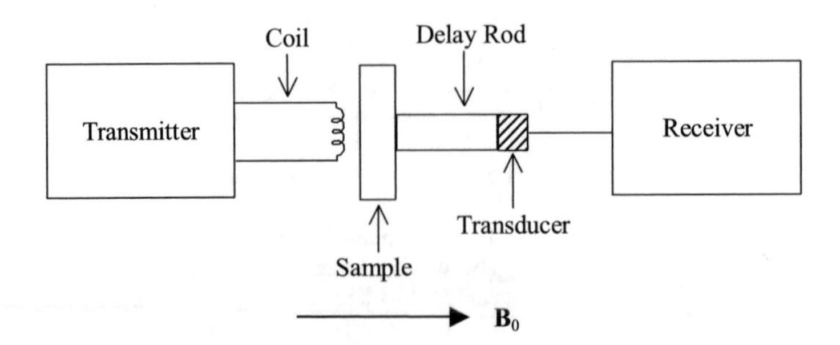

Fig. 1.4 The sound wave generated by an rf pulse in the coil propagates through the sample and the 3-cm-long *X*-cut quartz delay rod to the 10-MHz *AC*-cut quartz transducer. The experiment was performed at 4.2 K in a 50-kg superconducting solenoid. Reprinted with permission from Houck et al. 1967, Copyright (1967) by the American Physical Society. (http://dx.doi.org/10.1103/PhysRevLett.19.224)

these were size-controlled, they were sharp resonances and the acoustic resonator had a Q larger than 10^4." They revealed that the resonance occurs when the odd integer number of half-wavelength lies across the sample thickness.

The helicon-phonon interaction is not substantial in the present context, but it offered an opportunity for understanding what they called a direct electromagnetic generation of ultrasonic shear wave. Houck et al. (1967) presented "evidence that rf-ultrasonic coupling can occur near the surface of a metal in the presence of a magnetic field, independent of helicon propagation." Using the pulsing setup in Fig. 1.4, the signal amplitude was shown to be proportional to the magnetic strength. Betjemann et al. (1967) found that the helium temperature is not a necessary condition for the ultrasonic generation. Gaerttner et al. (1969) conducted an experiment on aluminum and tungsten crystals to observe the dependence of received shear wave amplitude on frequency (10–70 MHz) and temperature (4.2–300 K). Temperature varies the metal's conductivity and then the skin depth, which changes the generation efficiency with the Lorentz force mechanism. Good agreement with the theory was presented. During the same period, Abeles (1967) demonstrated the transduction in the absence of a static field. Indium film evaporated on germanium rod was excited by microwaves to generate a shear wave propagating in the rod. Converse effect was also observed that the shear wave echoes excited the indium film to radiate electromagnetic wave into the microwave cavity.

Using a similar configuration to Fig. 1.4 but at room temperature, Dobbs (1970) showed that the shear wave and longitudinal wave can be separately generated by orienting the static field to be normal and parallel to the specimen surface, respectively, and concluded that "the Lorentz force on the electronic current in the skin depth was responsible for the generation." He emphasized the usefulness for the nondestructive testing of metals and also measured the exponential decay of shear wave amplitude with the air gap up to 1.0 mm off the aluminum surface.

Concentrating on the practical nondestructive measurements, the former Soviet Union researchers achieved remarkable advancement with the EMAT technique. In a review paper, Butenko et al. (1972) asserted that the research had begun in 1959, but much work had been done in the late 1960s. They pointed out the advantages and disadvantages of EMATs just as we do forty years later: (i) noncontact aspect and tolerance to the surface conditions, (ii) easy generation and reception of shear wave, (iii) applications to high-temperature measurements up to 1000 °C (Chaps. 8 and 18), (iv) detection of anisotropic velocities with polarized shear waves (Chap. 12), (v) low transfer efficiency compared to the piezoelectric transducers, and (vi) inapplicability to nonmetallic materials. (The last point does not matter much, since depositing thin metal film allows coupling with nonconducting materials.) The Lorentz force was considered to be the coupling mechanism, but some experiments indicated that the magnetostriction participated. Instrumentations were done for the anisotropic velocity measurement using polarized shear waves, the thickness measurements for hot-rolled pipes with shear wave pulse echo technique, a mobile system for inspection of railway track, and an automatic flaw detector on seamless pipes. However, these pioneering researches were published only in the local community, and the progress was not known worldwide.

In the 1970s, the EMATs were studied with a distinct aim of developing noncontact flaw detection means. Frost (1979) gave an intensive review on the EMATs' principles and applications, including surface wave and bulk-wave transduction for both physical acoustic studies and industrial uses, that is, nondestructive testing of materials. He used the acronym "EMT" rather than EMAT. The EMAT designing and instrumentation were facilitated by the physical models, which were best described by Thompson (1973, 1990) and extended by Ogi (1997). Drawing much on his previous work starting in 1973, Thompson provided a general framework and analytical formalism covering both of the Lorentz force and magnetostrictive coupling mechanisms. The key papers contributing to establishing the principles of EMATs are cited there. Practical designing of EMATs and discussions on them can be found in Maxfield and Fortunko (1983) and Alers and Burns (1987), to name but a few. The bulk-wave EMAT (Sect. 3.1) and periodic permanent magnet (PPM) EMAT (Sect. 3.3) emerged from these studies, which are still of much value. Intensive studies have been done by many researchers in many countries over decades. However, the EMATs were not recognized as a generally useful technique and could not replace the piezoelectric transducers in flaw detection applications. This is solely because the received signals were too small to make reliable flaw detection on metal components.

As such, the EMATs did not gain much attention during the 1980s except for the limited applications including measurements on metals at elevated temperature (Sects. 18.4 and 18.5) and moving at moderate speeds (Sects. 11.4, 18.1, and 18.2). It is then not surprising to see that the EMAT technique was said to be "mature" in *Nondestructive Testing Handbook*, Volume 7, (McIntire 1991) in comparison with two other noncontact ultrasonic techniques, laser ultrasonic, and air-coupled transducer.

1.3 Electromagnetic Acoustic Resonance - EMAR

Electromagnetic acoustic resonance (EMAR) is the main subject in this monograph. The EMAR method is a combination of EMATs and resonance measurement. This noncontact method matched the recent upsurge of interest in nondestructive materials characterization, rather than flaw detection, and the distinctive capability has found many applications in both fundamental and industrial measurements.

Low transduction efficiency is only one weakness of EMATs. Once this weakness is overcome or compensated for, their unique and attractive features open novel ultrasonic methodology. Comprehending the coupling mechanism (Chap. 2) is important to make the most of the EMAT effects for individual measurements. Furthermore, there are several ideas to cope with the low transfer efficiency with EMATs. A simplest way is to drive an EMAT of a large aperture with low-frequency, high-power burst signals. For better performance, an intensification mechanism should incorporate along with the high-power excitation. Individual signals emitted by the EMATs are weak, but if they are superimposed in a coherent way, the resultant signal must be of a sufficient intensity. Low transfer efficiency is beneficial to make a large number of echoes superimposed (see Fig. 1.1b). This idea is mechanized by resonance measurement, that is, EMAR. Spectrometric approach gives rise to the ultrasonic resonance, and all echoes coherently overlap each other, improving the signal-to-noise ratio to a large extent. Combination of an EMAT with the ultrasonic resonance method successfully overcomes the disadvantage of an EMAT with the advantage of the resonance method and vice versa. The low transfer efficiency with an EMAT is improved by acquiring heavily and coherently overlapping signals at a resonant state. At the same time, the specimen is isolated from the transducer, thereby preventing other elements from participating into the measurement. Part II is devoted to describing the principles of EMAR; Part III provides the physical acoustic measurements using EMAR; and Part IV presents a number of the industrial applications. The EMATs were thought to be the final choice and adopted when the conventional transducers are useless (such as high temperature and speed of objects). But, being incorporated in EMAR, they demonstrate much higher accuracy and workability in ordinary measurements than the piezoelectric transducers. Many of these instances are provided in Parts III and IV for measurements of elastic constants, stress, texture, surface modification, grain size, aging caused by fatiguing and creeping, cracks, corrosion, and solid-liquid interface.

Resonant ultrasound spectroscopy (RUS) is an established methodology for studying a variety of physical phenomena through resonance frequency and resonance peak sharpness (Q^{-1} value), especially in small, simple-shaped specimens (Migliori and Sarrao 1997). The point-contacting piezoelectric transducers can be replaced with EMATs to realize free-vibration and mode-selective spectroscopy, which results in unmistakable determination of anisotropic elastic constants (Chap. 8).

The idea of EMAR is not new. Many of the early EMATs were employed in resonance to maximize the signal intensity. Adding to the studies mentioned above, Filimonov et al. (1971) and Nikiforenko et al. (1971) developed instrumentation for the electromagnetic excitation in the 1–20-MHz frequency range. They observed resonance peaks attributable to the longitudinal and shear modes of thickness oscillations of metal plates and splitting shear-mode peaks, which are now interpreted as a birefringence effect caused by the texture-induced elastic anisotropy in the rolled metal sheet (Chap. 11). They claimed the usefulness of this contactless method for measuring the elastic wave speeds and the anisotropic elastic constants. Twenty years later, Kawashima (1990) applied this technique to detect the rolling texture and estimate the formability (or *r*-value) in thin steel sheets. The available frequency range was up to 50 MHz and was later increased to 120 MHz for a 20-μm-thick aluminum foil (Kawashima and Wright 1992).

In the same period, Johnson et al. (1992, 1994), Clark et al. (1992), and Fukuoka et al. (1993) proposed to use EMAR for materials characterization. With the recent advances of the electronics and signal processing, EMAR has showed great potentials for the velocity and attenuation measurements. The difficulty of a precise velocity measurement is overcome by incorporating highly sensitive EMATs and employing a superheterodyne signal processing (Hirao et al. 1993; Petersen et al. 1994). Furthermore, it was found that EMAR can give an absolute value of attenuation owing to the noncontacting nature and low transfer efficiency (Ogi et al. 1995a, b). This is one of the attractive aspects of EMAR, because the absolute measurement of attenuation has never been realized with contact techniques.

Though less extensively, three more principles of coherent signal superposition and amplitude enhancement were implemented. One is line- and point-focusing of SV waves radiated from a meander-line coil for flaw detection (Sects. 3.10 and 14.4). The second is arrayed-coil EMAT, which is excited at given time intervals and polarity for productive interference of guided longitudinal waves into a direction along a long wire (Sect. 3.2). Incorporation of phased array technique also appears promising (Sect. 18.7.1). As the third of them, chirp pulse compression (Sect. 3.6) is sometimes useful to compensate for the saturation on high-power excitation. In a preliminary experiment, meander-line coils of variable spaces are shown to function in compressing a lengthy pulse of guided plate mode.

References

Abeles, B. (1967). Electromagnetic excitation of transverse microwave phonons in metals. *Physical Review Letters, 19*, 1181–1183.

Alers, G. A., & Burns, L. R. (1987). EMAT design for special applications. *Materials Evaluation, 45*, 1184–1189.

Betjemann, A. G., Bohm, H. V., Meredith, D. J., & Dobbs, E. R. (1967). RF ultrasonic wave generation in metals. *Physics Letters A, 25*, 753–755.

Buess, M. L., & Petersen, G. L. (1978). Acoustic ringing effects in pulsed nuclear magnetic-resonance probes. *Review of Scientific Instruments, 49*, 1151–1155.

Butenko, A. I., Ermolov, I. N., & Shkarlet, Y. M. (1972). Electromagneto-acoustic non-destructive testing in the Soviet Union. *Non-Destructive Testing, 5*, 154–159.

Clark, A. V., Fortunko, C. M., Lozev, M. G., Schaps, S. R., & Renken, M. C. (1992). Determination of sheet steel formability using wide band electromagnetic-acoustic transducers. *Research in Nondestructive Evaluation, 4*, 165–182.

Dobbs, E. R. (1970). Electromagnetic generation of ultrasonic waves in metals. *Journal of Physics and Chemistry of Solids, 31*, 1657–1667.

Filimonov, S. A., Budenkov, B. A., & Glukhov, N. A. (1971). Ultrasonic contactless resonance testing method. *Soviet Journal of Nondestructive Test* (*translated from Defektoskopiya*), (1), 102–104.

Frost, H. M. (1979). Electromagnetic-ultrasound transducers: Principles, practice, and applications. In *Physical Acoustics* (Vol. 14, pp. 179–275). New York: Academic Press.

Fukuoka, H., Hirao, M., Yamasaki, T., Ogi, H., Petersen, G. L., & Fortunko, C. M. (1993). Ultrasonic resonance method with EMAT for stress measurement in thin plate. In *Review of Progress in Quantitative Nondestructive Evaluation* (Vol. 12, pp. 2129–2136).

Gaerttner, M. R., Wallace, W. D., & Maxfield, B. W. (1969). Experiments relating to the theory of magnetic direct generation of ultrasound in metals. *Physical Review, 184*, 702–704.

Grimes, C. C., & Buchsbaum, S. J. (1964). Interaction between helicon waves and sound waves in potassium. *Physical Review Letters, 12*, 357–360.

Hirao, M., Ogi, H., & Fukuoka, H. (1993). Resonance EMAT system for acoustoelastic stress evaluation in sheet metals. *Review of Scientific Instruments, 64*, 3198–3205.

Houck, J. R., Bohm, H. V., Maxfield, B. W., & Wilkins, J. W. (1967). Direct electromagnetic generation of acoustic waves. *Physical Review Letters, 19*, 224–227.

Johnson, W., Auld, B. A., & Alers, G. A. (1994). Application of resonant modes of cylinders to case depth measurement. In *Review of Progress in Quantitative Nondestructive Evaluation* (Vol. 13, pp. 1603–1610).

Johnson, W., Norton, S., Bendec, F., & Pless, R. (1992). Ultrasonic spectroscopy of metallic spheres using electromagnetic-acoustic transduction. *The Journal of the Acoustical Society of America, 91*, 2637–2642.

Kawashima, K. (1990). Nondestructive characterization of texture and plastic strain ratio of metal sheets with electromagnetic acoustic transducers. *The Journal of the Acoustical Society of America, 87*, 681–690.

Kawashima, K., & Wright, O. B. (1992). Resonant electromagnetic excitation and detection of ultrasonic waves in thin sheets. *Journal of Applied Physics, 72*, 4830–4839.

Larsen, P. K., & Saermark, K. (1967). Helicon excitation of acoustic waves in aluminum. *Physics Letters A, 24*, 668–669.

Maxfield, B. W., & Fortunko, C. M. (1983). The design and use of electromagnetic acoustic wave transducers (EMATs). *Materials Evaluation, 41*, 1399–1408.

McIntire, P. (Ed.). (1991). *Ultrasonic Testing: Nondestructive Testing Handbook* (2nd ed., Vol. 7). ASNT.

Migliori, A., & Sarrao, J. (1997). *Resonant Ultrasound Spectroscopy*. New York: Wiley-Interscience.

Mould, P. R., & Johnson, T. E., Jr. (1973). Rapid assessment of drawability of cold-rolled low-carbon steel sheets. *Sheet Metal Industries, 50*, 328–333.

Nikiforenko, Z. G., Glukhov, N. A., & Averbukh, I. I. (1971). Measurement of the speed of elastic waves and acoustic anisotropy in plates. *Soviet Journal of Nondestructive Test* (*translated from Defektoskopiya*), (4), 427–432.

Ogi, H. (1997). Field dependence of coupling efficiency between electromagnetic field and ultrasonic bulk waves. *Journal of Applied Physics, 82*, 3940–3949.

Ogi, H., Hirao, M., & Honda, T. (1995a). Ultrasonic attenuation and grain size evaluation using electromagnetic acoustic resonance. *The Journal of the Acoustical Society of America, 98*, 458–464.

Ogi, H., Hirao, M., Honda, T., & Fukuoka, H. (1995b). Absolute measurement of ultrasonic attenuation by electromagnetic acoustic resonance. In *Review of Progress in Quantitative Nondestructive Evaluation* (Vol. 14, pp. 1601–1608).

Petersen, G. L., Chick, B. B., Fortunko, C. M., & Hirao, M. (1994). Resonance techniques and apparatus for elastic-wave velocity determination in thin metal plates. *Review of Scientific Instruments, 65*, 192–198.

Randall, R. H., Rose, F. C., & Zener, C. (1939). Intercrystalline thermal currents as a source of internal friction. *Physical Review, 56*, 343–348.

Thompson, R. B. (1973). A model for the electromagnetic generation and detection of Rayleigh and Lamb waves. *IEEE Transaction on Sonics and Ultrasonics, SU-20*, 340–346.

Thompson, R. B. (1990). Physical principles of measurements with EMAT transducers. In *Physical Acoustics* (Vol. 19, pp. 157–200). New York: Academic Press.

Wegel, R. L., & Walther, H. (1935). Internal dissipation in solids for small cyclic strains. *Journal of Applied Physics, 6*, 141–157.

Part I
Development of EMAT Techniques

Part I is intended to be a self-contained description of the basic elements of coupling mechanism along with practical designing of EMATs for various purposes. There are several implementations to compensate for the low transfer efficiency of the EMATs. Useful tips to make an EMAT are also presented.

Chapter 2
Coupling Mechanism

Abstract An EMAT consists of a coil to induce dynamic electromagnetic fields at the surface region of a conductive material, and permanent magnets (or electromagnets) to provide a biasing magnetic field. An EMAT configuration depends on the modes of elastic waves to be excited and detected. Optimum design of an EMAT requires understanding the coupling mechanism of energy transfer between the electromagnetic and elastic fields. This is a long-running topic and many studies appeared (Thompson (1977, 1978, 1990); Kawashima (1976, 1985); Il'in and Kharitonov (1981); Wilbrand (1983, 1987); Ogi (1997); Ogi et al. (2003); Ribichini et al. (2012)). This chapter presents the comprehensive analysis on physical principles of EMATs.

Keywords Eddy current · Liftoff · Lorentz force · Magnetostriction · Piezomagnetic constants · Theoretical calculation

2.1 Background

Previous studies revealed that three mechanisms contribute to the coupling: (i) Lorentz force mechanism caused by the interaction between eddy currents and the static magnetic flux density, (ii) magnetization force mechanism between the oscillating magnetic field and the magnetization, and (iii) magnetostriction mechanism by the piezomagnetic effect. The Lorentz force mechanism arises in all conducting materials, while other two appear only in ferromagnetic materials. For nonmagnetic metals, therefore, the Lorentz force mechanism explains the transfer with an EMAT (Gaerttner et al. 1969). The coupling is rather complicated for ferromagnetic materials. Thompson (1978) studied the field dependence of the guided-wave amplitude in ferromagnetic thin plates and derived a theoretical model

M. Hirao and H. Ogi, *Electromagnetic Acoustic Transducers*,
Springer Series in Measurement Science and Technology,
DOI 10.1007/978-4-431-56036-4_2

to explain the results. Il'in and Kharitonov (1981) calculated the efficiency of detecting Rayleigh waves radiated by a meander-line coil in a ferromagnetic metal. Wilbrand (1983, 1987) discussed bulk-wave detection involving the three mechanisms. Following them, Ogi (1997) improved Wilbrand's model to explain the field dependence of the bulk-wave excitation and detection in a ferromagnetic metal. Ribichini et al. (2012) analyzed EMAT phenomenon in more practical cases with numerical simulations.

2.2 Generation Mechanism

2.2.1 Governing Equations

When an alternating current is applied to an EMAT's coil element placed near a ferromagnetic specimen, the electromagnetic fields occur and penetrate through the material. The fields inside the material interact with the biasing magnetic field, cause body forces, and generate elastic waves. Thus, the wave generation analysis takes three steps: (i) calculation of the electromagnetic fields inside the material (Sect. 2.2.2), (ii) calculation of the body forces caused by the interactions between the electromagnetic and elastic fields (Sects. 2.2.3–2.2.5), and (iii) calculation of the acoustic fields caused by the body forces.

The governing equations are Maxwell's equations (Ampere's law and Faraday's law of induction)

$$\mathrm{rot}\mathbf{H} = \frac{\partial \mathbf{D}}{\partial t} + \mathbf{J}, \tag{2.1}$$

$$\mathrm{rot}\mathbf{E} = -\frac{\partial \mathbf{B}}{\partial t}, \tag{2.2}$$

Ohm's law

$$\mathbf{J} = \eta \mathbf{E}, \tag{2.3}$$

the constitutive relation

$$\mathbf{B} = \mu_0 \mathbf{H} + \mathbf{M}, \tag{2.4}$$

and the equation of motion

$$\rho \frac{\partial^2 u_i}{\partial t^2} = \frac{\partial \sigma_{ij}}{\partial x_j} + f_i. \tag{2.5}$$

Summation convention is implied in Eq. (2.5). Here, **H** (A/m) denotes the magnetic field, **D** (C/m^2) the electric flux density, **J** (A/m^2) the current density, **E** (V/m) the electric field, **B** (T) the magnetic flux density, **M** (T) the magnetization, **f** (N/m^3) the body force per unit volume, and **u** (m) the elastic displacement. These are all vector quantities. σ_{ij} is a component of the stress tensor. μ_0 (= $4\pi \times 10^{-7}$ H/m) and η (S/m) are the free-space permeability and the electrical conductivity, respectively.

For a polycrystalline ferromagnetic material subjected to a homogeneous biasing magnetic field, the resulting magnetization shows anisotropy about the polarization direction. For example, when the biasing field is applied along the z-axis, the relationship between **M** and **H** takes the form

$$\mathbf{M} = [\chi]\mathbf{H} = \begin{bmatrix} \chi_{xx} & 0 & 0 \\ 0 & \chi_{xx} & 0 \\ 0 & 0 & \chi_{zz} \end{bmatrix} \mathbf{H}, \tag{2.6}$$

where $[\chi]$ denotes the magnetic susceptibility tensor. The **B**-**H** relationship can then be written as

$$\mathbf{B} = \mu_0 \begin{bmatrix} \bar{\mu}_{xx} & 0 & 0 \\ 0 & \bar{\mu}_{xx} & 0 \\ 0 & 0 & \bar{\mu}_{zz} \end{bmatrix} \mathbf{H}, \tag{2.7}$$

with the normalized permeability tensor $[\bar{\mu}]$, that is, $\bar{\mu}_{ij} = 1 + \chi_{ij}/\mu_0$.

2.2.2 Dynamic Magnetic Fields in a Ferromagnetic Material

For simplifying the analysis, we use several approximations. (i) The electromagnetic and elastodynamic fields are variables in the two-dimensional space of the x-z plane (see Fig. 2.1): The half-space of $z > 0$ is filled with a ferromagnetic metal with the isotropic permeability $\mu_0\bar{\mu}$, in which the x-y plane defines the interface with a vacuum. (ii) The magnetostriction causes no volume change (isovolume). This is true for a nontextured polycrystalline ferromagnetic material, because randomly oriented easy axes average out the anisotropic magnetostriction of individual magnetic domains (Chikazumi 1964). (iii) The displacement current $\partial\mathbf{D}/\partial t$ in Eq. (2.1) is neglected, because EMATs usually use frequencies of megahertz orders. (iv) All time-dependent quantities cause harmonic oscillation and involve the $e^{j\omega t}$ factor, which we omit in equations. Quantities with subscript 0 are time-independent and homogeneous.

We consider two basic coils as shown in Fig. 2.1: (a) unidirectionally aligned coils with n terns per unit length in the x-axis and (b) a meander-line coil with

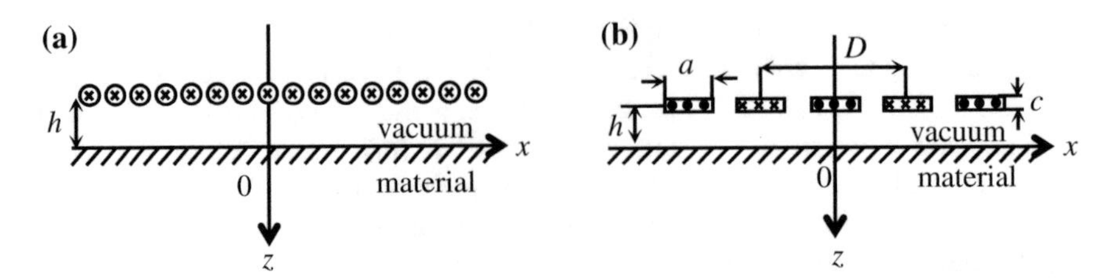

Fig. 2.1 Two-dimensional model of basic EMAT coils: **a** unidirectional coil and **b** meander-line coil

period D and width a. Current I flows in the coils. In this two-dimensional formulation, Eqs. (2.1)–(2.4) reduce to

$$\frac{\partial H_x}{\partial z} - \frac{\partial H_z}{\partial x} = J_y, \tag{2.8}$$

$$\frac{\partial E_y}{\partial z} = \mu_0 \bar{\mu} \frac{\partial H_x}{\partial t}, \tag{2.9}$$

$$\frac{\partial E_y}{\partial x} = -\mu_0 \bar{\mu} \frac{\partial H_z}{\partial t}, \tag{2.10}$$

$$J_y = \eta E_y. \tag{2.11}$$

Comparison between the x derivative of Eq. (2.9) and the z derivative of Eq. (2.10) leads to

$$\frac{\partial H_x}{\partial x} = -\frac{\partial H_z}{\partial z}. \tag{2.12}$$

Differentiating Eq. (2.8) with respect to z and using Eqs. (2.9), (2.11), and (2.12), we have

$$\left(\frac{\partial^2}{\partial x^2} + \frac{\partial^2}{\partial z^2}\right) H_x - j\omega\eta\mu_0\bar{\mu} H_x = 0. \tag{2.13}$$

In a vacuum, H_x satisfies Laplace equation

$$\left(\frac{\partial^2}{\partial x^2} + \frac{\partial^2}{\partial z^2}\right) H_x^V = 0. \tag{2.14}$$

For the unidirectional coil (Fig. 2.1a), a current element at x, $nI\mathrm{d}x$, provides the tangential magnetic field at the origin, which is obtained by superimposing the fields caused by the original element $nI\mathrm{d}x$ and its reflection image $nI\mathrm{d}x(\bar{\mu}-1)/(\bar{\mu}+1)$ (Hammond and Sykluski 1994)

$$dH_x^V = \frac{nI\mathrm{d}x}{2\pi(x^2+h^2)} \cdot \frac{2h}{\bar{\mu}+1}.$$

Thus, the total tangential field at $z = 0$ caused by the sheet currents is given by

$$H_x^V = 2\int_0^\infty \mathrm{d}H_x^V = 2 \times \frac{nI}{\pi(\bar{\mu}+1)} \int_0^\infty \frac{h}{x^2+h^2}\mathrm{d}x = \frac{nI}{\bar{\mu}+1}, \tag{2.15}$$

using a formula

$$\int_0^\infty \frac{h}{x^2+h^2}\mathrm{d}x = \frac{\pi}{2}.$$

The magnetic field in the material, H_x^M, must satisfy Eq. (2.13) and the electromagnetic boundary condition at the interface

$$H_x^V(x,0) = H_x^M(x,0) = \frac{nI}{\bar{\mu}+1}. \tag{2.16}$$

Because $\partial/\partial x = 0$ for all quantities, Eq. (2.13) can be written as

$$\frac{\partial^2 H_x^M}{\partial z^2} - q^2 H_x^M = 0, \tag{2.17}$$

where

$$q \equiv -\frac{1}{\delta}(1+j), \tag{2.18}$$

$$\delta = \sqrt{\frac{2}{\omega\eta\mu_0\bar{\mu}}}. \tag{2.19}$$

The solution of Eq. (2.17), which does not diverge at $z \to \infty$ and satisfies the boundary condition of Eq. (2.16), is

$$H_x^M = \frac{nI}{\bar{\mu}+1}e^{qz} = \frac{nI}{\bar{\mu}+1}e^{-\frac{z}{\delta}}e^{-j\frac{z}{\delta}}. \tag{2.20}$$

The magnetic field decays exponentially in the z direction depending on a factor $1/\delta$. The penetration depth is represented by δ and is called the *electromagnetic skin depth*. For copper at 1 MHz, $\delta \approx 0.07$ mm. For steels, $\delta \approx 0.01$ mm. Thus, δ is much smaller than the ultrasonic wavelength used in EMAT measurements, and it can be assumed that the electromagnetic fields occur only at the material surfaces.

Concerning the meander-line coil (Fig. 2.1b), the calculation of the fields inside the material is explained in detail in Thompson's monograph (Thompson 1990). The current per unit length along the x-axis of the meander-line coil is given by

$$\frac{I}{a}s(x),\quad \text{where}\quad s(x)=\begin{cases}1, & -a/2<x<a/2\\ 0, & -(D-a)/2<x<-a/2\ \text{or}\ a/2<x<(D-a)/2\\ -1, & -D/2<x<-(D-a)/2\ \text{or}\ (D-a)/2<x<D/2\end{cases}. \tag{2.21}$$

$s(x)$ can be expressed by a Fourier series,

$$s(x)=\sum_m A_m\cos k_m x, \tag{2.22}$$

where

$$A_m=\frac{4}{(2m+1)\pi}\sin\left\{\frac{a}{D}(2m+1)\pi\right\}, \tag{2.23}$$

$$k_m=\frac{2\pi}{D}(2m+1). \tag{2.24}$$

The integrated form of Eq. (2.1) gives the relationship that the contour integral of the magnetic field along the closed curve surrounding the meander-line current equals the total current passing through the cross section. This gives the magnetic field beneath the meander-line coil:

$$H_x^V(x,-h)=\sum_m\frac{I}{2a}A_m\cos k_m x. \tag{2.25}$$

Supposing $\partial^2/\partial x^2=-k_m^2$, the solution of Eq. (2.14) that satisfies the boundary condition of Eq. (2.25) and does not diverge at $z\to\infty$ becomes

$$H_x^V=\sum_m\frac{I}{2a}A_m e^{-k_m h}e^{-k_m z}\cos k_m x. \tag{2.26}$$

Equation (2.12) leads to

$$H_z^V=-\sum_m\frac{I}{2a}A_m e^{-k_m h}e^{-k_m z}\sin k_m x. \tag{2.27}$$

Solutions (2.26) and (2.27) express the magnetic fields below the meander-line coil in vacuum. They involve the factor $e^{-k_m h}$, which rapidly decays with increasing m. For simplicity, only the first term (m = 0) is considered:

$$\begin{cases} H_x^V = \dfrac{I}{2a} A_0 e^{-k_0 h} e^{-k_0 z} \cos k_0 x, \\ H_z^V = -\dfrac{I}{2a} A_0 e^{-k_0 h} e^{-k_0 z} \sin k_0 x. \end{cases} \tag{2.28}$$

The image method is then applied to calculate the magnetic field with the presence of the half-space conductor:

$$\begin{cases} H_x^M = IaA_0 e^{-k_0 h} e^{qz} \cos k_0 x, \\ H_z^M = k_0 \delta \sqrt{2} IaA_0 e^{-j\frac{\pi}{4}} e^{-k_0 h} e^{qz} \sin k_0 x. \end{cases} \tag{2.29}$$

where $q^2 = k_0^2 + 2j/\delta^2$ ($\mathrm{Re}(q) < 0$). We used a reasonable approximation that $k\delta \ll 1$ and then $|H_x^M| \gg |H_z^M|$. The presence of the gap h between the coil and material surface, called *liftoff*, exponentially decreases the magnitude of the electromagnetic fields in the material.

The magnetic fields induced by EMAT coils in the material take the form

$$\begin{cases} H_x^M = f(x) e^{qz} \\ H_z^M = g(x) e^{qz} \end{cases}, \tag{2.30}$$

where f and g are functions of only x. Because the variable changes in the z-axis are much larger than in the x-axis ($k_0\delta \ll 1$), the following relation usually holds

$$\left| \frac{\partial H_x^M}{\partial z} \right| \gg \left| \frac{\partial H_x^M}{\partial x} \right| \approx \left| \frac{\partial H_z^M}{\partial z} \right| \gg \left| \frac{\partial H_z^M}{\partial x} \right|. \tag{2.31}$$

2.2.3 Lorentz Force

The Lorentz force mechanism was studied by many researchers (Gaerttne et al. 1969; Beissner 1976; Kawashima 1976; Maxfield and Fortunko 1983; Thompson 1973, 1990). We summarize their studies here.

When an electric field $\mathbf{E}$ is applied to a conducting material, the Coulomb force $-e\mathbf{E}$ occurs on individual electrons. In the presence of the biasing magnetic field $\mathbf{B}_0$, the Lorentz force $e\mathbf{v}_e \times \mathbf{B}_0$ appears. $\mathbf{v}_e$ denotes the mean electron velocity. Thus, the equation of motion for an electron is

$$m\dot{\mathbf{v}}_e = -e(\mathbf{E} + \mathbf{v}_e \times \mathbf{B}_0) - \frac{m\mathbf{v}_e}{\tau}. \tag{2.32}$$

Here, m denotes the electron mass and e denotes the electron charge. τ denotes the mean time of the electron-ion collision and is of the order of 10^{-14} s for common metals at room temperature. Supposing harmonic oscillation of the electric field with the angular frequency ω and $\omega\tau \ll 1$, Eq. (2.32) reduces to

$$n_e \frac{m\mathbf{v}_e}{\tau} = -n_e e(\mathbf{E} + \mathbf{v}_e \times \mathbf{B}_0), \tag{2.33}$$

where n_e is the electron density. This momentum is transferred to the ions via the collisions. Thus, the body forces applied to the ions are approximated as

$$\mathbf{f} = NZ_e(\mathbf{E} + \dot{\mathbf{u}} \times \mathbf{B}_0) + n_e \frac{m\mathbf{v}_e}{\tau}, \tag{2.34}$$

where N denotes the ion density, Z_e the ion charge, and $\mathbf{u}$ the ion displacement. Because $n_e e = NZ_e$ and $\mathbf{v}_e \gg \dot{\mathbf{u}}$, the force per unit volume on ions reduces to

$$\mathbf{f} = -n_e e \mathbf{v}_e \times \mathbf{B}_0 = \mathbf{J}_e \times \mathbf{B}_0 \equiv \mathbf{f}^{(L)}. \tag{2.35}$$

$\mathbf{J}_e = -n_e e \mathbf{v}_e$ is the electron eddy current density. The Lorentz force in Eq. (2.35) can cause an acoustic vibration.

From Eq. (2.8), we have

$$J_e = \frac{\partial H_x^M}{\partial z} - \frac{\partial H_z^M}{\partial x}. \tag{2.36}$$

According to Eq. (2.31), the second term in the right-hand side is negligible compared with the first term. Then, the Lorentz forces are

$$\begin{cases} f_x^{(L)} = B_{0z} \dfrac{\partial H_x^M}{\partial z}, \\ f_z^{(L)} = -B_{0x} \dfrac{\partial H_x^M}{\partial z}. \end{cases} \tag{2.37}$$

When we consider the contribution of the dynamic magnetic field $\mathbf{B}$ caused by the EMAT coil as well as the static field, the Lorentz forces are expressed by

$$\begin{cases} f_x^{(L)} = (B_{0z} + \mu_0 \bar{\mu} H_z^M) \dfrac{\partial H_x^M}{\partial z}, \\ f_z^{(L)} = -(B_{0x} + \mu_0 \bar{\mu} H_x^M) \dfrac{\partial H_x^M}{\partial z}. \end{cases} \tag{2.38}$$

Equation (2.7) is used. Thus, the Lorentz forces caused by the static field (first terms of the right-hand sides) are proportional to the driving current I and vibrate with the same frequency ω as the driving current. Whereas those caused by the dynamic field (second terms) are proportional to I^2 and possess the doubled frequency component of 2ω. Because the magnitude of the static field is usually much larger than that of the dynamic field, we can neglect the second terms. However, when a large driving current is in use, the second terms exceed the first terms to give rise to the double-frequency ultrasound. Such an excitation requires no static field and makes the EMAT compact as will be shown in Sect. 18.4.2.

2.2.4 *Magnetization Force*

The force acting in the volume and on the surface of a magnetic material because of the presence of the magnetization **M** are summed up to give (Brown 1966; Moon 1984)

$$\mathbf{F} = \int_V \mathbf{M} \cdot \nabla \mathbf{H} \, \mathrm{d}V + \frac{1}{2}\mu_0 \int_S \mathbf{n} M_n^2 \mathrm{d}S. \tag{2.39}$$

n is a unit vector normal to the material surface and M_n the normal component of the magnetization at the surface. The second term in Eq. (2.39) appears because of a steep change of the electromagnetic fields at the surface and disappears inside the material. The integrand in the first term of Eq. (2.39) consists of gradients of dynamic magnetic field, involving some components of Eq. (2.35). In this book, we remain the definition for the Lorentz force mechanism in Eq. (2.35) and regard the body force per unit volume

$$\mathbf{f}^{(M)} = (\nabla \mathbf{H}) \cdot \mathbf{M}_0, \tag{2.40}$$

as the magnetization force. When the biasing magnetic field is applied, Eq. (2.40) becomes

$$\begin{cases} f_x^{(M)} = M_{0x} \dfrac{\partial H_x^M}{\partial x} + M_{0z} \dfrac{\partial H_z^M}{\partial x}, \\ f_z^{(M)} = M_{0x} \dfrac{\partial H_x^M}{\partial z} + M_{0z} \dfrac{\partial H_z^M}{\partial z}. \end{cases} \tag{2.41}$$

The second terms are negligible because of $\left|H_x^M\right| \gg \left|H_z^M\right|$. Combining with the Lorentz forces in Eq. (2.37) and using Eq. (2.4), we obtain

$$\begin{cases} f_x^{(M)} + f_x^{(L)} = B_{0z} \dfrac{\partial H_x^M}{\partial z} + M_{0x} \dfrac{\partial H_x^M}{\partial x}, \\ f_z^{(M)} + f_z^{(L)} = (M_{0x} - B_{0x}) \dfrac{\partial H_x^M}{\partial z} = -\mu_0 H_{0x} \dfrac{\partial H_x^M}{\partial z}. \end{cases} \tag{2.42}$$

The biasing magnetic field in EMAT phenomena is normally less than a few Tesla and $|\mu_0 \mathbf{H}_0| \ll |\mathbf{M}_0|$. It is important to note that, in the case of a tangential biasing magnetic field ($H_{0z} = 0$), $f_z^{(L)}$ and $f_z^{(M)}$ act in opposite directions and the magnetization force cancels a major part of the Lorentz force in the z direction. Thus, the

Lorentz and magnetization forces are ineffective in generating longitudinal-mode waves in ferromagnetic materials.

The magnetizaiton force defined in Eq. (2.40) applies in most cases, but it fails to explain the wave generation in some particular cases, for which the total magnetic forces in Eq. (2.39) have to be considered (Seher and Nagy 2016).

2.2.5 Magnetostriction Force

When an external magnetic field is applied to a ferromagnetic material, a dimensional change arises depending on the magnitude and direction of the field. The normalized dimensional change is called *magnetostriction*. It occurs because, in iron for instance, the external magnetic field affects the 3*d* subshell and changes the shape and size of the 3*d*-electron orbits to minimize energy in the presence of the field. This can be intuitively understood by considering a dimensional change in a magnetic domain caused by the rotation of magnets connected by elastic springs (Fig. 2.2). The magnets simulate atomic spins. There is an elastic strain to make a balance between the spring and magnetic forces in equilibrium even without an external field, which is the spontaneous magnetostriction and is equivalent to an eigenstrain in individual magnetic domains. In iron, a positive spontaneous magnetization exists in domain's magnetization directions oriented to easy axes <100>.

In polycrystalline materials, the magnetostriction response to the external field is more complicated. In polycrystalline iron, the dimensional change occurs in two steps. First, the domains, whose magnetizations are oriented near along the external field, expand in volume, causing a positive dimensional change (elongation) along the field because of the positive spontaneous magnetostriction (Fig. 2.3b). After this domain's rearrangement, the magnetization rotates about the easy axis within the domain, which reduces the dimension due to the rotation of the spins. Thus, the magnetostriction ε^{M} along the applied field is a function of the external field H and

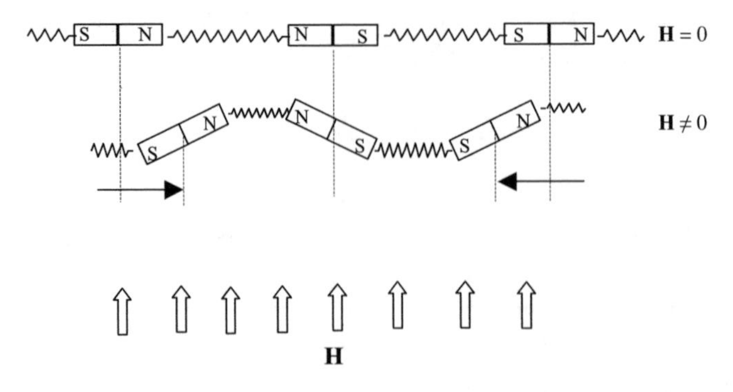

Fig. 2.2 Intuitive understanding of magnetostriction in a magnetic domain. Dimensional change caused by the external field on magnets connected to each other by elastic springs corresponds to magnetostriction

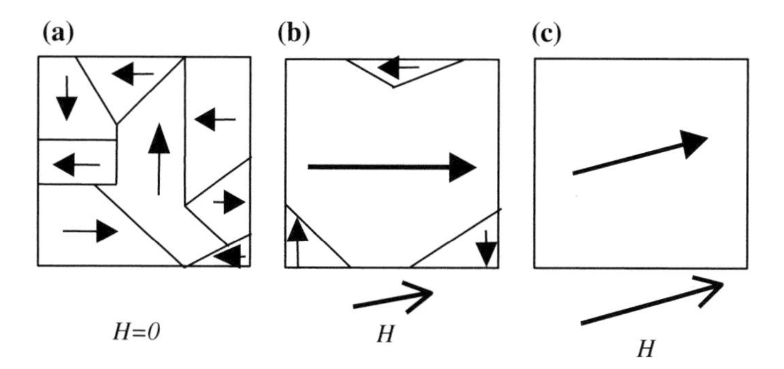

Fig. 2.3 Evolution of domains and rotation of magnetization with the increase of the external field in a polycrystalline (or multi-domain) material. **a** Randomly oriented domains at $H = 0$. **b** Volume increase in the domain oriented nearly parallel to the field. **c** Rotation of the magnetization to the field direction

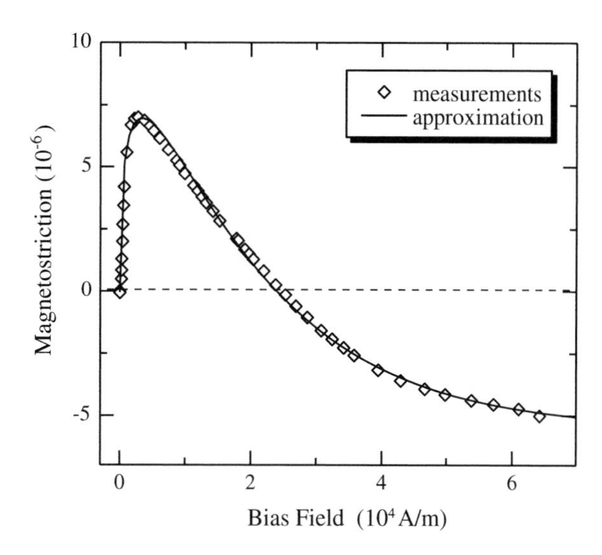

Fig. 2.4 Measured (*open symbols*) and approximated (*solid line*) magnetostriction curve for a low-carbon steel

then the ε-H curve, or the magnetostriction curve, shows a maximum. Figure 2.4 exemplifies the measured magnetization curve for a low-carbon steel. The magnitude of maximum magnetostriction is of the order of 10^{-6} for steels and 10^{-5} for nickel.

Coupling between the elastic and magnetic fields for such a ferromagnetic (or piezomagnetic) material can be assumed to take a similar form as that for a piezoelectric material:

$$S_I = d_{Ij}^{(MS)} H_j + s_{IJ}^H \sigma_J. \quad (I, J = 1, 2, \ldots, 6;\ j = x, y, z) \tag{2.43}$$

Here, S_I is a component of the engineering strain, s_{IJ}^{H} a component of the compliance matrix at a constant field, and σ_J the stress component in contracted notation (Auld 1973). $d_{Ij}^{(MS)}$ can be written as

$$d_{Ij}^{(MS)} = \left(\frac{\partial S_I}{\partial H_j}\right)_{|\sigma}, \tag{2.44}$$

which indicate the piezomagnetic strain coefficients. When we apply a magnetic field to a ferromagnetic material at the stress-free state, the strain $S_I = d_{Ij}^{(MS)} H_j$ will appear. The same strain field occurs with the stress field $\sigma_K = c_{KI}^{H} S_I$ without the magnetic field, where c_{KI}^{H} is the elastic-stiffness coefficient at a constant field. The equivalent stress to cause the magnetostriction is $\sigma_K = c_{KI}^{H} d_{Ij}^{(MS)} H_j$. If the applied magnetic field changes quickly or vibrates with a high frequency, the strain will fail to respond simultaneously with the field and the stress field $-\sigma_K$ will occur inside the material, which is the magnetostriction stress. We can thus define the magnetostriction stress as

$$\sigma_I^{(MS)} = -c_{IJ}^{H} d_{Jj}^{(MS)} H_j = -e_{Ij}^{(MS)} H_j, \tag{2.45}$$

with the converse piezomagnetic stress coefficients

$$e_{Ij}^{(MS)} = c_{IJ}^{H} d_{Jj}^{(MS)} = -\left(\frac{\partial \sigma_I^{(MS)}}{\partial H_j}\right)_{|S}. \tag{2.46}$$

The constitutive equation among the stress, strain, and field is then given by

$$\sigma_I = -e_{Ij}^{(MS)} H_j + c_{IJ}^{H} S_J, \tag{2.47}$$

Equation (2.47) is similar to that of the piezoelectric stress equation. The body forces caused by the magnetostriction stress are

$$\begin{cases} f_x^{(MS)} = \dfrac{\partial \sigma_1^{(MS)}}{\partial x} + \dfrac{\partial \sigma_6^{(MS)}}{\partial y} + \dfrac{\partial \sigma_5^{(MS)}}{\partial z}, \\ f_y^{(MS)} = \dfrac{\partial \sigma_6^{(MS)}}{\partial x} + \dfrac{\partial \sigma_2^{(MS)}}{\partial y} + \dfrac{\partial \sigma_4^{(MS)}}{\partial z}, \\ f_z^{(MS)} = \dfrac{\partial \sigma_5^{(MS)}}{\partial x} + \dfrac{\partial \sigma_4^{(MS)}}{\partial y} + \dfrac{\partial \sigma_3^{(MS)}}{\partial z}. \end{cases} \tag{2.48}$$

The acoustic fields generated by the magnetostriction forces can be calculated using Eqs. (2.5) and (2.48) with boundary conditions when the piezomagnetic coefficients $\mathbf{d}^{(MS)}$ and $\mathbf{e}^{(MS)}$ are known. They depend highly on the magnitude and direction of the applied field, but they are estimated from the magnetostriction curve as exemplified in Fig. 2.4. Following are the examples for calculating the piezomagnetic coefficients.

(a) Bias field normal to the surface ($\mathbf{H}_0 = (0, 0, H_{0z})$)

When a homogenous static magnetic field H_{0z} is applied in the z direction (normal to the specimen surface) in a ferromagnetic material in the stress-free state, the longitudinal magnetostriction $\varepsilon(H_{0z})$ appears along the field. The magnetostriction perpendicular to the field will be $-\varepsilon(H_{0z})/2$ because of the isovolume dimensional change. Thus, the time-independent strain field $\mathbf{S}^0$ caused by the applied field is

$$S_3^0 = \varepsilon(H_{0z}), \quad S_1^0 = S_2^0 = -\frac{1}{2}\varepsilon(H_{0z}), \quad S_4^0 = S_5^0 = S_6^0 = 0. \tag{2.49}$$

The dynamic field applied in the z direction, H_z, disturbs the strain field to cause the dynamic strain $\mathbf{S}$. When the dynamic field is much smaller in magnitude than the static field, we can approximate as

$$\begin{aligned} &S_3 = \left(\frac{\partial S_3}{\partial H_z}\right) H_z, \quad S_1 = S_2 = -\frac{1}{2}\left(\frac{\partial S_3}{\partial H_z}\right) H_z, \\ &S_4 = S_5 = S_6 = 0. \end{aligned} \tag{2.50}$$

The piezomagnetic strain coefficients related to the field in the z direction are therefore given by

$$\begin{cases} d_{3z}^{(MS)} = \left(\dfrac{\partial S_3}{\partial H_z}\right)_{|\sigma} \equiv \gamma, \quad d_{1z}^{(MS)} = d_{2z}^{(MS)} = -\frac{1}{2}\gamma, \\ d_{4z}^{(MS)} = d_{5z}^{(MS)} = d_{6z}^{(MS)} = 0, \end{cases} \tag{2.51}$$

which indicates that the shearing deformation is not caused by H_z. γ denotes the slope of the magnetostriction curve and is measurable.

Components of the coefficients related to the x direction dynamic field can be calculated as follows. When the dynamic field H_x is added, the total field occurs in the direction inclined by θ about the z-axis and the principal strains arise in the directions parallel and normal to the total field. Thus, in the coordinate system where the z'-axis is along the total field (Fig. 2.5), the strains can be expressed as

$$S_3' = \varepsilon(H_t) \equiv \varepsilon_t, \quad S_1' = S_2' = -\frac{1}{2}\varepsilon(H_t) = -\frac{1}{2}\varepsilon_t, \tag{2.52}$$

where ε_t denotes the magnetostriction along the total field $\mathbf{H}_t$. Note that no shearing deformation occurs in the x'-y'-z' coordinate system. The strain field in the original coordinate system takes the form

$$\begin{cases} S_1 = S_1' \cos^2\theta + S_3' \sin^2\theta, \\ S_2 = S_2', \\ S_3 = S_3' \cos^2\theta + S_1' \sin^2\theta, \\ S_5 = (S_3' - S_1')\sin 2\theta = \frac{3}{2}\varepsilon_t \sin 2\theta. \end{cases} \tag{2.53}$$

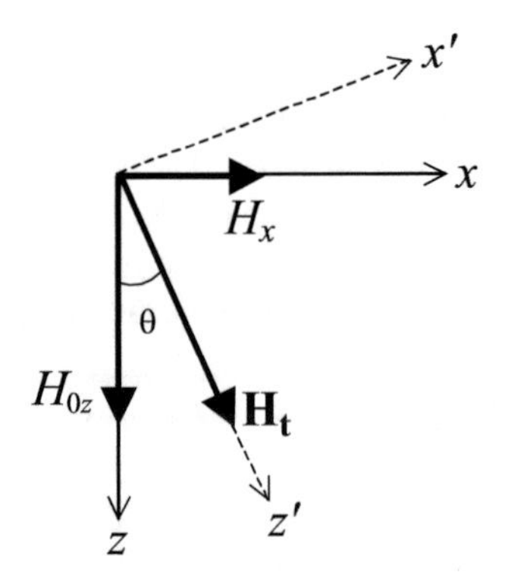

Fig. 2.5 Rotation of the total field about the static field direction

It is clear that $d_{4x}^{(MS)} = d_{6x}^{(MS)} = 0$ because S_4 and S_6 equal zero. Then, we have

$$\begin{cases} d_{1x}^{(MS)} = \left(\dfrac{\partial S_1}{\partial H_x}\right)_{|\sigma} = \dfrac{\partial}{\partial H_x}(S'_1 \cos^2\theta + S'_3 \sin^2\theta) \\ \quad = \dfrac{3\varepsilon_t}{H_{0z}}\cos^3\theta \sin\theta + \gamma \sin\theta\left(-\dfrac{1}{2}\cos^2\theta + \sin^2\theta\right), \\ d_{2x}^{(MS)} = -\dfrac{1}{2}\gamma \sin\theta, \\ d_{3x}^{(MS)} = -\dfrac{3\varepsilon_t}{H_{0z}}\cos^3\theta \sin\theta + \gamma \sin\theta\left(-\dfrac{1}{2}\sin^2\theta + \cos^2\theta\right), \\ d_{5x}^{(MS)} = \dfrac{3\gamma}{2}\sin 2\theta \sin\theta + \dfrac{3\varepsilon_t}{H_{0z}}\cos^2\theta \cos 2\theta. \end{cases} \tag{2.54}$$

Analogous calculation results in the coefficients relating to H_y:

$$\begin{cases} d_{1y}^{(MS)} = d_{2x}^{(MS)},\ d_{2y}^{(MS)} = d_{1x}^{(MS)},\ d_{3y}^{(MS)} = d_{3x}^{(MS)},\ d_{4y}^{(MS)} = d_{5x}^{(MS)}, \\ d_{5y}^{(MS)} = d_{6y}^{(MS)} = 0. \end{cases} \tag{2.55}$$

Thus, the matrix $\mathbf{d}^{(MS)}$ takes the following form for the normal bias field,

$$\left[d_{Ij}^{(MS)}\right] = \begin{bmatrix} d_{1x} & d_{2x} & d_{1z} \\ d_{2x} & d_{1x} & d_{1z} \\ d_{3x} & d_{3x} & d_{3z} \\ 0 & d_{5x} & 0 \\ d_{5x} & 0 & 0 \\ 0 & 0 & 0 \end{bmatrix}. \tag{2.56}$$

EMATs normally use a static magnetic field much larger than the dynamic fields, that is, $H_{0z} \gg H_x$ and H_z. Such a high-field approximation reduces Eq. (2.56) to

$$\left[d_{Ij}^{(MS)}\right] = \begin{bmatrix} 0 & 0 & -\frac{\gamma}{2} \\ 0 & 0 & -\frac{\gamma}{2} \\ 0 & 0 & \gamma \\ 0 & \frac{3\varepsilon_t}{H_{0z}} & 0 \\ \frac{3\varepsilon_t}{H_{0z}} & 0 & 0 \\ 0 & 0 & 0 \end{bmatrix}. \tag{2.57}$$

Equation (2.57) is comparable with the piezoelectric strain coefficients of crystals, which belong to a hexagonal *6mm* point group.

Equations (2.48) and (2.57) are combined to derive the two-dimensional magnetostriction forces as

$$f_x^{(MS)} = -\frac{3c_{55}\varepsilon_t}{H_{0z}}\frac{\partial H_x}{\partial z} + \frac{\gamma}{2}(c_{11} - c_{12})\frac{\partial H_z}{\partial x}, \tag{2.58}$$

$$f_z^{(MS)} = -\frac{3c_{55}\varepsilon_t}{H_{0z}}\frac{\partial H_x}{\partial x} - \gamma(c_{11} - c_{12})\frac{\partial H_z}{\partial z}. \tag{2.59}$$

Equation (2.31) allows us to neglect the second term on the right-hand side of Eq. (2.58) and to predict $\left|f_x^{(MS)}\right| \gg \left|f_z^{(MS)}\right|$, indicating that the shear wave polarized in the x direction can be generated more efficiently than the longitudinal wave with the magnetostriction forces.

It is important to note that the stress-free condition at the surface needs the apparent traction force, which markedly cancels the magnetostriction force $f_x^{(MS)}$ (Ribichini et al. 2012). The contribution of the magnetostriction mechanism to the wave generation is comparable with that of the Lorentz force mechanism in lower magnetic field region ($\lesssim$1000 A/m). However, it becomes minor at higher field (Ogi 2012).

For example, consider the shear plane wave generation by the sheet current as shown in Fig. 2.1a and a homogenous normal magnetic field B_{0z}. The wave equation takes the form

$$\rho\frac{\partial^2 u_x}{\partial t^2} = \frac{\partial \sigma_5}{\partial z} + B_{0z}\frac{\partial H_x^M}{\partial z}, \tag{2.60}$$

where

$$\sigma_5 = G\frac{\partial u_x}{\partial z} - e_{5x}^{(MS)} H_x^M. \tag{2.61}$$

G denotes the shear modulus, and H_x^M is given by Eq. (2.20). Assuming the harmonic vibration with ω, we have

$$\frac{\partial^2 u_x}{\partial z^2} + K^2 u_x = \frac{Pq}{G}\left(e_{5x}^{(MS)} - B_{0z}\right)e^{qz}e^{j\omega t}. \quad (2.62)$$

Here, $K \equiv \omega\sqrt{\rho/G}$ and $P \equiv nI/(1+\bar{\mu})$. The general solution takes the form $u_x = (A^* e^{-jKz} + B^* e^{qz})e^{j\omega t}$, where A^* and B^* denote complex coefficients: The first term represents the homogeneous solution, allowing long-distance propagation, and is important for EMAT applications. The second term indicates a particular solution, which decays rapidly with propagation. Substituting this form into Eq. (2.62) yields

$$B^* = \frac{P}{G}\left(e_{5x}^{(MS)} - B_{0z}\right)\frac{q}{q^2 + K^2}. \quad (2.63)$$

From the boundary condition ($\sigma_5 = 0$ at $z = 0$) and Eq. (2.63), we have

$$A^* \approx \frac{P}{KG}\left(\frac{e_{5x}^{(MS)}}{2}(K\delta)^2 + jB_{0z}\right), \quad (2.64)$$

with $|K\delta| \ll 1$. The first and second terms show contributions of the magnetostriction and Lorentz-force mechanisms, respectively. Their ratio, $(K\delta)^2 e_{5x}^{(MS)}/2B_{0z}$, then indicates dominant mechanism. Assuming $e_{5x}^{(MS)} = 3G\varepsilon_t/H_{0z}$ with $G = 80$ GPa and $\varepsilon_t = 2 \times 10^{-6}$, frequency of 5 MHz, conductivity of 5×10^6 S/m, $B_{0z} = 0.3$ T, and $\bar{\mu} = 1000$, this ratio becomes about unity when $H_{0z} = 1000$ A/m. It becomes less than 0.1 for $H_{0z} = 10{,}000$ A/m, indicating the dominant contribution of the Lorentz force mechanism at higher magnetic field.

(b) Bias field parallel to the surface in the y-axis ($\mathbf{H}_0 = (0, H_{0y}, 0)$)

A similar approach to the above yields an expression for the piezomagnetic coefficients for a bias field in the y direction;

$$\left[d_{Ij}^{(MS)}\right] = \begin{bmatrix} d_{1x} & d_{1y} & d_{3x} \\ d_{2x} & d_{2y} & d_{2x} \\ d_{3x} & d_{1y} & d_{1x} \\ 0 & 0 & d_{6x} \\ 0 & 0 & 0 \\ d_{6x} & 0 & 0 \end{bmatrix}. \quad (2.65)$$

For example,

$$d_{6x}^{(MS)} = \frac{3\gamma}{2}\sin 2\theta \sin\theta + \frac{3\varepsilon_t}{H_{0y}}\cos^2\theta\cos 2\theta. \quad (2.66)$$

At high fields ($H_{0y} \gg H_x, H_z$), Eq. (2.65) reduces to

$$\left[d_{Ij}^{(MS)}\right] = \begin{bmatrix} 0 & -\frac{\gamma}{2} & 0 \\ 0 & \gamma & 0 \\ 0 & -\frac{\gamma}{2} & 0 \\ 0 & 0 & \frac{3\varepsilon_t}{H_{0y}} \\ 0 & 0 & 0 \\ \frac{3\varepsilon_t}{H_{0y}} & 0 & 0 \end{bmatrix} \tag{2.67}$$

The nonzero body force occurs only in the y direction in this case:

$$f_y^{(MS)} = \frac{\partial \sigma_6^{(MS)}}{\partial x} + \frac{\partial \sigma_4^{(MS)}}{\partial z} = -\frac{3c_{66}\varepsilon_t}{H_{0y}}\frac{\partial H_x}{\partial x} - \frac{3c_{44}\varepsilon_t}{H_{0y}}\frac{\partial H_z}{\partial z}, \tag{2.68}$$

where we neglected the terms proportional to $\partial H_z/\partial x$. The first term in the right-hand side of Eq. (2.68) can generate the surface shear waves or plate shear waves propagating in the x direction with the y polarization. The second term generates the bulk shear wave propagating in the thickness direction with the y polarization.

2.3 Receiving Mechanisms

A dynamic deformation caused by an acoustic wave creates dynamic electromagnetic fields in a conductive material exposed to a steady magnetic field. The dynamic fields pass the material/vacuum boundary and can be detected by an EMAT coil. The analysis of EMAT's receiving mechanism then includes three factors: (i) the electromagnetic fields within the material caused by elastic waves, (ii) moving boundary at the material surface, through which the fields pass into a vacuum, and (iii) the electromagnetic fields in vacuum, where the EMAT coil is located.

The dynamic electric field induced by the deformation in a conducting material takes the form

$$\frac{\partial \mathbf{u}}{\partial t} \times \mathbf{B}_0. \tag{2.69}$$

This is the reversed Lorentz force mechanism. The induced current density in the material can be of the form

$$\mathbf{J} = \eta\left(\mathbf{E} + \frac{\partial \mathbf{u}}{\partial t} \times \mathbf{B}_0\right). \tag{2.70}$$

In a ferromagnetic material, the elastic deformation disturbs the steady magnetization state, because it varies a particular magnetic domain volume or affects the electron's subshell orbits, resulting in an additional flux magnetic density. This is the reversed magnetostriction mechanism (piezomagnetic effect). The constitutive relation will take the form

$$B_i = \mu_0\mu_{ij}^S H_j + \tilde{e}_{iJ}^{(MS)} S_J, \quad (i,j = x,y,z;\ J = 1,2,\ldots,6) \tag{2.71}$$

Here, μ_{ij}^S is the normalized permeability tensor at a constant strain. The piezomagnetic coefficients $\tilde{e}_{iJ}^{(MS)}$ relate the induced magnetic flux density to the applied strain field. Equation (2.71) is similar to the constitutive equation for expressing the piezoelectric effect (Auld 1973). We know from the thermodynamics of solids that the piezoelectric strain matrix is a transposed matrix of the converse piezoelectric stress matrix. An analogous relationship is expected for the piezomagnetic coefficients:

$$\tilde{e}_{iJ}^{(MS)} = e_{Ji}^{(MS)}. \tag{2.72}$$

Experiments indicate that Eq. (2.72) holds roughly, but not exactly.

Returning to the two-dimensional approximation, we obtain a differential equation for E_y by substituting Eq. (2.70) into Eq. (2.1) and eliminating H_x and H_z using Eqs. (2.2) and (2.71). It gives a general solution of the electric field in the material. (Also, it applies to obtain the electric field in a vacuum using zero conductivity and zero elastic displacement.) The general solutions include arbitrary constants. The boundary conditions at the moving surface are (Il'in and Kharitonov 1981)

$$\mathbf{n}_0 \times (\mathbf{E}^V - \mathbf{E}) = \frac{\partial u_3}{\partial t}(\mathbf{B}_0^V - \mathbf{B}_0), \tag{2.73}$$

$$\mathbf{n}_0 \times (\mathbf{H} - \mathbf{H}^V) = \mathbf{n}' \times (\mathbf{H}_0^V - \mathbf{H}_0). \tag{2.74}$$

These allow us to determine the constants and obtain the electric field in vacuum, which can be detected by an EMAT coil. Here, $\mathbf{n}_0$ is an outward unit vector normal to the surface and $\mathbf{n}'$ is the one when the surface is perturbed by the ultrasonic wave. The detailed calculation procedure was given by Ogi (1997).

2.4 Comparison with Measurements

The Lorentz forces are proportional to the applied magnetic field. The magnetization forces are proportional to the magnetization magnitude. Both of them increase monotonically with the applied field. The magnetostriction forces, however, are complicated functions of the field depending on material's magnetostriction response to the field. The above theoretical model can be confirmed by measuring the magnetic field dependence of ultrasonic wave amplitude, because the body forces produce acoustic waves through Eq. (2.5) and are proportional to the generated waves.

2.4.1 SH Plate Wave

Figure 2.6 shows an EMAT configuration to generate a shear-horizontal (SH) plate wave via the magnetostriction mechanism. It consists of a meander-line coil and a static magnetic field along the straight parts of the coil. Neither a Lorentz force nor a magnetization force occurs with this configuration. This type of EMAT was reported first by Thompson (1979). Considering the two-dimensional model in Fig. 2.1b and the stress-free condition $\sigma_4 = 0$ at the surface, it is easy to find that only the magnetostriction force

$$f_y^{(MS)} = \frac{\partial \sigma_6^{(MS)}}{\partial x} = -e_{6x}^{(MS)} \frac{\partial H_x}{\partial x}, \tag{2.75}$$

contributes to the SH wave generation. The field dependence of the wave amplitude is governed by that of the coefficient $e_{6x}^{(MS)} = c_{66} d_{6x}^{(MS)}$. $d_{6x}^{(MS)}$ is given by Eq. (2.66) and c_{66}, a shear modulus, is known; and the coefficient can be calculated as a function of the field using the measurable magnetostriction curve (ε_t-H_0 curve).

For an interstitial-free steel, the field dependence of the SH wave amplitude was measured. As shown in Fig. 2.6, the SH wave was launched by the EMAT and

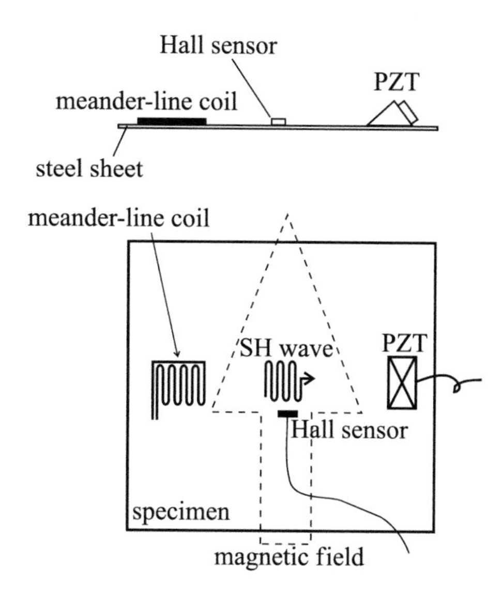

Fig. 2.6 Generation of SH plate wave with a meander-line-coil EMAT and detection of it with a PZT transducer. The tangential biasing field was applied using an electromagnet

detected by a wedge-mounted piezoelectric transducer. A Hall sensor detected the static field. The magnetostriction along the field was measured by a semiconductor strain gauge. We found that the magnetostriction curve can be well approximated by the following function as shown by the solid line in Fig. 2.7a

$$\varepsilon_t = 34H^{0.2} \times 1.48^{-1.5(H+1)} - 5.5, \tag{2.76}$$

where ε_t is in microstrain and H in 10^4 A/m.

Figure 2.7b compares the measurements with the calculated coefficient. They are principally consistent, showing the peak near the maximum magnetostriction, a minimum at the zero magnetostriction at the field of $\sim 2.4 \times 10^4$ A/m, and a gradual increase to high fields. Thus, the theory essentially explains the measurements. There are, however, discrepancies in magnitude between the theory and measurements; the measurements are lower than the calculations in low fields and larger in high fields. This may be attributed to the electromagnetic losses, including the hysteresis loss due to irreversible movement of magnetic domains and the eddy current loss (or the Joule-heating loss). They occur when the ferromagnetic material is dynamically magnetized and are closely related to the susceptibility of the material (Chikazumi 1964): high susceptibility causes high losses. Susceptibility is higher and the losses are more remarkable in the low-field region than in the high-field region.

When the bias magnetic field is applied in an angled direction by ϕ about the straight parts of the meander-line coil (in the y-axis) as shown in Fig. 2.8, similar expressions for the piezomagnetic coefficients to Eqs. (2.65) and (2.66) are

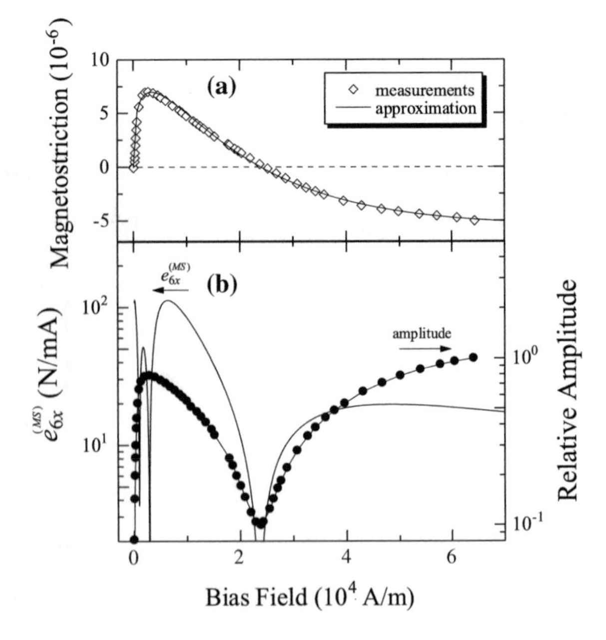

Fig. 2.7 **a** Magnetostriction curve of an interstitial-free steel. *Open marks* denote measurements and the *solid line* is an approximation given by Eq. (2.76). **b** Comparison between the field dependence of the SH plate wave amplitude generated by the EMAT and the calculated piezomagnetic coefficient, $e_{6x}^{(MS)}$

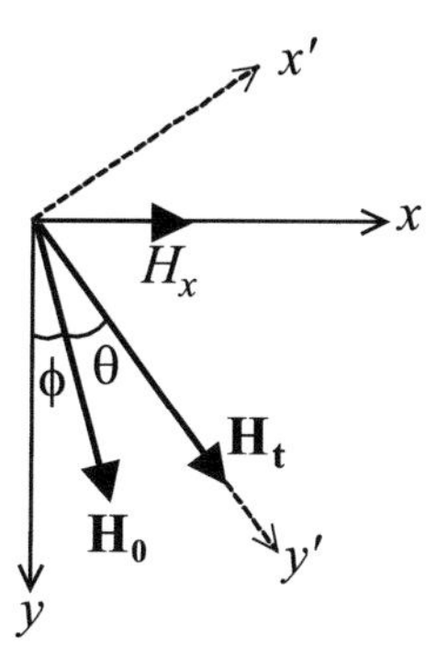

Fig. 2.8 Oblique bias magnetic field in the *y*-axis

obtained by replacing θ with θ + ϕ, H_{0y} with $H_{0y}\cos\phi$, and H_x with $H_x + H_0\sin\phi$. For example,

$$d_{6x}^{(MS)} = \frac{3\gamma}{2}\sin 2(\theta+\phi)\sin(\theta+\phi) + \frac{3\varepsilon_t}{H_{0y}\cos\phi}\cos^2(\theta+\phi)\cos 2(\theta+\phi). \quad (2.77)$$

The resulting SH wave amplitude is proportional to $e_{6x}^{(MS)}$ and its dependence on the inclined angle of the bias magnetic field, ϕ, can be calculated. Figure 2.9a presents such calculations of coefficient $e_{6x}^{(MS)}$, showing a maximum in the range of ϕ = 45°–60°, which depends on the relative magnitude between ε_t and γ. The corresponding measurements are shown in Fig. 2.9b, which supports the calculation. Thus, an

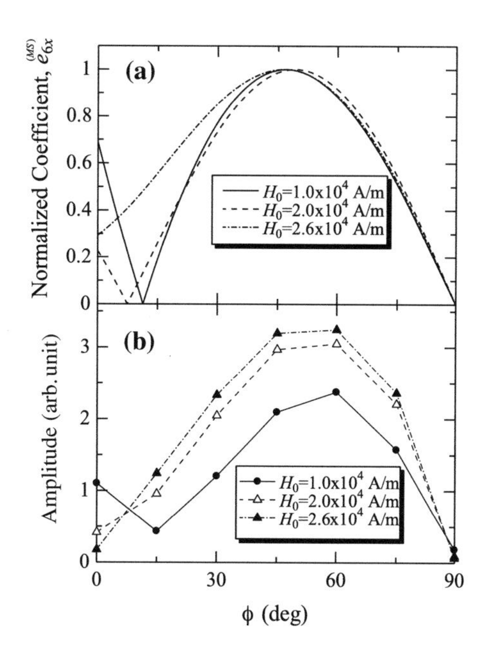

Fig. 2.9 Dependence of **a** the piezomagnetic coefficient $e_{6x}^{(MS)}$ and **b** the SH wave amplitude on the direction of the bias magnetic field. (After Ogi et al. 2003)

oblique static magnetic field is quite effective to increase the efficiency of this type of EMAT as shown in Fig. 3.13.

2.4.2 Bulk Shear Wave

A bulk shear wave can be excited when the tangential field is applied normal to the straight parts of the meander-line coil (i.e., in the x direction in Fig. 2.1b). It propagates perpendicular to the specimen surface with a polarization parallel to the surface. The electromagnetic ultrasonic resonance (EMAR) technique described in Chap. 5 was used to measure the normalized amplitude of the bulk shear wave as a function of the tangential field for a low-carbon steel. The result is shown in Fig. 2.10.

The Lorentz force mechanism will not contribute to the shear wave generation. (It can generate the longitudinal wave, however.) The magnetostriction force that contributes to the bulk shear wave generation is

$$f_x^{(MS)} = \frac{\partial \sigma_5^{(MS)}}{\partial z} = -e_{5z}^{(MS)} \frac{\partial H_z}{\partial z}. \tag{2.78}$$

Thus, the normal component of the dynamic field, H_z, which occurs between the straight parts of the coil, plays an important role for the generation as illustrated in Fig. 2.11. The coefficient $e_{5z}^{(MS)}$ equals $C_{55} d_{5z}^{(MS)}$ and the form of $d_{5z}^{(MS)}$ is essentially the same as that of $d_{6x}^{(MS)}$ in Eq. (2.66), where H_{0y} has to be replaced by H_{0x}. Figure 2.10 shows the calculation of the coefficient using the material's magnetostriction response. Again, the measurements and calculation are compared favorably with each other.

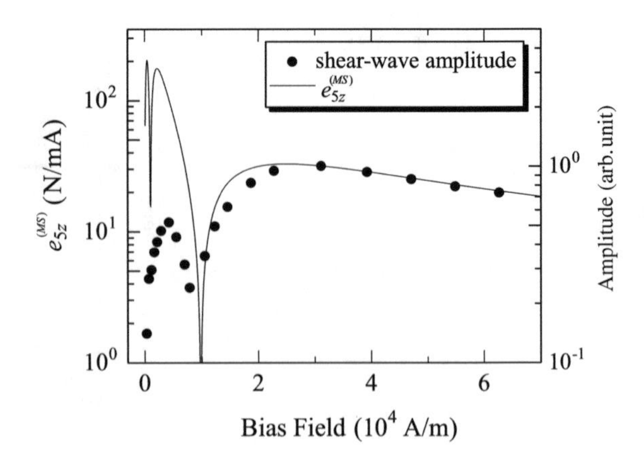

Fig. 2.10 Field dependence of the shear wave amplitude and that of the piezomagnetic coefficient, $e_{5z}^{(MS)}$

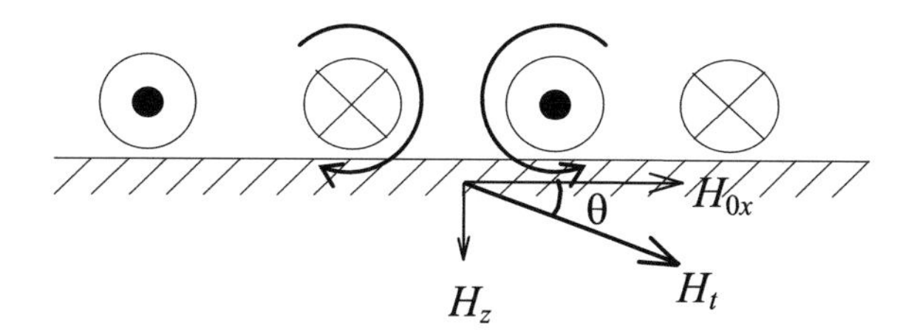

Fig. 2.11 Inclined total magnetic field about the static field direction by θ to cause shear wave traveling in the *z* direction

References

Auld, A. B. (1973). *Acoustic Fields and Waves in Solids*. New York: Wiley.

Beissner, R. E. (1976). Electromagnetic-acoustic transducers: A study of the state-of-the-art. *Southwest Research Institute*, Report NTIAC-76–1.

Brown, W. F., Jr. (1966). *Magnetoelastic Interactions* (p. 57). Berlin: Springer.

Chikazumi, S. (1964). *Physics of Magnetism*. New York: Wiley-Interscience.

Gaerttner, M. R., Wallace, W. D., & Maxfield, B. W. (1969). Experiments relating to the theory of magnetic direct generation of ultrasound in metals. *Physical Review, 184*, 702–704.

Hammond, P., & Sykluski, J. K. (1994). *Engineering Electromagnetism Physical Processes and Computation* (pp. 155–165). Oxford: Oxford Science Publications.

Il'in, I. V., & Kharitonov, A. V. (1981). Theory of the EMA method of detecting Rayleigh waves for ferromagnetic and ferrimagnetic materials. *Soviet Journal of Nondestructive Testing-USSR, 16*, 549–554.

Kawashima, K. (1976). Experiments with two types of electromagnetic ultrasonic transducers. *The Journal of the Acoustical of America, 60*, 365–373.

Kawashima, K. (1985). Electromagnetic acoustic wave source and measurement and calculation of vertical and horizontal displacements of surface waves. *IEEE Transaction on Sonics and Ultrasonics, SU-32*, 514–522.

Maxfield, B. W., & Fortunko, C. M. (1983). The design and use of electromagnetic acoustic wave transducers (EMATs). *Materials Evaluation, 41*, 1399–1408.

Moon, F. C. (1984). *Magneto-Solid Mechanics*. New York: Wiley-Interscience.

Ogi, H. (1997). Field dependence of coupling efficiency between electromagnetic field and ultrasonic bulk waves. *Journal of Applied Physics, 82*, 3940–3949.

Ogi, H., Goda, E., & Hirao, M. (2003). Increase of efficiency of magnetostriction SH-wave EMAT by angled bias field: Piezomagnetic theory and measurement. *Japanese Journal of Applied Physics, 42*, 3020–3024.

Ogi, H. (2012). Erratum: "Field dependence of coupling efficiency between electromagnetic field and ultrasonic bulk waves" [*J. Appl. Phys*. **82**, 3940 (1997)]. *Journal of Applied Physics, 112*, 059901.

Ribichini, R., Nagy, P. B., & Ogi, H. (2012). The impact of magnetostriction on the transduction of normal bias field EMATs. *NDT & E International, 51*, 8–15.

Seher, M., & Nagy, P. B. (2016). On the separation of Lorentz and magnetization forces in the transduction mechanism of Electromagnetic Acoustic Transducers (EMATs). *NDT & E International, 84*, 1–10.

Thompson, R. B. (1973). A model for the electromagnetic generation and detection of Rayleigh and Lamb waves. *IEEE Transaction on Sonics and Ultrasonics, SU-20*, 340–346.

Thompson, R. B. (1977). Mechanism of electromagnetic acoustic generation and detection of ultrasonic Lamb waves in iron-nickel alloy polycrystals. *Journal of Applied Physics, 48*, 4942–4950.

Thompson, R. B. (1978). A model for the electromagnetic generation of ultrasonic guided waves in ferromagnetic metal polycrystals. *IEEE Transaction on Sonics and Ultrasonics, SU-25*, 7–15.

Thompson, R. B. (1979). Generation of horizontally polarized shear-waves in ferromagnetic materials using magnetostrictively coupled meander-coil electromagnetic transducers. *Applied Physics Letters, 34*, 175–177.

Thompson, R. B. (1990). Physical principles of measurements with EMAT transducers. In *Physical Acoustics* (Vol. 19, pp. 157–200). New York: Academic Press.

Wilbrand, A. (1983). EMUS-probes for bulk waves and Rayleigh waves. Model for sound field and efficiency calculations. In *New Procedures in Nondestructive Testing* (pp. 71–80). Berlin: Springer.

Wilbrand, A. (1987). Quantitative modeling and experimental analysis of the physical properties of electromagnetic-ultrasonic transducers. In *Review of Progress in Quantitative Nondestructive Evaluation* (Vol. 7, pp. 671–680).

Chapter 3
Available EMATs

Abstract Many EMATs have been developed for bulk waves, surface waves, guided waves, wave focusing, and applications to materials in move and at elevated temperatures. An antenna transmission technique for piezoelectric materials is also proposed. This chapter presents various EMAT structures and their operating principles.

Keywords Antenna transmission · Bulk-wave EMAT · Guided wave · High-temperature measurement · Line-/point-focusing EMAT · Lorentz force · Magnetostriction force · Meander-line coil · Piezoelectric material · PPM EMAT · SV wave

3.1 Bulk-Wave EMATs

Figure 3.1 shows the configuration of a bulk-wave EMAT, which simultaneously generates and detects a shear wave with a polarization parallel to the specimen surface and a longitudinal wave. Both bulk waves propagate perpendicular to the specimen surface. The compact structure, broad bandwidth (0.05–50 MHz), and high transfer efficiency promote its frequent use in many scientific and practical applications. The permanent magnets produce biasing fields normal to the surface under the unidirectional coil elements and the tangential fields at the center and edges of the coil elements. This EMAT functions for exciting both the shear and longitudinal waves in a nonmagnetic material because of the Lorentz force mechanism (Eq. 2.37). However, it generates only the shear wave in a ferromagnetic material; the longitudinal wave is normally too weak to be observed. This is because the Lorentz force for the longitudinal wave generation is almost canceled by the magnetization force (Sect. 2.2.4), and the magnetostriction force for the longitudinal wave generation is much smaller than that for the shear wave (Sect. 2.2.5). (In such cases, this EMAT will be referred as shear-wave EMAT.) Figure 3.2 explains schematically the bulk-wave generation in nonmagnetic and ferromagnetic materials.

M. Hirao and H. Ogi, *Electromagnetic Acoustic Transducers*,
Springer Series in Measurement Science and Technology,
DOI 10.1007/978-4-431-56036-4_3

Fig. 3.1 Appearance (*left*) and constitution (*right*) of a bulk-wave EMAT

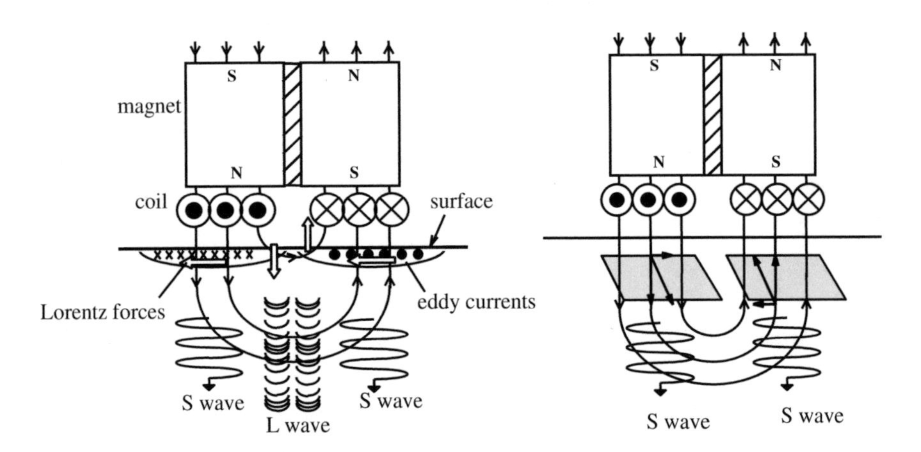

Fig. 3.2 Mechanisms of the bulk-wave generation by the Lorentz forces (*left*) and magnetostriction (*right*)

Figure 3.3 shows appearance and generation mechanism with Lorentz forces of another bulk-wave EMAT. It consists of a single-cylindrical permanent magnet and a flat pancake coil. The static field from the magnet has the radial and the normal components, which interact with the eddy currents and induce the Lorentz forces in the normal and radial directions, ($f_z^{(L)}$ and $f_r^{(L)}$), respectively. (For a ferromagnetic material, the magnetostriction forces appear as well.) These forces generate the longitudinal wave and the radially polarized shear wave both propagating in the thickness direction at the same time, for which this is called dual-mode EMAT. If the metal has an orthorhombic elastic anisotropy, due to the texture and stress, the shear waves decompose into two polarizations in the two principal directions producing the split resonance peaks (see Fig. 12.17). This EMAT is convenient for the measurement of in-plane anisotropy in rolled steel sheets (Chap. 11) and the birefringent acoustoelastic stress measurement (Chap. 12).

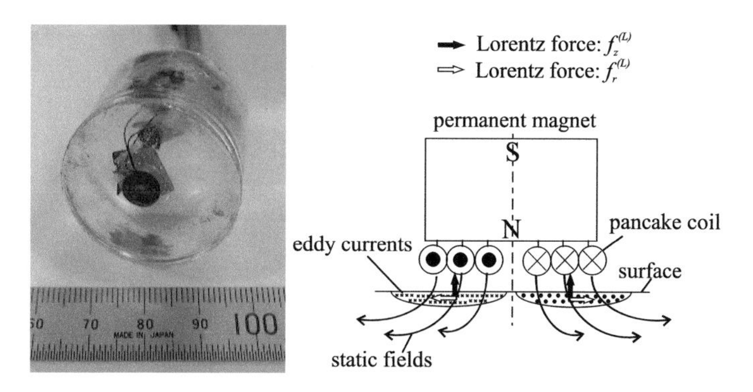

Fig. 3.3 Appearance (*left*) and the Lorentz force mechanism of a dual-mode EMAT (*right*) with pancake coil and a single magnet

For more detailed analysis, a two-dimensional FEM computation was performed to map out the body-force distributions in materials (Ogi et al. 1996b). The region shown in Fig. 3.4 was divided into 46,400 elements by the first-order triangle meshes with 0.01–0.1 mm sides. The magnetization inside the permanent magnets was fixed to 0.5 T. The driving current density was 1.6×10^8 A/m^2, and the frequency was 1 MHz. Conventional discrete method was applied to the diffusion equation for the magnetic vector potential (Segerlind 1984; Ludwig and Dai 1990, 1991). The analysis contained the magnetization and magnetostriction responses of the material to the static field.

Figure 3.5 shows typical results for the Lorentz forces and magnetization forces arising in a low-carbon steel. The z direction magnetization force $f_z^{(M)}$ cancels much of the z direction Lorentz force $f_z^{(L)}$, giving little contribution to the longitudinal wave generation. This is consistent with Eq. (2.42).

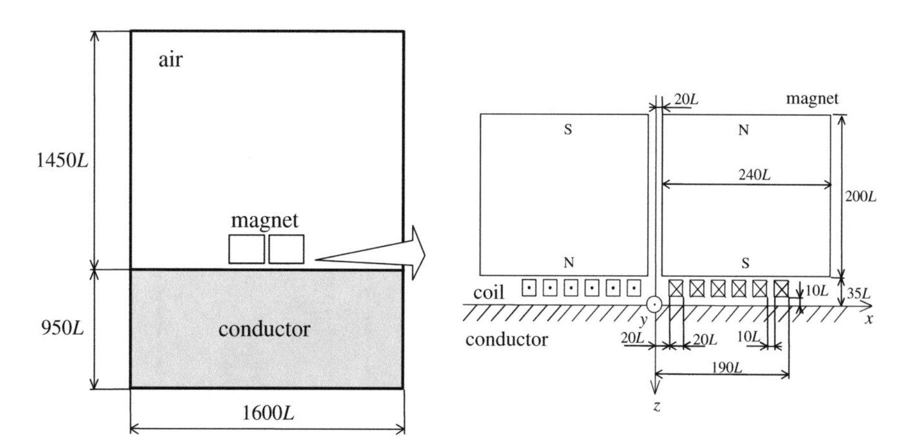

Fig. 3.4 Two-dimensional FEM meshes for dividing the bulk-wave EMAT, vacuum region, and conductive region. L (=0.01 mm) denotes the unit dimension of the first-order triangle mesh

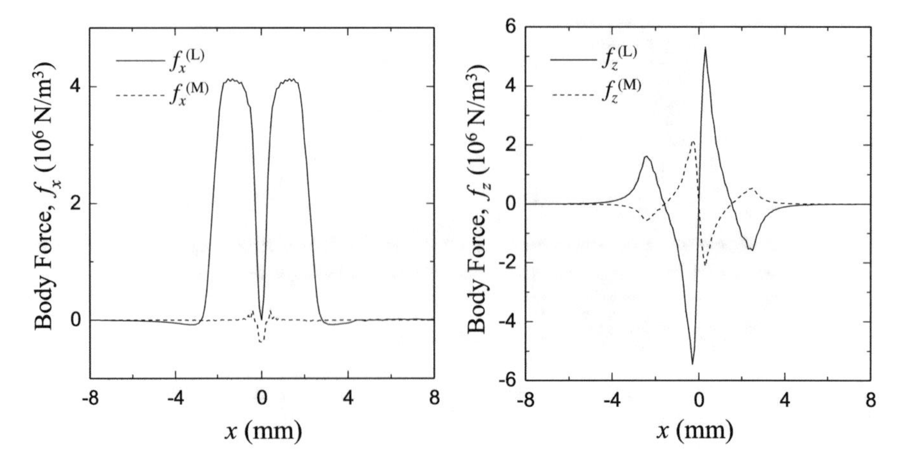

Fig. 3.5 Distributions of the Lorentz forces and magnetization forces in the x direction at $z = 0.005$ mm and $t = 760$ ns

3.2 Longitudinal-Guided-Wave EMAT for Wires and Pipes

Figure 3.6 shows an EMAT to generate and detect longitudinal guided waves along a rod or pipe specimen (Mason 1958; Tzannes 1966; Kwun and Teller 1994; Kwun and Bartels 1998; Yamasaki et al. 1996, 1998). It consists of an electromagnet or permanent magnet to give the biasing field along the specimen's axis and a solenoid coil to cause the dynamic field superimposed on the biasing field. The magnetostriction mechanism is dominant. Taking the x-axis in the radial direction and the z-axis in the longitudinal direction of the specimen, the piezomagnetic coefficients take the same form as Eq. (2.57), and the governing magnetostrictive force is obtained as follows:

$$f_x^{(MS)} = \frac{\gamma}{2}(c_{11} - c_{12})\frac{\partial H_z}{\partial x}, \quad (3.1)$$

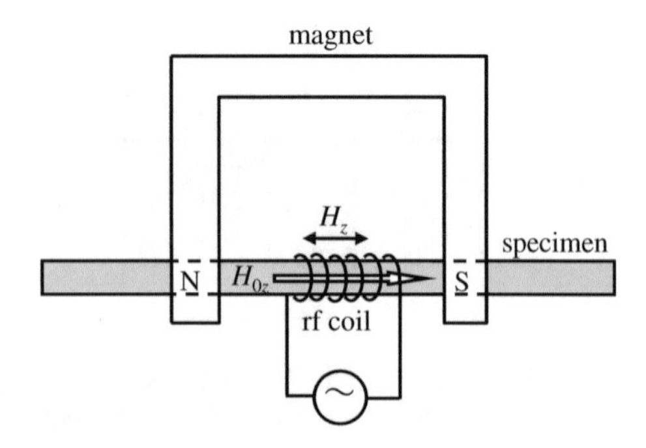

Fig. 3.6 Magnetostrictive EMAT for longitudinal waves along wires and pipes

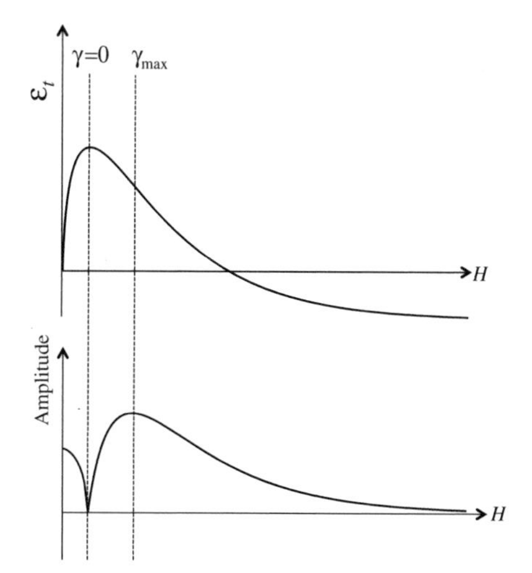

Fig. 3.7 Magnetostriction curve for a steel (*upper*) and excited longitudinal wave amplitude (*below*)

$$f_z^{(MS)} = -\gamma(c_{11} - c_{12})\frac{\partial H_z}{\partial z}. \tag{3.2}$$

The generation efficiency is thus determined by γ, the slope of the magnetostriction curve, as demonstrated by Yamasaki et al. (1998) and illustrated in Fig. 3.7. The maximum amplitude of the launched wave occurs at the maximum absolute value of γ, and the minimum amplitude occurs at the maximum magnetostriction.

The longitudinal wave generated by this EMAT travels along the wire as long as 100 m or more. Low damping rate stems from the low operating frequencies in the submegahertz range and the guided wave nature that the wave energy is confined to within the wire. The generated signal is made up mainly with the (nondispersive) fundamental longitudinal mode with the nominal speed of $\sqrt{E/\rho}$, where E is Young's modulus and ρ is the mass density. Long-range inspection will then be possible to detect large precipitation particles and flaws in wires in a production circumstance. This EMAT, however, generates the waves symmetrically in both sides, which makes the signal interpretation complicated. Unidirectional generation with amplification is possible by arranging an array of generation coils with prescribed spaces (Yamasaki et al. 1998). Experimental setup of Fig. 3.8 mechanizes such an idea, which consists of an array of transmitter coils, a single receiver coil, and a multiple-delayed pulser. The steel wire is 2 m long and 5.6 mm in diameter. Top trace of Fig. 3.9 is the received signal when channel 2 is excited by a square pulse. It is composed with the direct signal to the receiver and a signal propagated first to the left and reflected at the end. Middle is the trace when channel 1 is excited with the same pulse, but of reversed polarity and a delay. These two waveforms are overlapped in the bottom trace to cancel out the left-propagating signal and amplify the right-propagating signal. This is accomplished by adjusting the transmitter coil

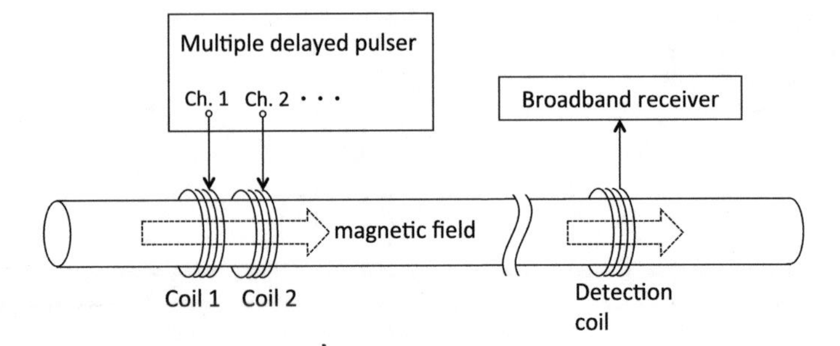

Fig. 3.8 Arrangement for unidirectional generation of longitudinal wave

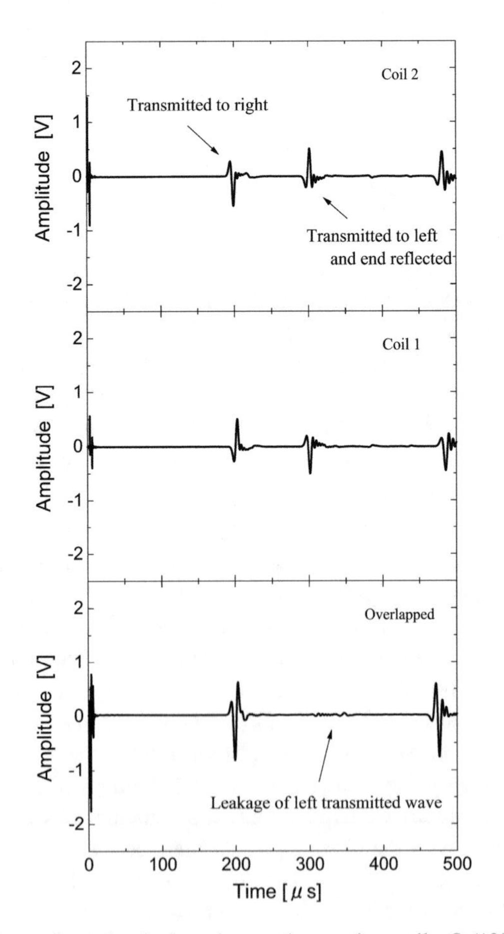

Fig. 3.9 Received waveforms by single and arrayed transmitter coils. © (1998) IEEE. Reprinted, with permission, from Yamasaki et al. (1998)

space to a quarter wavelength of the excited longitudinal wave and delaying channel 1 excitation by a quarter period relative to channel 2 excitation. The rear half of channel 2 signal is superimposed in phase by the leading half of channel 1 signal. We observe some residue, which is attributable to the asymmetric magnetostrictive response to the opposite polarities of exciting pulses at rather high voltages. Multiplying the number of transmitter coil pairs intensifies the longitudinal wave amplitude, leading to the long-range inspection at high S/N ratio.

3.3 PPM EMAT

Figure 3.10 illustrates a periodic permanent magnet EMAT (PPM EMAT) for generating shear waves polarized parallel to the specimen surface in both sides. The shear waves are launched not only along the surface but also into the material (Thompson 1990). The EMAT has been often used for measurements with the surface or plate SH waves.

The polarity of the permanent magnets is alternately varied to cause the tangential Lorentz force and magnetostriction force whose directions alternately change with the same period of the magnet array. Thinner magnets are required to generate higher-frequency SH waves, which are preferable for nondestructive inspection of surface flaws. The maximum transfer efficiency occurs when the wavelength is tuned to the spacing of magnets. However, the efficiency drastically decreases with the magnet thinness, and magnets thicker than 1 mm are normally used; that is, the available shear wave frequency is lower than 1.6 MHz for common steels. Thus, a disadvantage of this EMAT is its operation limited to low frequencies.

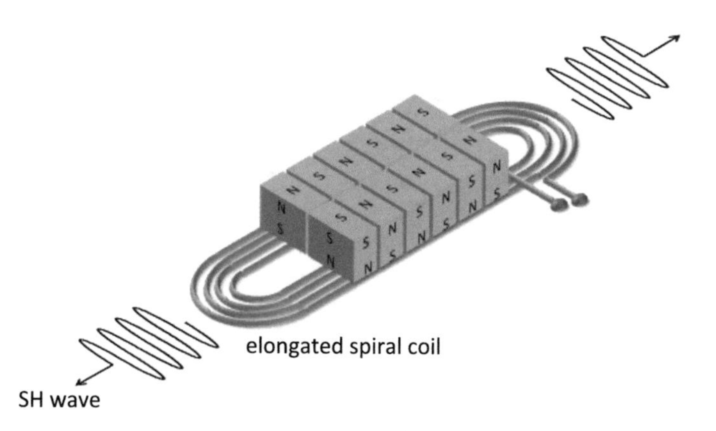

Fig. 3.10 Appearance of the periodic permanent magnet EMAT (PPM EMAT) for SH wave

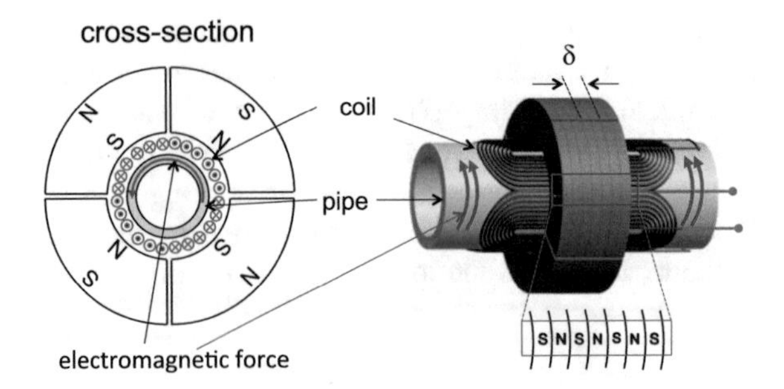

Fig. 3.11 Appearance of the torsional-wave EMAT for pipes and rods

3.4 EMAT for Torsional Wave along Pipes

The above geometry of PPM EMAT can be modified to generate and receive torsional guided waves propagating along pipes and rods as shown in Fig. 3.11 (Nurmalia et al. 2013). It consists of four elongated spiral coils and the surrounding periodic magnets so that the electromagnetic forces arise in the circumferential direction on the cross section for generating the torsional waves. The permanent magnets are arranged so as to indicate the periodically alternating magnetization normal to the specimen surface. The period of the magnet array is equal to the wavelength of the generated torsional waves, δ.

3.5 Meander-Line-Coil SH-Wave EMAT

Tangential biasing magnetic field to surface and a meander-line coil work for the surface SH waves as discussed in Sect. 2.4.1. This is called meander-line coil SH-wave EMAT (Fig. 3.12). The SH-wave frequency is governed by the meander-line period,

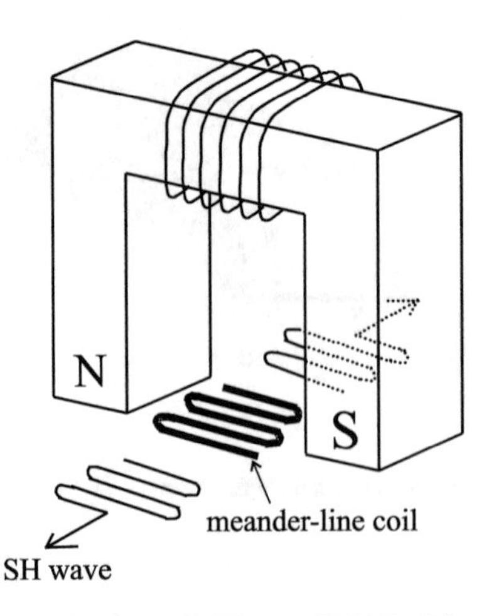

Fig. 3.12 Structure of the meander-line coil SH wave EMAT. (After Ogi et al. 2003a)

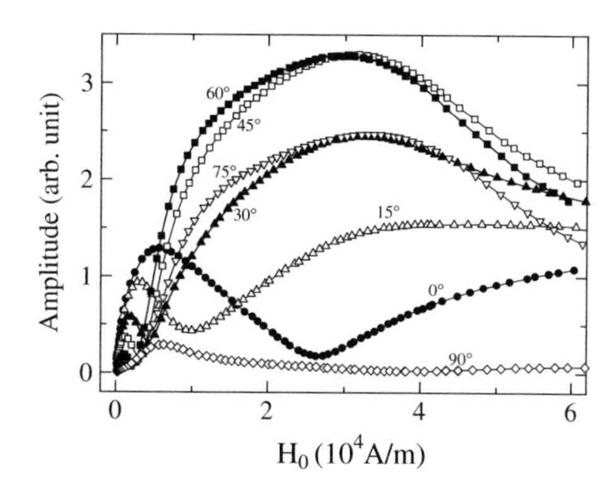

Fig. 3.13 Field dependence of the SH wave amplitude with oblique bias magnetic fields relative to the meandering coil. (After Ogi et al. 2003a)

which can be controlled by fabricating the coil by a printed circuit technique as will be described in Sect. 4.1. The advantage is then capability of generating surface SH waves of higher frequencies.

A disadvantage is the necessity of very large biasing magnetic field when the static magnetic field is applied along the straight lines of the meander-line coil. However, using an oblique biasing magnetic field, the efficiency can be improved as demonstrated in Sect. 2.4.1. Figure 3.13 shows the measurements of the field dependence of the SH wave amplitude with various bias field directions ϕ for the interstitial-free steel used in Sect. 2.4.1. By applying an oblique bias field, the SH wave amplitude increases to a large extent even at low fields (Ogi et al. 2003a). The most efficient generation was achieved by the bias field angled by $\phi = 45°–60°$ and with magnitude around 3×10^4 A/m; the maximum amplification by a factor 2.5 is then possible relative to the aligned field of $\phi = 0$ (classical case).

3.6 SH-Wave EMAT for Chirp Pulse Compression

A potential application of the meander-line coil SH-wave EMAT (Fig. 3.12) is the flaw detection in ferromagnetic metal sheets. Like other EMATs, it can operate on moving sheets for online inspection. But one will face a dilemma when he/she intends to enhance the signal intensity and the temporal resolution at the same time. Being excited by an impulse, the EMAT launches a wave train of constant amplitude. The carrier wavelength equals the meander-line period. The same EMAT receives the reflected SH wave. The received signal has a trapezoidal envelope after a coherent superposition of the currents induced at individual straight parts of the coil. The maximum amplitude is then proportional to the number of turns in the meander-line coil. Long coil of many turns produces high amplitude, but it results in the elongated received signal, deteriorating the temporal resolution.

The dilemma can be solved by generating the SH wave with a meander-line coil, whose period changes linearly in the propagation direction, and by detecting the reflected SH wave with another meander-line coil having the reversed period

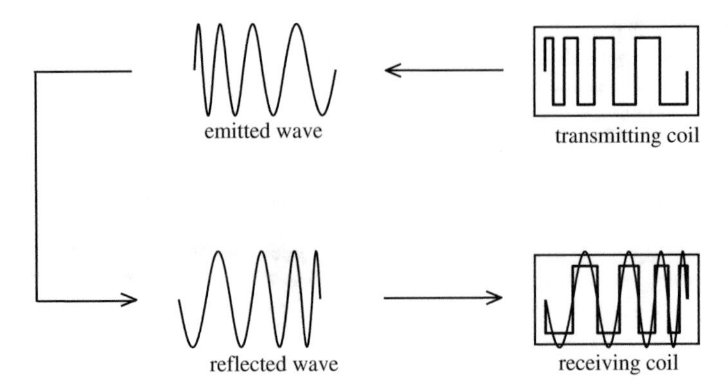

Fig. 3.14 Mechanism of chirp pulse compression with meander-line coils of variable period

variation of the generating coil. This configuration results in a compressed pulse with high amplitude. Operation mechanism is based on the selective sensitivity that the meander-line period equals the wavelength of the excited SH wave, and also, the coil picks up the SH wave having the same wavelength as the period. When an impulse is fed to the coil, the generator coil emits an SH wave package, in which the wavelength varies continuously from the head to the tail, that is, a *chirp* signal (see Fig. 3.14). Narrow spacing generates higher frequencies, and wide spacing generates lower frequencies. After the signal is reflected at the discontinuity (flaw, edge of the sheet, etc.), the echo heads back to the receiver coil. Unless the wave is dispersive, all the frequency components arrive at the matching coil elements with a common delay time. In other words, the coincidence between the wavelength and the coil period occurs only at a time over the whole length of the receiving coil. At this moment, a current is induced in the coil, making a sharp pulse.

Figure 3.15 shows such a meander-line coil, where the period varies linearly from 0.38 to 1.5 mm over 57 turns of 65 mm length. Two identical meander-line wires are precisely put together at reversed positions using the printing-circuit technique, one for excitation and the other for reception. Insulator polyimide sheets are placed on the top, bottom, and between them. Figure 3.16a is a linearly swept FM-chirped signal generated by one of these meander-line coils and received by a conventional wedge-mounted PZT transducer for 1-mm-thick nickel plate. A square impulse of 0.5 μs width was used for excitation. We observe the lower frequencies at the leading part and higher frequencies at the rear part; the duration is 18 μs. Figure 3.16b is the reflected signal from the sheet edge, for which the coils of variable period were used for generation and reception. The received signal is compressed to 5-μs duration.

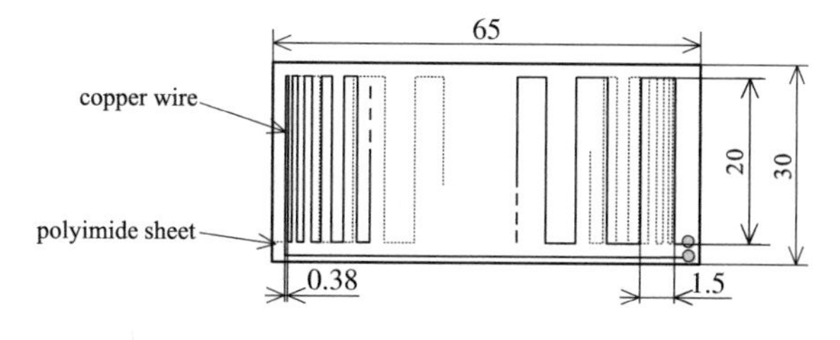

Fig. 3.15 Meander-line coils of variable period for making chirp signal of SH wave

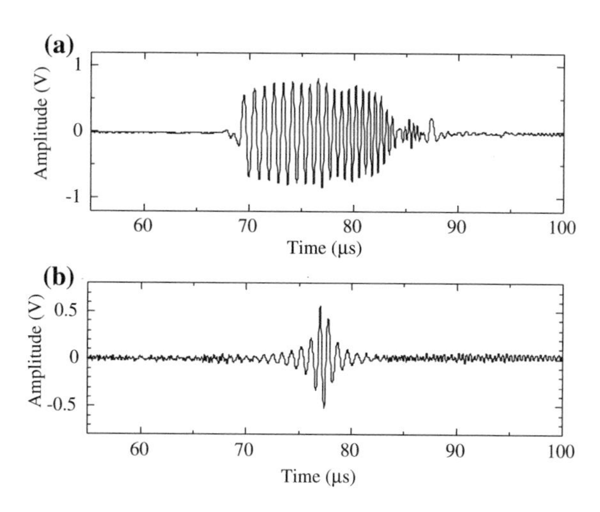

Fig. 3.16 Reflected signals of FM-chirped SH_0-mode received by **a** PZT transducer and **b** variable period coil EMAT

Care should be taken to avoid the influence of the higher modes of SH plate modes (Auld 1973). The chirp signal contains frequency components of certain bandwidth. If the higher modes take place, they become a noise to disturb the inspection. It is then advisable to set the bandwidth below or near the cutoff frequency of the first higher mode of SH wave in a sheet.

The technique of chirp pulse compression is a standard signal processing and has been utilized in many cases of sensing and imaging (Kino 1987; Murray et al. 1997; Iizuka 1998; Iizuka and Awajiya 2014; Anastasi and Madaras 1999). Incorporation in the EMAT measurement, however, has not been explored much, although such a compression processing is required to compensate for the low transfer efficiency.

3.7 Axial-Shear-Wave EMAT

An axially polarized shear wave, called *axial shear wave*, travels in the circumferential direction along a cylindrical surface of a circular rod or pipe specimen. Figure 3.17 shows the magnetostrictively coupled EMAT designed for the axial shear wave in ferromagnetic materials (Ogi et al. 1996a, 2001). It consists of a solenoid coil to supply the biasing magnetic field in the axial direction and a meander-line coil surrounding the cylindrical surface to induce the dynamic field in the circumferential direction. The total field oscillates about the axial direction at the same frequency as the driving currents and produces a shearing vibration through the magnetostriction effect, resulting in the axial shear wave excitation. The same coil works as a receiver through the reverse mechanism.

For a nonmagnetic material, the axial shear wave can be generated by the Lorentz force mechanism using permanent magnets arranged with radial polarity of alternating sign from one magnet to the next and a solenoid coil surrounding the cylindrical surface as shown in Fig. 3.18 (Johnson et al. 1994).

Axial shear wave is useful for resonance measurements focused on the outer regions of cylinders (Johnson et al. 1994; Hirao et al. 2001). Driving the meander-line coil with long RF bursts causes interference among the axial shear waves traveling around the surface. A frequency scan can pick up the resonance frequencies at which all of them overlap coherently and large amplitudes occur. Theoretical resonance frequencies can be calculated by solving the equation of motion with the traction-free boundary condition on a cylindrical surface. For a rod of radius R, the characteristic equation takes the form (Johnson et al. 1994)

$$nJ_n(kR) - kRJ_{n+1}(kR) = 0. \tag{3.3}$$

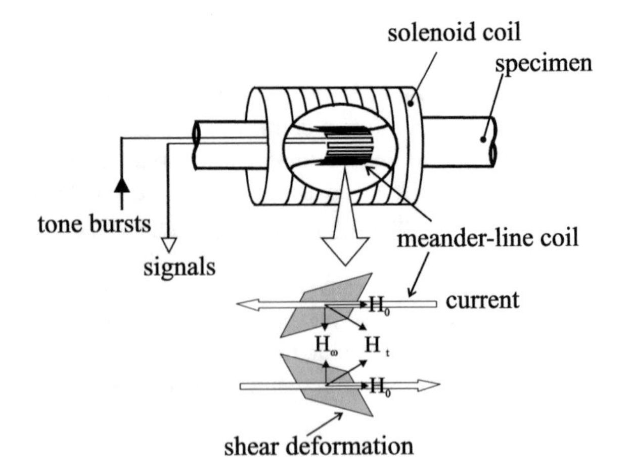

Fig. 3.17 Axial-shear-wave EMAT consisting of a solenoid coil and a meander-line coil surrounding the cylindrical surface. The magnetostriction mechanism causes the axially polarized surface SH wave. Reprinted with permission from Ogi et al. (2001). Copyright (2001), AIP Publishing LLC

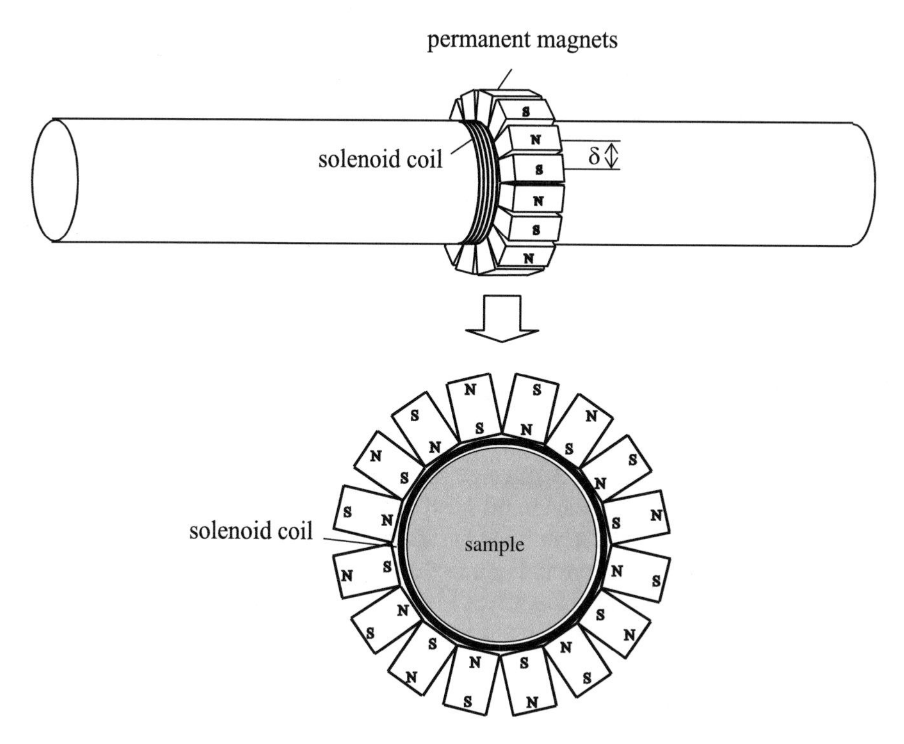

Fig. 3.18 Axial-shear-wave EMAT consisting of periodic permanent magnets with the alternating radial polarity and a solenoid coil surrounding the cylindrical surface. The Lorentz force mechanism causes an axially polarized surface SH wave

Here, J_n denotes the nth Bessel function of the first kind, $k = \omega/c$ the wavenumber, ω the angular frequency, and c the shear wave velocity. Integer n is the number of turns around the circumference determined by

$$n \approx 2\pi R/\delta, \tag{3.4}$$

with the meander-line period δ. Equation (3.3) provides a series of resonance frequencies, $f_m^{(n)}$, with another integer m to assign overtones ($m = 1, 2, 3, \ldots$). Figure 3.19 shows the calculated ultrasonic power for the first three resonance modes when $n = 49$ and $R = 7$ mm. The wave energy of the fundamental mode is concentrated to the outer surface region, and as the mode becomes a higher overtone, the peak amplitude moves to the inside. It is then possible to evaluate the radial gradient using different resonance modes as demonstrated in Chaps. 13 and 16.

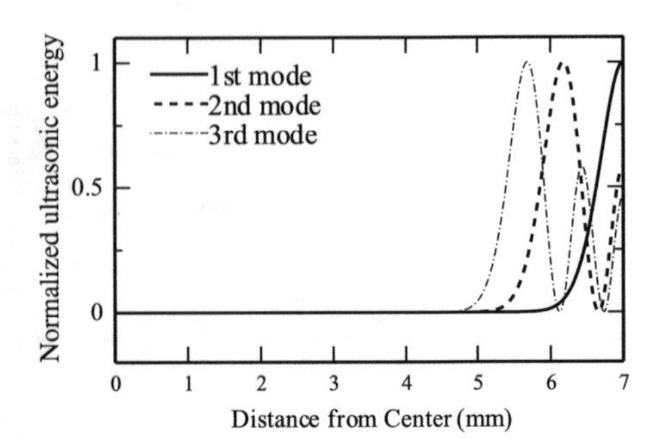

Fig. 3.19 Radial distribution of the axial shear wave energy for the lowest three modes at $n = 49$ in a steel rod with 14 mm diameter. (After Ogi et al. 2000)

3.8 SH-Wave EMAT for Resonance in Bolt Head

The EMAT geometry of Fig. 3.18 can also be modified for generation and reception of the axial shear resonance in the head of a bolt (Fig. 3.20). It consists of thin permanent magnets arrayed to provide the alternating bias fields normal to the side faces, and the solenoid coil wound in a hexagonal shape. When the driving current is applied to the coil, the shearing forces in the axial direction arise periodically and generate the shear waves propagating in the circumferential direction with the polarization along the bolt axis.

When the specimen is a cylindrical rod, the resonance modes are easily determined by Eq. (3.3) and it is revealed that the shear-wave propagates near the outer surface region at a lower resonance mode, and as the mode becomes higher, the vibration penetrates inside. This property essentially holds in the case of the hexagonal rod, although the resonance modes cannot be explicitly obtained with a simple frequency equation. A two-dimensional FEM calculation was then made to determine the resonance frequencies and the vibration distribution for individual modes (Ogi and Hirao 1998). Details will be given in Sect. 12.5.3.

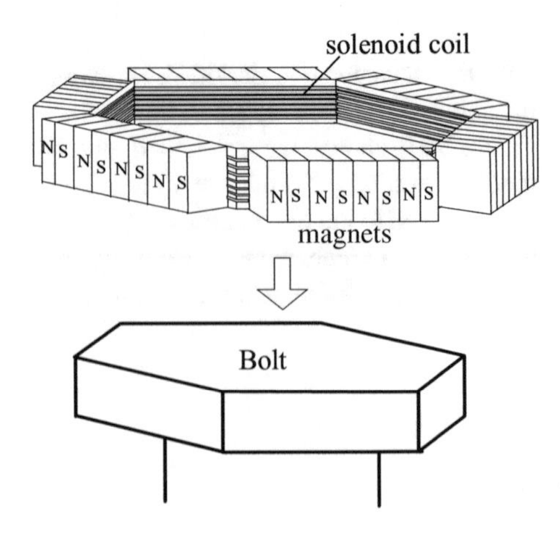

Fig. 3.20 Perspective view of the EMAT for generating axial shear waves in a bolt head

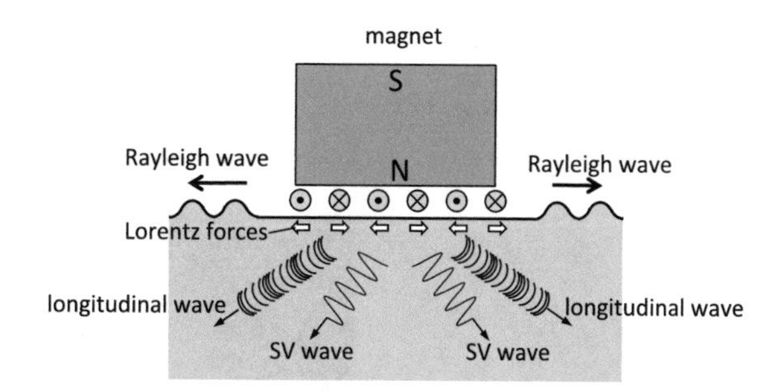

Fig. 3.21 Rayleigh-wave EMAT consisting of a permanent magnet and a meander-line coil

3.9 Rayleigh-Wave EMAT

A normal biasing field and a meander-line coil as shown in Fig. 3.21 induce the Lorentz force (and magnetostriction force) parallel to the surface, whose directions change alternately with the meandering period. Such a body-force distribution generates Rayleigh waves traveling along the surface; and simultaneously, the longitudinal waves and shear-vertical (SV) waves traveling obliquely into the specimen. Thus, a disadvantage of this EMAT is difficulty of the mode identification in the observed echoes. This type of EMAT was used to detect cracks in railroad wheels (Sect. 18.2).

3.10 Line- and Point-Focusing EMATs

The elastic waves excited by some EMATs propagate in nearly all possible directions in a solid, which is undesirable for flaw detection purposes. Typical case occurs in using the EMAT shown in Fig. 3.21, which generates the longitudinal wave and the SV wave propagating obliquely in the solid and the Rayleigh wave along the surface. They are simultaneously generated for both sides with broad directivity patterns and will be reflected not only by flaws but also by the sample edges to overlap on a flaw echo. It is often difficult to distinguish the flaw signal from others. Moreover, the broad radiation decreases the ultrasonic energy to be concentrated on the target flaw. To cope with this, a specially designed array of the meander-line elements allows focusing the SV waves on a line inside the material. This is called line-focusing EMAT (LF EMAT) (Ogi et al. 1998, 1999).

Design of an LF EMAT coil makes use of the sharp radiation pattern of the SV wave radiated from a single line source on the surface. Thus, we can concentrate the SV wave beams having incident angles centered around this angle to a focal line. Figure 3.22 shows a configuration of the LF EMAT, which consists of a permanent magnet block to supply the bias field normal to the surface and a meander-line coil to induce eddy currents and the dynamic fields in the surface region. The meander-line spacing is not uniform (Figs. 3.22 and 3.23). Shearing forces arise

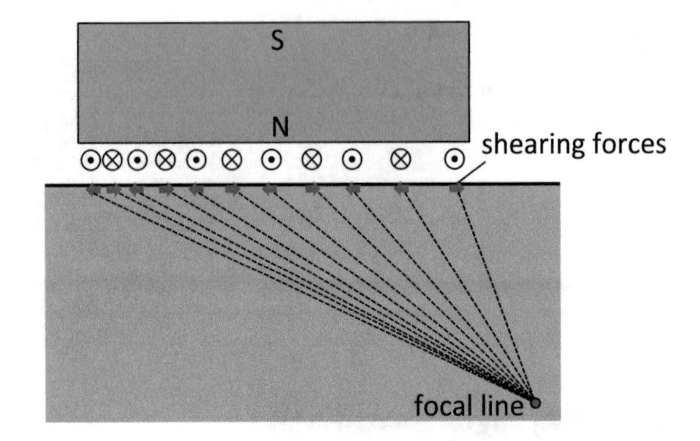

Fig. 3.22 Focusing of the SV waves on a focal line in solid by the line-focusing EMAT

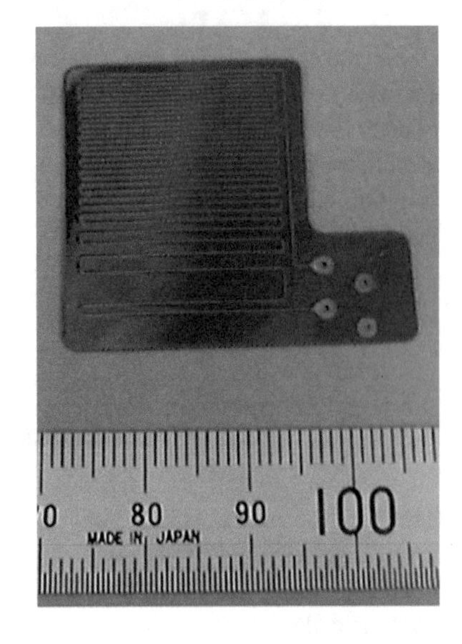

Fig. 3.23 Meander-line coil with nonuniform spacing fabricated by a printed circuit technique

parallel to the surface under the coil segments, which are caused by the Lorentz force mechanism in a nonmagnetic metal and also by the magnetostriction force mechanism in a ferromagnetic metal.

A single line source on the surface, oscillating parallel to the surface and normal to the line itself, radiates the SV and longitudinal waves traveling obliquely in the two-dimensional half space. The amplitude directivity of the SV wave has a sharp peak at the angle around 30°, depending on Poisson's ratio, from the normal direction to the surface as shown in Fig. 3.24, whereas that of the longitudinal wave is rather broad (Miller and Pursey 1954). This angle corresponds to the critical angle of the longitudinal wave, at which all the excitation energy is directed toward the SV-wave generation. The SV-wave amplitude drops rapidly beyond this angle. We can create an intense energy concentration on a line, *focal line*, in the sample by placing the line sources approximately at 30° on the surface so that all the radiated SV waves arrive on this line with the same phase.

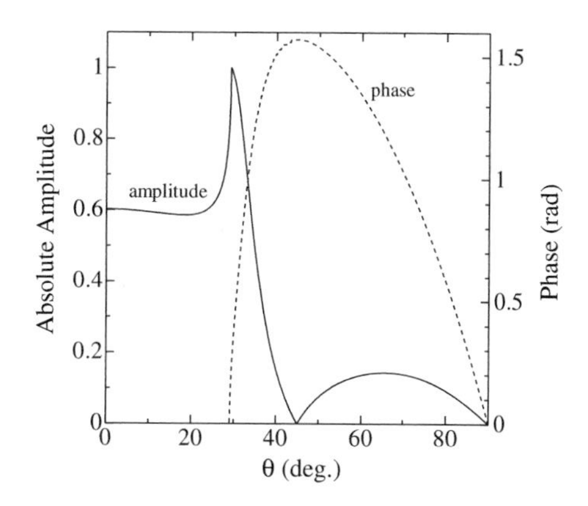

Fig. 3.24 Amplitude and phase directivity of the SV wave launched by a single line source oscillating parallel to the surface and normal to the line. Aluminum is assumed for this calculation.

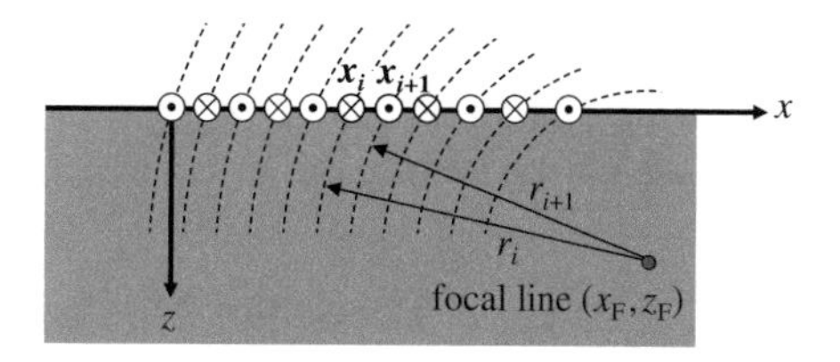

Fig. 3.25 Location of the coil segments to focus the SV waves on the focal line

Figure 3.25 shows locations of the coil segments (Ogi et al. 1999). For a given set of parameters (the frequency f, the location of the focal line (x_F, z_F), and the shear wave velocity c), the location of (i + 1)th segment, x_{i+1}, is determined from the relationship that the difference between propagation paths from the ith and (i + 1)th sources to the focal line is equal to a half wavelength:

$$r_i - r_{i+1} = \frac{c}{2f}. \tag{3.5}$$

This requires the coil segments to be located at the intersections of the surface and the concentric circles centered at the focal line, whose radii are stepped at every half wavelength. Such a coil can be fabricated by printed circuit technique with 1-μm accuracy (Fig. 3.23). Details will be given in Chap. 14.

The same focusing principle in Fig. 3.25 and Eq. (3.5) is also applicable for a curved meander-line coil, leading to the point-focusing EMAT (PF-EMAT). Figure 3.26a shows a pair of PF-EMAT coils for generation and detection. Providing the normal magnetic field to the surface with permanent magnets, the SV wave generated by each partial-arc source converges to the focal point inside the material (Fig. 3.26b). When a defect is present there, the scattered SV waves are coherently received by the detection coil, giving rise to a signal of high signal-to-noise ratio. This PF-EMAT allows us to detect very small defects (≲50 μm height) in an austenitic stainless steel as demonstrated in Sect. 14.4.2.

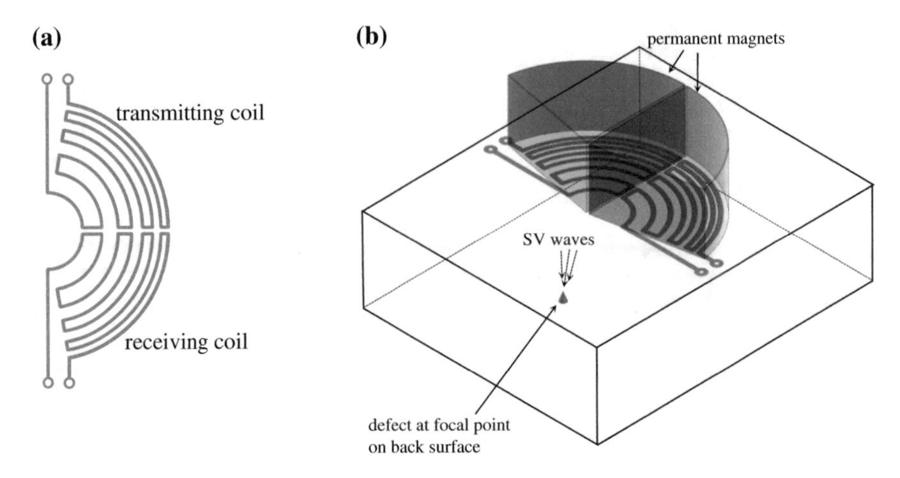

Fig. 3.26 **a** Arched meander-line coils for generation and detection of SV waves for the point-focusing EMAT (PF-EMAT). **b** A pair of permanent magnets provide the normal magnetic field, leading to the focusing SV waves at the focal point. (After Takishita et al. 2015)

3.11 Trapped-Torsional-Mode EMAT

Trapping of vibrational energy in a region of a solid has been a subject of considerable importance in two fields: (i) the designing of electronic oscillators and (ii) the study of internal friction of materials. Electronic resonators must be mechanically supported at their edges, and if vibrational amplitudes are significant at the points of support, there is a leakage of acoustic energy into the surrounding structure, which causes a decrease in Q of the device. Studies of internal friction seek to obtain information on the dynamics and symmetries of anelasticity of a material such as point defects, dislocations, and magnetic domains. In such studies, the energy loss arising from mechanical support is often so large that it obscures the internal friction of the material. Johnson et al. (1996) presented an EMAT to trap torsional vibration energy at a central region of a cylinder that has a slightly larger diameter over the central region as shown in Fig. 3.27. They also presented an approximate theory to explain the measurements.

The EMAT consists of magnets arranged with radial polarity and alternating sign from one magnet to the next and a meander-line coil surrounding the cylinder as illustrated in Fig. 3.28. Alternating currents in the coil result in Lorentz forces in the circumferential direction at the cylinder surface, thus coupling to torsional vibrations of the cylinder.

The circumferential displacement u_θ of a torsional mode in an isotropic cylinder rod with a constant diameter takes the form, in a cylindrical coordinate system (r, θ, z) (Eringen and Şuhubi 1975),

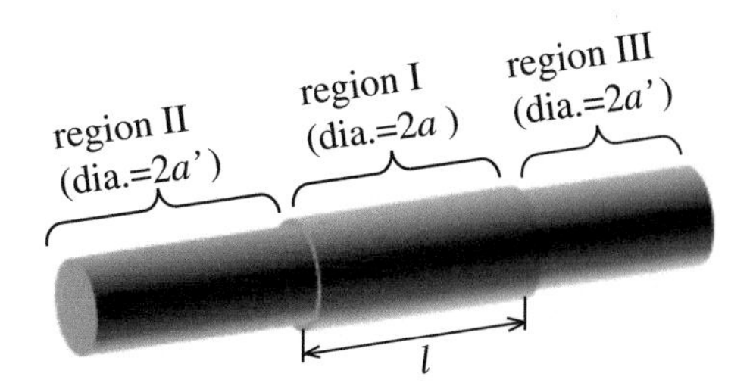

Fig. 3.27 Solid circular cylinder of total length L with radius a over a central region of length l (*region I*) and with radius a' elsewhere (*regions II* and *III*)

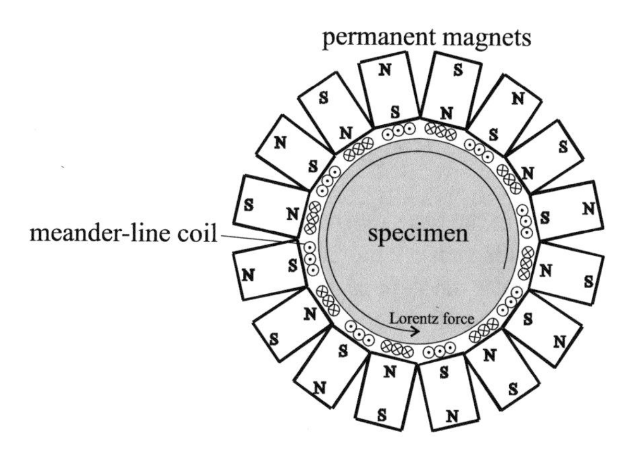

Fig. 3.28 Schematic cross section of the torsional-mode EMAT consisting of permanent magnets with alternating radial polarity and a meander-line coil. Reprinted with permission from Johnson et al. (1996). Copyright 1996, Acoustical Society of America

$$u_\theta = \frac{a}{\eta_0} J_1(\eta_0 r/a)\left(C_1 e^{j\kappa z/a} + C_2 e^{-j\kappa z/a}\right) e^{-j\omega t}, \tag{3.6}$$

where η_0 and κ are dimensionless wavenumbers, a denotes the rod radius, J_1 denotes the Bessel function of the first order, and C_1 and C_2 are constants. The boundary condition $\sigma_{r\theta} = 0$ at $r = a$ requires that $J_2(\eta_0) = 0$, and this equation has an infinite number of solutions: $\eta_0 = 0$, 5.136, 8.417, 11.62,... The dispersion relation for a given κ and η_0 is as follows:

$$\kappa^2 = \left(\frac{\omega a}{v_s}\right)^2 - \eta_0^2, \tag{3.7}$$

or for nonzero values of η_0

$$\kappa^2 = \eta_0^2\left[\left(\frac{\omega}{\omega_c}\right)^2 - 1\right]. \tag{3.8}$$

Here, v_s denotes the plane-wave shear velocity and $\omega_c = v_s\eta_0/a$ is the cutoff frequency. (Frequencies below ω_c provide imaginary values of κ and nonpropagating modes, which decay with distance from the origin along the z-axis.) The traction force on surfaces perpendicular to the z-axis is given by

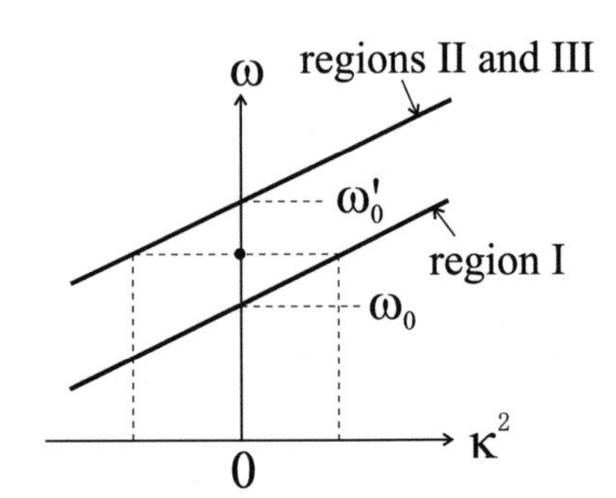

Fig. 3.29 Dispersion curves for *regions I, II,* and *III* near their respective cutoff frequencies ω_0 and ω_0'. Reprinted with permission from Johnson et al. (1996). Copyright 1996, Acoustical Society of America

$$F_z = \rho v_s^2 \frac{\partial u_\theta}{\partial z} = j\rho v_s^2 \frac{\kappa}{\eta_0} J_1(\eta_0 r/a)\left(C_1 e^{j\kappa z/a} - C_2 e^{-j\kappa z/a}\right)e^{-j\omega t}, \tag{3.9}$$

where ρ is the mass density. For a rod of finite length, the requirement of zero force at the ends restricts the allowed values of κ.

Consider the stepped rod shown in Fig. 3.27. Assuming a common η_0, the cutoff frequencies will differ from each other at regions I and II. Let them be ω_0 and ω_0', respectively. Figure 3.29 shows the dispersion curves near cutoff for the two regions. Note that, for small axial wavenumbers, the frequencies between ω_0 and ω_0' will provide real values of κ in region I and imaginary values of κ in regions II and III, indicating that the displacement of such a resonance mode decreases exponentially with distance from the center in regions II and III to remain finite at $z \to \pm\infty$. Thus, the resonance vibration must be trapped at region I. The displacement distribution can be approximately calculated using Eqs. (3.6) and (3.9) as a basis in each of the three regions and imposing the boundary condition of continuity at the interfaces and zero stress on the end faces (Johnson et al. 1996). Figure 3.30 shows the calculated displacement of the lowest mode at $\eta_0 = 5.136$ along the z-axis for an aluminum cylindrical rod with dimensions a = 11.562 mm,

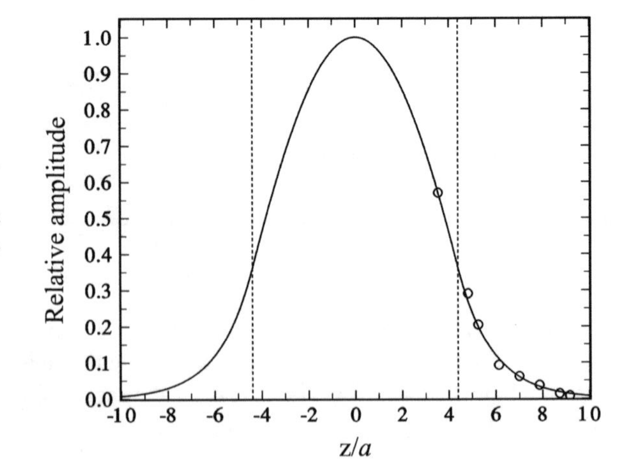

Fig. 3.30 Measured and theoretical displacements of the lowest-frequency trapped mode in an aluminum rod. The frequency is 222.6 kHz. The origin was located at the center of the rod. Reprinted with permission from Johnson et al. (1996). Copyright 1996, Acoustical Society of America

$a' = 11.441$ mm, and $l/a = 8.750$, and total length equals $19.81a$. The resonance frequency was 222.6 kHz. The vertically dashed lines indicate the positions of the steps in the rod, $z/a = \pm 4.4$. The figure also shows the relative displacements measured with a mobile EMAT along the z-axis, which show good agreement with the theoretical values. The vibrations are strongly localized in and near region I, dropping in amplitude by two orders of magnitude as z/a increases from 3.5 to 9.2.

3.12 EMATs for High-Temperature Measurements

Figure 3.31 shows an EMAT for measuring elastic constants and internal friction of small specimens at high temperatures (Ogi et al. 2003b, 2004). The specimen is inserted in a solenoid coil located within a stainless steel cylindrical chamber. The solenoid coil is made with a Ni-alloy wire, and its shape is held by a ceramic cement (see Fig. 3.32) so as to operate at higher temperatures above 1000 °C. A cantal-line heater supplies the heat to the coil and specimen. Combination of rotary and turbomolecular pumps evacuates the air in the chamber from the bottom and keeps the pressure below 10^{-4} Torr for the specimen not to be oxidized. A pair of permanent magnets located outside the chamber provides the static magnetic field to the specimen for the electromagnetic excitation and detection of free vibration via the Lorentz force mechanism and the magnetostriction force mechanism. The permanent magnet assembly is mounted on casters, which facilitated the

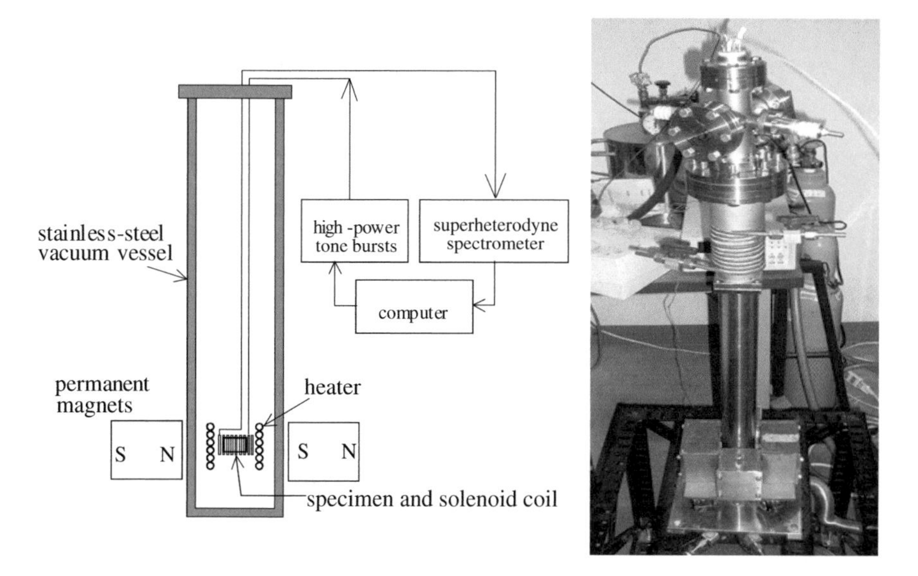

Fig. 3.31 Details of an EMAT for high-temperature acoustic study (*left*) and its appearance (*right*). Reprinted from Ogi et al. (2004). Copyright (2004), with permission from Elsevier

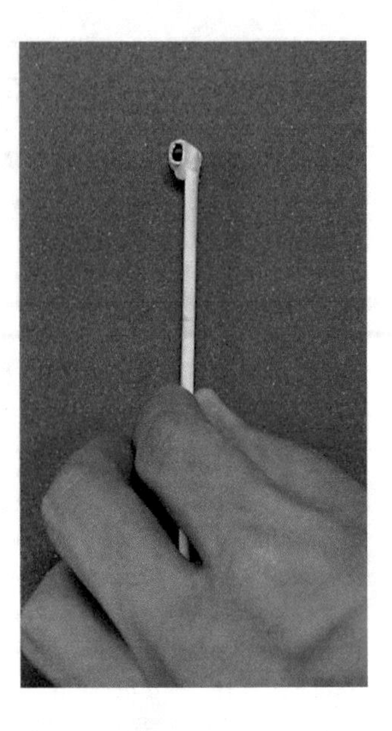

Fig. 3.32 Appearance of the solenoid coil for high temperatures

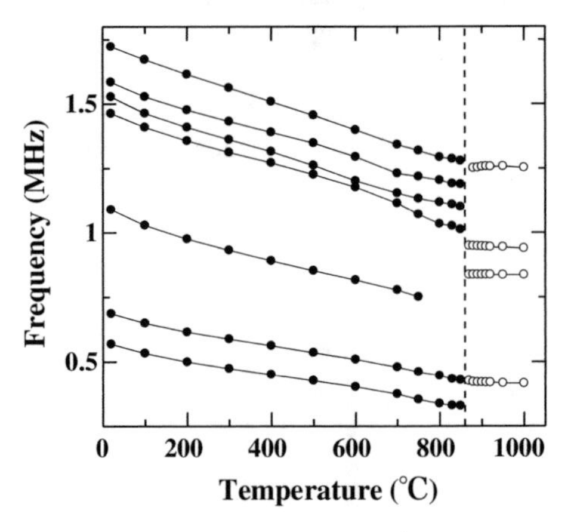

Fig. 3.33 Temperature dependence of the free-vibration resonance frequencies of a Ti monocrystal specimen. Discontinuity at 882 °C implies the $\alpha \rightarrow \beta$ phase transformation

rotation of field direction about the cylinder axis and selection of the detectable vibration modes (see Sect. 8.4.6).

Figure 3.33 demonstrates usefulness of this type of EMAT for physical acoustic study. It shows measurements of the resonance frequencies of several free vibrations of a pure titanium single crystal during a heating process from 20 to 1000 °C. They keep linearly decreasing until the $\alpha \rightarrow \beta$ (hcp $\rightarrow$ bcc) phase transformation at 882 °C gives rise to totally different temperature dependences of resonance frequencies.

Air- or water-cooled EMAT retains compact structure and permits oneside access to a specimen even at high temperatures. Figure 3.34 shows cross section of such a high-temperature EMAT developed by Burns et al. (1988). A magnetizing

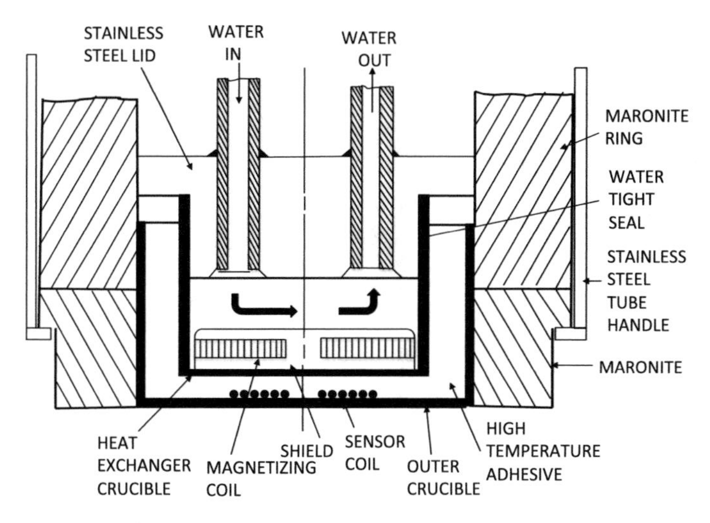

Fig. 3.34 Cross-section of a water-cooled high-temperature EMAT consisting of a magnetizing coil and an RF sensor coil. The outer diameter of the EMAT probe is about 5 cm. (After Burns et al. 1988)

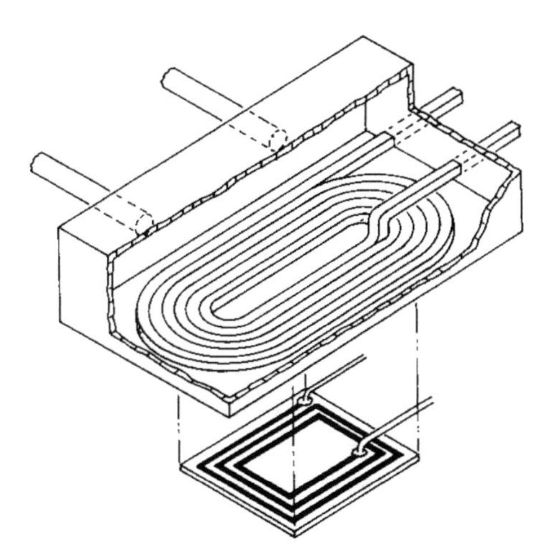

Fig. 3.35 Schematic diagram of the magnetizing coil, heat exchanger, and sensor coil. (After Burns et al. 1988)

coil shown in Fig. 3.35 is used instead of a permanent magnet. A short but very strong current produces high enough tangential magnetic field at the surface region of specimen to achieve an electromagnetic coupling to a longitudinal wave. In fact, classical formulae predict that 1 T can be observed at the surface of a 2-mm-diameter wire carrying 5000 A. Such a current is obtained for a short period of time by discharging a capacitor bank into the wire. Coupling to EMAT phenomena finishes within a short time, and then, such a short-life magnetic field is indeed practical.

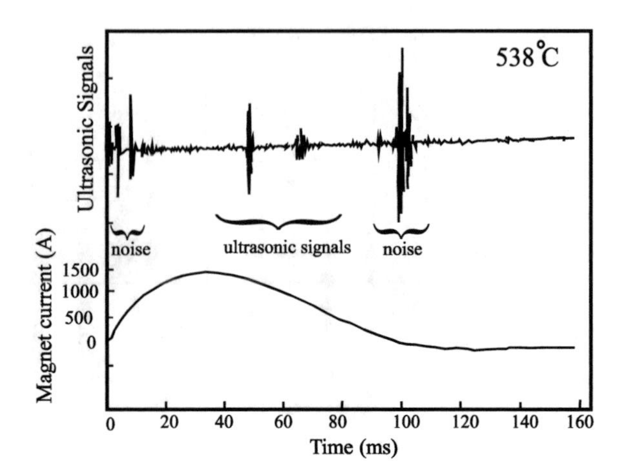

Fig. 3.36 Ultrasonic signals generated by a pulsed laser in hot steel as detected by the high-temperature EMAT whose magnetic field was applied by the pulsed current waveform at the *bottom* of the figure. (After Burns et al. 1988)

The magnetizing coil is placed in the flowing water in an alumina crucible whose bottom thickness is about 0.5 mm. The top of the crucible is closed by a stainless steel lid, which bears two stainless steel tubes for delivering and removing water. On the bottom of this heat exchanger crucible, the EMAT sensor coil is attached with high-temperature cement and protected from the hot specimen by a second alumina crucible with a thin bottom.

This EMAT was tested on a 250-mm-thick stainless steel slab as it sat in a tube furnace at a temperature of 538 °C. It received the longitudinal wave signal excited by the laser on the other side of the plate. Figure 3.36 shows the results in the form of two traces after triggering the pulsed magnet. The bottom trace plots the magnet current; it reached a maximum value of 1400 A and 30 μs after being triggered. Separate calibration measurements made at room temperature with this maximum current showed approximately 0.8 T at the position of the steel surface. The top trace shows the detected signals; large electromagnetic noise appears both early and late in the time sequence when the time rate of change of the magnetic field is at its maximum and when there are switching transients in the timing circuits. Acoustical signals were recognized because it arrived at the correct time for a longitudinal wave produced by the laser impulse.

Another test was made on an 89-mm-thick aluminum plate. Figure 3.37 shows the results obtained with the EMAT acting both as a transmitter and receiver when the specimen was heated to near the melting point of 950 °F (510 °C). Plotted is the longitudinal wave signal amplitude as it was reflected from the back surface of the slab as a function of temperature. Since these data were obtained without any coupling fluid and with the probe simply resting on the slab, it is clear that the transducer system has enough sensitivity to overcome the attenuation of the sound as the melting temperature is approached. Four more examples of high-temperature EMATs can be found in Sects. 18.4 and 18.5.

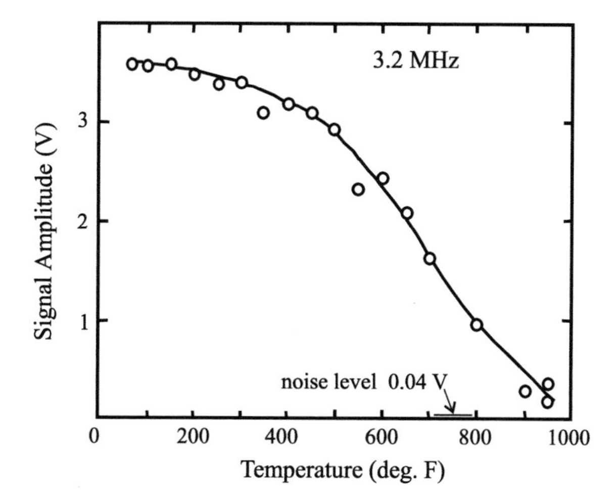

Fig. 3.37 Variation of the ultrasonic echo amplitude from the back surface of an 89-mm-thick aluminum slab as a function of temperature. (After Burns et al. 1988)

3.13 Antenna Transmission Technique for Piezoelectric Materials

An RF coil in an EMAT generates electromagnetic fields, and the magnetic field participates in the wave generation for metallic and magnetic materials. The electric field similarly allows the noncontacting coupling for piezoelectric materials. The phenomenon is, however, not based on the rotational component of the electric field $\mathbf{E}_\omega$, which is expressed by $\mathbf{E}_\omega = -\partial \mathbf{B}/\partial t$, but the nonrotational component or the quasistatic electric field is $\mathbf{E}_{qs}$ (= gradφ, where φ denotes the electric potential). This was made clear with an α-TeO_2 plate, which shows tetragonal symmetry and possesses only one independent piezoelectric coefficient (Ogi et al. 2006). Figure 3.38 shows the α-TeO_2 plate, where x'- , y'-, and z'-axes are set in the $\langle 110\rangle$, $\langle 110\rangle$, and $\langle 001\rangle$ crystallographic directions. The dimensions are 0.318 (thickness), 7, and 5 mm along the x', y', and z' axes, respectively. In this coordinate system, the piezoelectric coefficient matrix takes the form

$$[e'_{ij}] = \begin{bmatrix} 0 & 0 & 0 & e'_{14} & 0 & 0 \\ 0 & 0 & 0 & 0 & -e'_{14} & 0 \\ 0 & 0 & 0 & 0 & 0 & 0 \end{bmatrix}. \tag{3.10}$$

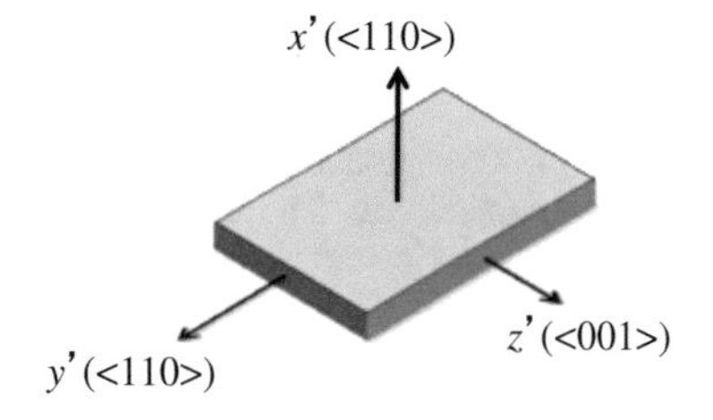

Fig. 3.38 α-TeO_2 crystal plate for clarifying the coupling mechanism. Reprinted with permission from Ogi et al. (2006). Copyright (2006) by American Chemical Society

Thus, no coupling occurs between the electric and acoustic fields with the electric field along the z'-axis ($\langle 001 \rangle$ direction), and the shear wave polarized in the z' direction will propagate in the x' direction by applying the electric field along the y'-axis. Therefore, by placing the α-TeO_2 plate on an elongated spiral coil, the governing coupling mechanism (either $\mathbf{E}_\omega$ or $\mathbf{E}_{qs}$) will be clarified, because $\mathbf{E}_\omega$ appears parallel to the current direction, whereas the direction of $\mathbf{E}_{qs}$ is perpendicular to the current direction.

Figure 3.39 shows resonance spectra measured using an elongated spiral coil with the two measurement setups. In the measurement setup in Fig. 3.39a, the direction of $\mathbf{E}_\omega$ is parallel to the $\langle 001 \rangle$ direction, and no acoustic resonance would have been caused if the rotational electric field had been the principal factor. However, the resonance peaks of the first, third, and fifth overtones appear clearly. On the other hand, the measurement setup in Fig. 3.39b caused no resonance peaks, regardless of the parallel direction of $\mathbf{E}_\omega$ to $\langle 110 \rangle$. Considering $\mathbf{E}_{qs}$ as the dominant factor, however, these observations can be explained consistently.

Figure 3.40 shows the resonance spectra of the same α-TeO_2 plate measured by the line antenna, which produces the quasistatic electric field more efficiently. In this case, the quasistatic electric field appears in the direction perpendicular to the straight line parts of the antenna. The results agree well with those observed by the

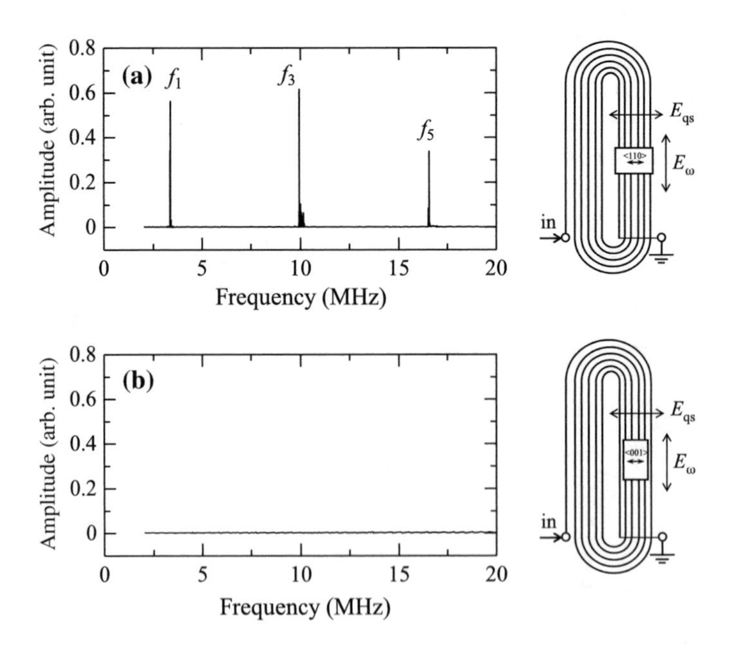

Fig. 3.39 Through-thickness shear wave resonance spectra of the α-TeO_2 plate measured by an elongated spiral coil. The α-TeO_2 plate was located on the straight parts of the coil. **a** The quasistatic electric field is generated in the $\langle 110 \rangle$ direction, whereas **b** it is generated in the $\langle 001 \rangle$ direction. Reprinted with permission from Ogi et al. (2006). Copyright (2006) by American Chemical Society

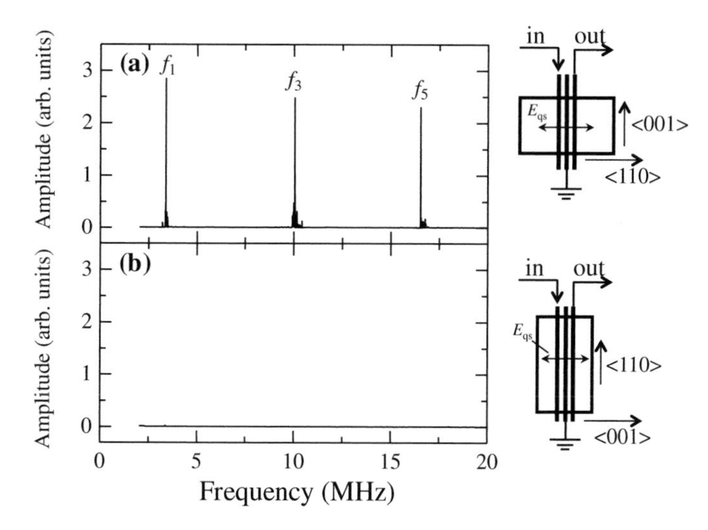

Fig. 3.40 Through-thickness shear wave resonance spectra of the α-TeO_2 plate measured by the line antenna. The α-TeO_2 plate was located on the straight parts of the antenna. **a** The quasistatic electric field is generated in the ⟨110⟩ direction. **b** It is generated in the ⟨001⟩ direction. Reprinted from Ogi et al. (2006). Copyright (2006) by American Chemical Society

elongated spiral coil. Furthermore, the resonance peak height increased by a factor of 5, indicating higher coupling efficiency with this configuration. This antenna transmission resonance method has been used for studying piezoelectric materials (Sect. 8.9) and the resonant ultrasound microscopy (Chap. 9).

References

Anastasi, R. F., & Madaras, E. I. (1999). Pulse compression techniques for laser generated ultrasound. In *Proceedings of the 1999 IEEE Ultrasonics Symposium* (pp. 813–817).

Auld, A. B. (1973). *Acoustic Felds and Waves in Solids*. New York: Wiley.

Burns, L. R., Alers, G., & MacLauchlan, D. T. (1988). A compact EMAT receiver for ultrasonic testing at elevated temperatures. In *Review of Progress in Quantitative Nondestructive Evaluation* (Vol. 7, pp. 1677–1683).

Eringen, A. C., & Şuhubi, E. S. (1975). *Elastodynamics*. New York: Academic Press.

Hirao, M., Ogi, H., & Minami, Y. (2001). Contactless measurement of induction-hardening depth by an axial-shear-wave EMAT. In *Nondestructive Characterization of Materials* (Vol. 10, pp. 379–386).

Iizuka, Y. (1998). High signal-to-noise ratio ultrasonic testing system using chirp pulse compression. *Insight, 40*, 282–285.

Iizuka, Y., & Awajiya, Y. (2014). High sensitivity EMAT system using chirp pulse compression and its application to crater end detection in continuous casting. *Journal of Physics: Conference Series, 520*, 012011.

Johnson, W., Auld, B. A., & Alers, G. A. (1994). Application of resonant modes of cylinders to case depth measurement. In *Review of Progress in Quantitative Nondestructive Evaluation* (Vol. 13, pp. 1603–1610).

Johnson, W., Auld, B. A., Segal, E., & Passarelli, F. (1996). Trapped torsional modes in solid cylinders. *The Journal of the Acoustical Society of America, 100*, 285–293.

Kino, G. S. (1987). *Acoustic Wave: Devices, Imaging, and Analog Signal Processing*. Englewood Cliffs: Prentice-Hall.

Kwun, H., & Bartels, K. A. (1998). Magnetostrictive sensor technology and its applications. *Ultrasonics, 36*, 171–178.

Kwun, H., & Teller, C. M. (1994). Detection of fractured wire in steel cables using magnetostrictive sensors. *Materials Evaluation, 52*, 503–507.

Ludwig, R., & Dai, X.-W. (1990). Numerical and analytical modeling of pulsed eddy currents in a conducting half-space. *IEEE Transactions on Magnetics, 26*, 299–307.

Ludwig, R., & Dai, X.-W. (1991). Numerical simulation of electromagnetic acoustic transducer in the time domain. *Journal of Applied Physics, 69*, 89–98.

Mason, W. P. (1958). *Physical Acoustics and Properties of Solids*. Princeton: Van Nostrand.

Miller, G. F., & Pursey, H. (1954). The field and radiation impedance of mechanical radiators on the free surface of a semi-infinite isotropic solid. In *Proceedings of the Royal Society of London A: Mathematical, Physical and Engineering Sciences* (Vol. 223, pp. 521–541).

Murray, T. W., Baldwin, K. C., & Wagner, J. W. (1997). Laser ultrasonic chirp sources for low damage and high detectability without loss of temporal resolution. *The Journal of the Acoustical Society of America, 102*, 2742–2746.

Nurmalia, Nakamura, N., Ogi, H., & Hirao, M. (2013). Mode conversion and total reflection of torsional waves for pipe inspection. *Japanese Journal of Applied Physics, 52*, 07HC14.

Ogi, H., Goda, E., & Hirao, M. (2003a). Increase of efficiency of magnetostriction SH-wave EMAT by angled bias field: Piezomagnetic theory and measurement. *Japanese Journal of Applied Physics, 42*, 3020–3024.

Ogi, H., Hamaguchi, T., & Hirao, M. (2000). Ultrasonic attenuation peak in steel and aluminum alloy during rotating bending fatigue. *Metallurgical and Materials Transactions A, 31*, 1121–1128.

Ogi, H., & Hirao, M. (1998). Electromagnetic acoustic spectroscopy in the bolt head for evaluating the axial stress. In *Nondestructive Characterization of Materials* (Vol. 8, pp. 671–676).

Ogi, H., Hirao, M., & Aoki, S. (2001). Noncontact monitoring of surface-wave nonlinearity for predicting the remaining life of fatigued steels. *Journal of Applied Physics, 90*, 438–442.

Ogi, H., Hirao, M., & Ohtani, H. (1998). Line-focusing of ultrasonic SV wave by electromagnetic acoustic transducer. *The Journal of the Acoustical Society of America, 103*, 2411–2415.

Ogi, H., Hirao, M., & Ohtani, T. (1999). Line-focusing electromagnetic acoustic transducers for detection of slit defects. *IEEE Transactions on Ultrasonics, Ferroelectrics, and Frequency Control, UFFC-46*, 341–346.

Ogi, H., Hirao, M., & Minoura, K. (1996a). Generation of axial shear acoustic resonance by magnetostrictively coupled EMAT. In *Review of Progress in Quantitative Nondestructive Evalution* (Vol. 15, pp. 1939–1944).

Ogi, H., Kai, S., Ichitsubo, T., Hirao, M., & Takashima, K. (2003b). Elastic-stiffness coefficients of a silicon-carbide fiber at elevated temperatures: Acoustic spectroscopy and micromechanics modeling. *Philosophical Magazine A, 83*, 503–512.

Ogi, H., Kai, S., Ledbetter, H., Tarumi, R., Hirao, M., & Takashima, K. (2004). Titanium's high-temperature elastic constants through the hcp-bcc phase transformation. *Acta Materialia, 52*, 2075–2080.

Ogi, H., Minoura, K., & Hirao, M. (1996b). Bulk-wave generation in a ferromagnetic metal by electromagnetic acoustic transduction. *Transactions of the Japan Society of Mechanical Engineers Part A, 62*, 1955–1962 (in Japanese).

Ogi, H., Motohisa, K., Matsumoto, T., Hatanaka, K., & Hirao, M. (2006). Isolated electrodeless high-frequency quartz crystal microbalance for immunosensors. *Analytical Chemistry, 78*, 6903–6909.

Segerlind, L. J. (1984). *Applied Finite Element Analysis*. New York: Wiley-Interscience.

Takishita, T., Ashida, K., Nakamura, N., Ogi, H., & Hirao, M. (2015). Development of shear-vertical-wave point-focusing electromagnetic acoustic transducer. *Japanese Journal of Applied Physics, 54*, 07HC04.

Thompson, R. B. (1990). Physical principles of measurements with EMAT transducers. In *Physical Acoustics* (Vol. 19, pp. 157–200). New York: Academic Press.

Tzannes, N. S. (1966). Joule and Wiedemann effects-The simultaneous generation of longitudinal and torsional stress pulses in magnetostrictive materials. *IEEE Transactions on Sonics and Ultrasonics, SU-13*, 33–41.

Yamasaki, T., Tamai, S., & Hirao, M. (1996). Electromagnetic acoustic transducers for flaw detection in steel pipes and wires. In *Proceedings of the 1st US-Japan Symposium on Advances in NDT* (pp. 117–122).

Yamasaki, T., Tamai, S., & Hirao, M. (1998). Arrayed-coil EMAT for longitudinal waves in steel wire. In *Proceedings of the 1998 IEEE Ultrasonics Symposium* (pp. 789–792).

Chapter 4
Brief Instruction to Build EMATs

Abstract Some EMAT coils can be made by handwork. A more sophisticated method is to use the printed circuit technique. An EMAT is regarded as a coil element in the electric circuit, and impedance matching is required for improving EMAT's transduction efficiency. This chapter describes how to fabricate EMAT with examples. Some designs for impedance-matching circuit are also given.

Keywords Meander-line coil · Neodymium-iron-boron (Nd-Fe-B) magnet · Polyimide sheet · Samarium-cobalt (Sm-Co) magnet · Spiral elongated coil

4.1 Coil

The high-frequency coil is the most important component of an EMAT. Its geometrical configuration determines the modes of generated and detected elastic waves. The liftoff (or air gap) between the coil and specimen surface must be minimized as much as possible because it exponentially lowers the efficiency of the EMAT as indicated by Eq. (2.29). Moreover, the coil should be flexible for bending and twisting to accommodate curved specimen surfaces. There are two fundamental and useful geometries of EMAT coils: spiral elongated coil and meander-line coil. The former is usually used for bulk-wave EMATs. The latter is for surface SH wave EMATs, Rayleigh wave EMATs, Lamb wave EMATs, etc. Coils for the focusing and chirp EMATs are rather complicated, requiring a high accuracy, and a handmade approach is impracticable.

The best way to fabricate an EMAT coil is to use a printed circuit technique. Recent advancement of the technique permits us to print any type of flat coil of arbitrary pattern with high accuracy, typically within one micrometer. Figure 4.1 shows a cross section of an EMAT coil fabricated by such a technique. It consists of five layers; top, middle, and bottom layers are polyimide resin sheets about 20 μm thick for sandwiching and insulating the copper coils between them. Patterning of the copper coils is made by the photoresist coating and etching methods. Two copper coils are used for each of generation and detection. This separation is often superior to using a single coil for both generation and detection because individual impedance matching

M. Hirao and H. Ogi, *Electromagnetic Acoustic Transducers*,
Springer Series in Measurement Science and Technology,
DOI 10.1007/978-4-431-56036-4_4

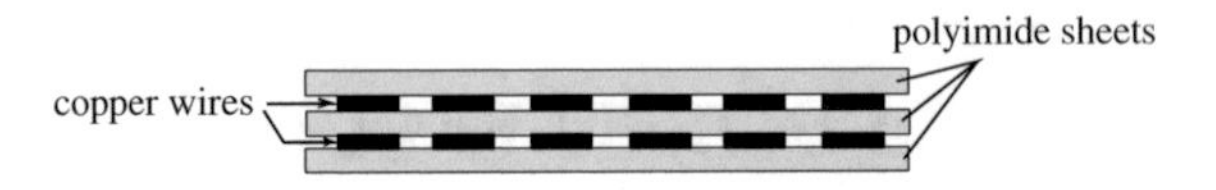

Fig. 4.1 Cross section of a layered sheet coil fabricated by a printed circuit technique

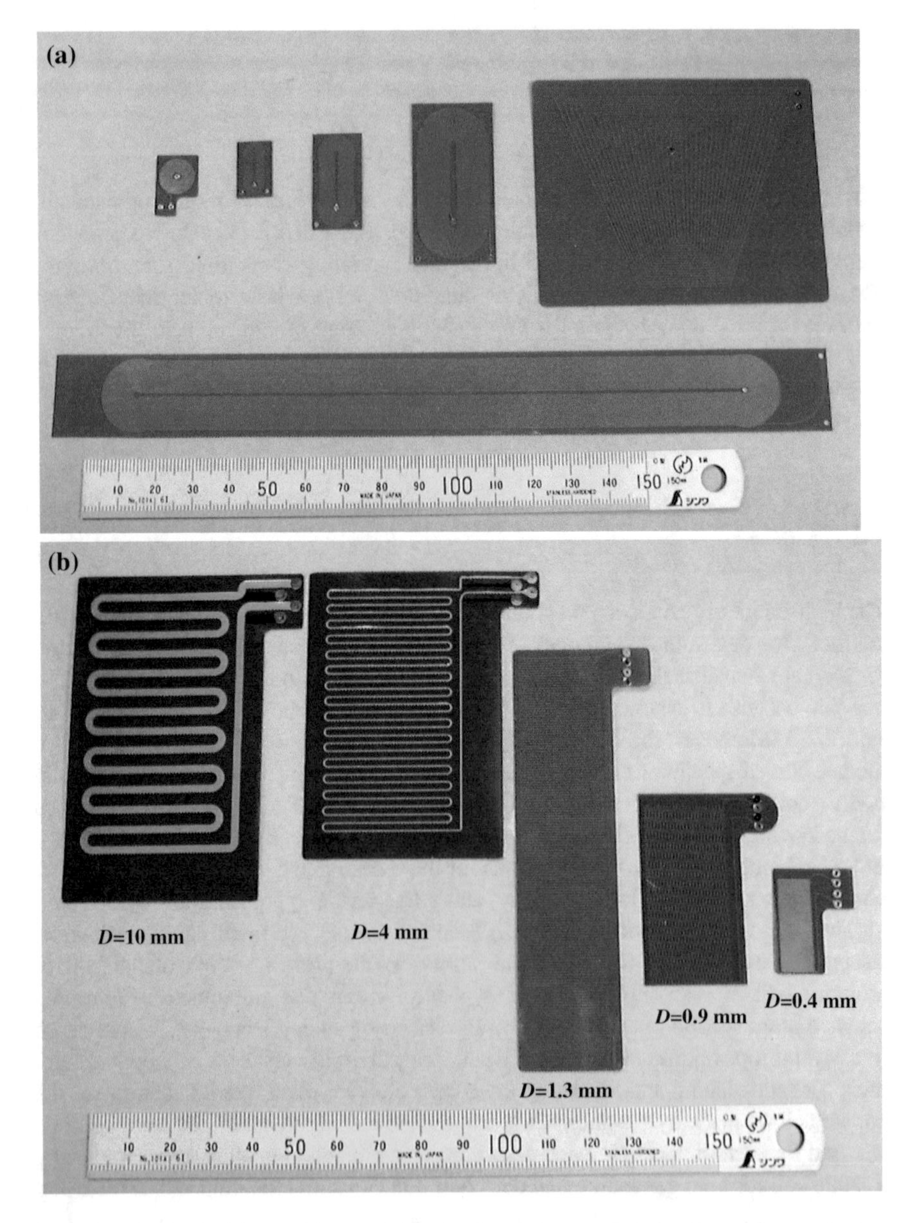

Fig. 4.2 Sheet coils fabricated by a printed circuit technique. **a** Spiral coils and **b** meander-line coils. *D* denotes the meander-line spacing

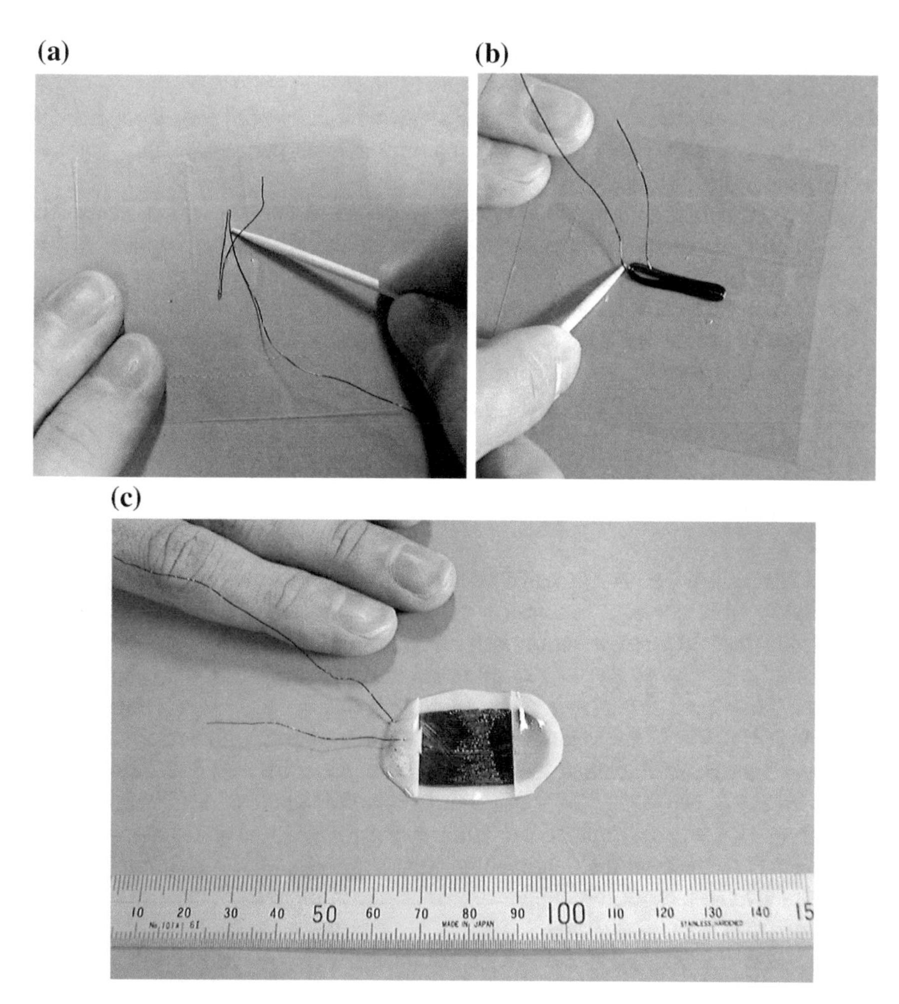

Fig. 4.3 Fabrication of a spiral elongated coil. **a** Making the coil core, **b** winding the enameled copper wire tightly using a toothpick, and **c** completing coiling by fastening with adhesive

for each coil is made possible. Total thickness of the sheet coil is about 100 μm. The liftoff equals the polyimide sheet thickness. Advantage of printed coil is multifold. Besides the accurate patterning and less liftoff, they are flexible and tolerant for friction, usable at elevated temperatures up to 600 K, and replaceable. Figure 4.2 shows various printed sheet coils fabricated by this technique.

Such high technology is sometimes costly and not easily accessible. Being apart from it, handmade coils are as good as the printed coils and useful enough for many purposes. Spiral coils are especially easy to make with low technology and low cost (see Fig. 3.1). Making a spiral elongated coil proceeds as follows:

(i) Prepare enameled copper wire of about 0.2 mm diameter.
(ii) Affix a double-faced adhesive tape on a flat platform.

(iii) Make a single-turn core by bending the wire and attach it onto the tape (Fig. 4.3a).
(iv) Use a toothpick or something similar to wind the wire, being parallel to the core so closely each other that a large turn number per unit width is obtained (Fig. 4.3b).
(v) Lay an epoxy adhesion bond over the coil so as to keep its flat shape. The bond must be removed at the region where the permanent magnets will be mounted (Fig. 4.3c).
(vi) Connect BNC cables and put the magnets on the top of the coil. Molding in polymer fixes the structure and facilitates uses. This finishes a *handmade* bulk-wave EMAT shown in Fig. 3.1.

4.2 Magnets

Many EMATs use permanent magnets because they can apply high enough magnetic fields with compact structures without power supply. Disadvantages are uncontrollable field strength and intolerance to high temperatures because of their low Curie points. Most EMATs adopt neodymium-iron-boron (Nd-Fe-B) sintered magnets. They exhibit the highest remanent flux density B_r ($\sim$1.4 T) and energy product $[BH]_{\max}$ ($\sim$47 megagauss oersted (MGOe)) among commercially available magnets. Because of their low Curie points, they cannot be used at high temperatures, normally below 80 °C. Thus, high-temperature applications need water- or air-cooling equipments. Sintered samarium-cobalt (Sm-Co) magnets are inferior to the Nd-Fe-B magnets in the magnetic properties ($[BH]_{\max} < 30$ MGOe), but more stable at elevated temperatures. Their maximum working temperatures are between 250 and 300 °C.

These permanent magnets are conductive and show magnetostriction. The EMAT coil then causes dynamic fields not only in the specimen but also in the magnets, and high-amplitude acoustic waves may occur within the magnets from the interactions between the magnetic field or magnetostriction within the magnets and the dynamic fields caused by the coil. This disturbs the signals from the specimen. Scratching the magnets with a grinder minimizes the influence, because the acoustic waves in the magnets are randomly scattered on reflections and cancel out quickly.

Another candidate is to use an electromagnet, which can control the strength of the biasing magnetic field and can be used at elevated temperatures with heat insulator and/or cooling system. Higher field is available with electromagnets, but this entails heating at the coils. Pulse current, being synchronous to the excitation current to the EMAT coil, can be used to minimize the Joule heating (as shown in Sect. 3.12 and Chap. 18). After all, EMATs with electromagnets occupy much more space than those with permanent magnets.

4.3 Impedance Matching

4.3.1 L-Matching Network

Transfer efficiency between electric energy and sound energy with an EMAT is much lower than that of a piezoelectric transducer, and every effort should be made to maximize it. There is usually large impedance mismatch between an EMAT and a pulse generator (or gated amplifier) that drives it, causing inefficient transfer into acoustic waves. The pulse generator itself possesses impedance of the form

$$Z_i = R_i + jX_i. \tag{4.1}$$

The impedance of the EMAT transducer connected to the pulse generator can be written as

$$Z_E = R_E + jX_E. \tag{4.2}$$

Here, R_i, R_E, X_i, and X_E are real quantities. The condition to maximize the power spent at the transducer is

$$R_i = R_E \quad \text{and} \quad X_i = -X_E. \tag{4.3}$$

Most pulse generators are 50 Ω resistive or less, while most EMATs are much less resistive. Therefore, Eq. (4.3) is usually not satisfied with the EMAT alone. Probably, the most useful LC matching network is the "L-matching network." Two reactive elements (coil or capacitor) with impedances jX_a and jX_b, respectively, are placed between the pulse generator and the EMAT in an effort to maximize the power transfer to the EMAT (Fig. 4.4). This requires that the impedance of the EMAT and the output impedance of the pulse generator be known. The impedance of the pulse generator can be assumed as R_i = 50 Ω and X_i = 0. The EMAT impedance must be measured or estimated.

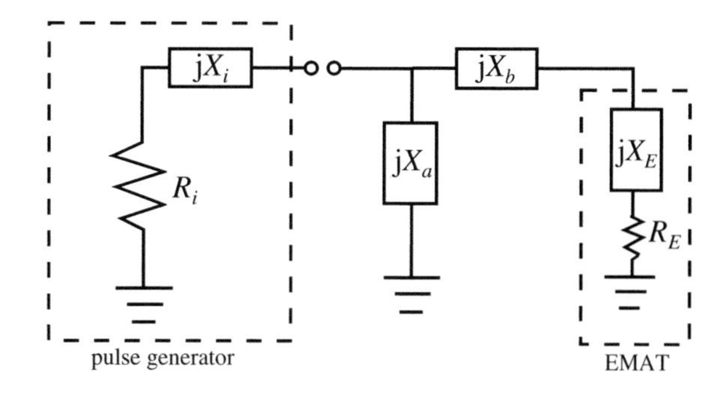

Fig. 4.4 L-matching network

The applicable equations for obtaining the matching impedances are

$$X_a = -\frac{R_i^2 + X_i^2}{X_i + QR_i}, \tag{4.4}$$

$$X_b = QR_E - X_E. \tag{4.5}$$

Here, Q is defined by

$$Q = \pm\sqrt{\frac{R_i}{R_E}\left[1 + \left(\frac{X_i}{R_i}\right)^2\right] - 1}. \tag{4.6}$$

If the transducer impedance is given in parallel form (Fig. 4.5), it can be converted to the series form using the following relationships:

$$R_s = R_p\left[1 + \left(\frac{R_p}{X_p}\right)^2\right]^{-1}, \tag{4.7}$$

$$X_s = X_p\left(\frac{R_p}{X_p}\right)^2\left[1 + \left(\frac{R_p}{X_p}\right)^2\right]^{-1}. \tag{4.8}$$

Conversions from series to parallel are also given by

$$R_p = R_s\left[1 + \left(\frac{X_s}{R_s}\right)^2\right], \tag{4.9}$$

$$X_p = X_s\frac{1 + \left(\frac{X_s}{R_s}\right)^2}{\left(\frac{X_s}{R_s}\right)^2}. \tag{4.10}$$

Note that the formalism Eqs. (4.4)–(4.6) assumes that the shunt element of the *L-matching circuit* is in parallel with the pulse generator. If one exchanges the

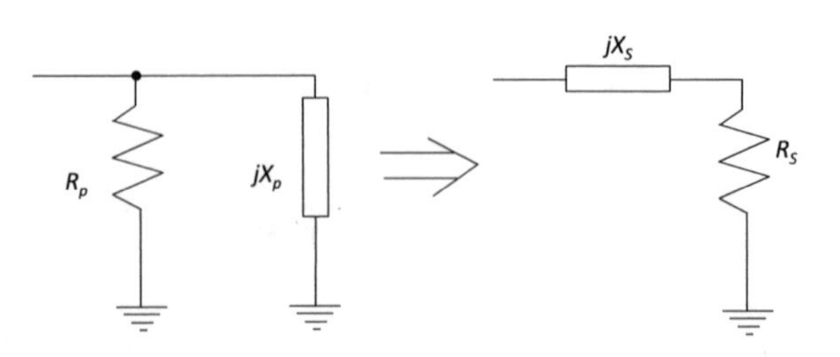

Fig. 4.5 Conversion from parallel form to series of transducer impedance

values of Z_i for those of Z_E and vice versa, another solution is obtained, in which the shunt element is in parallel with the EMAT. One also obtains separate solutions for positive and negative values of Q, giving a total of four possible matching networks. This may seem confusing, but in many real situations, the number of possibilities may easily be reduced. Q may be imaginary for one of the positions of the "L" network, and therefore, in this case, two of the solutions are physically meaningless. Also, some of the solutions are physically unrealizable because one of the components cannot be obtained or constructed easily.

4.3.2 Voltage Step-up Ratio

It is always interesting to know the voltage step-up or step-down from the output of the pulse generator to the EMAT. If the matching network is properly designed, the power output of the gated amplifier is given by

$$P = \frac{V_{\text{out}}^2}{R_i}. \tag{4.11}$$

The power applied to the EMAT becomes

$$P = \frac{R_E V_{\text{EMAT}}^2}{|Z_E|^2}, \tag{4.12}$$

where V_{out} and V_{EMAT} denote the rms voltage levels at the pulse generator output and at the EMAT contacts. $|Z_E|$ denotes the absolute magnitude of the complex EMAT impedance. When Eqs. (4.11) and (4.12) are equated, the solution for the ratio of the EMAT and pulse generator voltages (step-up ratio) is given by

$$\frac{V_{\text{EMAT}}}{V_{\text{out}}} = \frac{|Z_E|}{\sqrt{R_i R_E}}. \tag{4.13}$$

4.3.3 Example 1: Meander-Line-Coil EMAT

The impedance of the EMAT described in Sect. 3.5 was measured at 1.5 MHz with a vector impedance meter as

$$R_E = 0.512\,\Omega \quad \text{and} \quad X_E = 0.844\,\Omega.$$

The matching network components are calculated using Eqs. (4.4)–(4.6):

$$R_i = 50\,\Omega,\ X_i = 0,$$
$$Q = \pm\sqrt{\frac{50}{0.512} - 1} = \pm 9.831.$$

Using $Q = +9.831$,

$$X_a = -\frac{50^2}{9.831 \times 50} = -5.086\,\Omega,$$
$$X_b = 9.831 \times 0.512 - 0.844 = 4.189\,\Omega.$$

Thus, we can use a capacitor with $C_a = 0.0208$ μF and a coil with $L_b = 0.444$ μH for the two matching elements. The matching network circuit and the output response are shown in Fig. 4.6a.

Using $Q = -9.831$,

$$X_a = -\frac{50^2}{-9.831 \times 50} = 5.086\,\Omega,$$
$$X_b = -9.831 \times 0.512 - 0.844 = -5.877\,\Omega,$$

and we have $L_a = 0.540$ μH and $C_b = 0.0181$ μF. The circuit and response are shown in Fig. 4.6b.

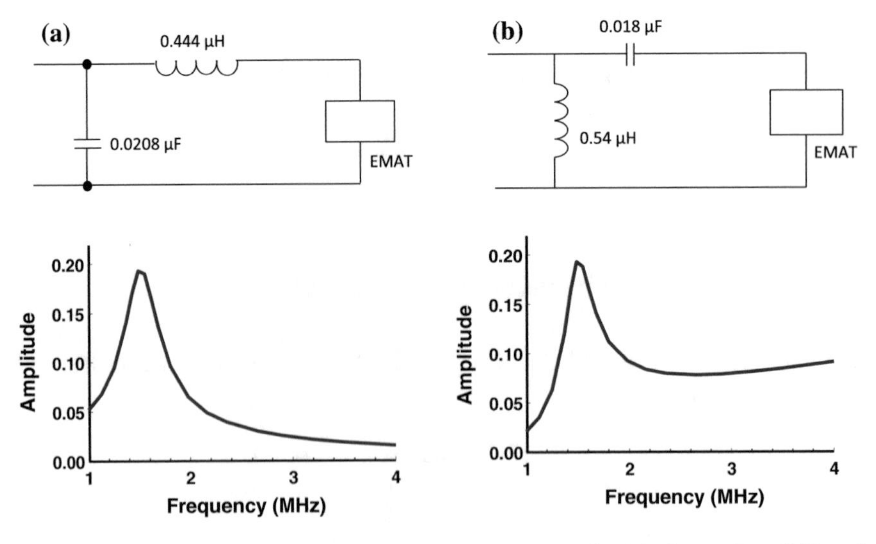

Fig. 4.6 Matching circuit (*upper*) and the output response (*lower*) for **a** $Q = +9.83$ and **b** $Q = -9.83$ for a meander-line-coil EMAT

We now reverse the positions of Z_i and Z_E and attempt to find two more solutions:

$$R_i = 0.512\,\Omega, X_i = 0.844\,\Omega, R_E = 50\,\Omega, \quad \text{and} \quad X_E = 0\,\Omega.$$

This case leads to $Q = \pm\sqrt{-0.96}$. Thus, these solutions are nonphysical, and only two solutions are available for this EMAT at 1.5 MHz.

The step-up ratio becomes

$$\frac{V_{\mathrm{EMAT}}}{V_{\mathrm{out}}} = \frac{|0.512 + j0.844|}{\sqrt{50 \times 0.512}} = 0.195. \tag{4.14}$$

This low step-up ratio is expected because low impedance is improved.

4.3.4 Example 2: Bulk-Wave EMAT with a Spiral Coil

The impedance of the EMAT described in Sect. 3.1 was measured at 1.5 MHz with a vector impedance meter as

$$R_E = 1.549\,\Omega \quad \text{and} \quad X_E = 9.295\,\Omega.$$

Then, we have

$$R_i = 50\,\Omega, X_i = 0,$$
$$Q = \pm\sqrt{\frac{50}{1.549} - 1} = \pm 5.593.$$

Using $Q = +5.593$,

$$X_a = -8.940\,\Omega,$$
$$X_b = -0.631\,\Omega.$$

Thus, the matching elements are capacitors with $C_a = 0.0119$ μF and $C_b = 0.168$ μF. The circuit diagram and response are given in Fig. 4.7a.

Using $Q = -5.593$,

$$X_a = +8.940\,\Omega,$$
$$X_b = -17.96\,\Omega.$$

and we have $L_a = 0.948$ μH and $C_b = 0.00591$ μF. The circuit diagram and response are shown in Fig. 4.7b.

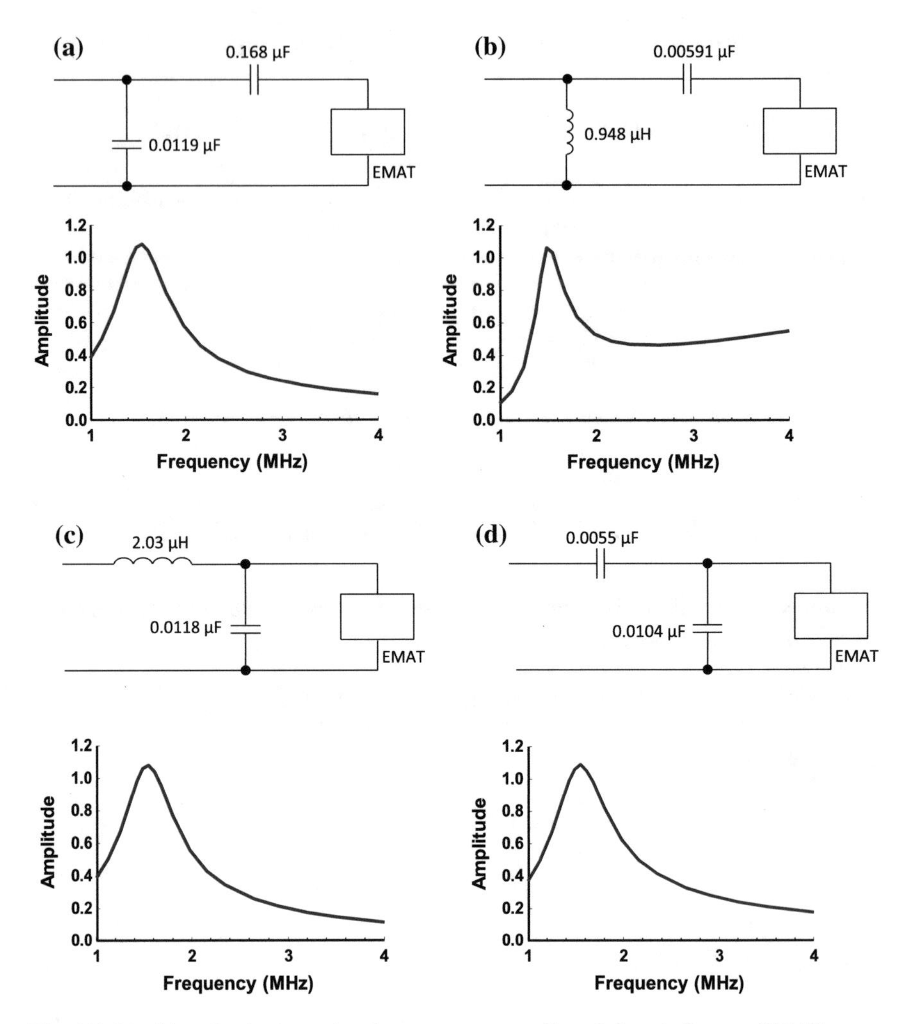

Fig. 4.7 Matching circuits (*upper*) and output responses (*lower*) for a bulk-wave EMAT

After reversing the positions of pulse generator and EMAT, the parameters for two more matching networks are calculated:

$$R_i = 1.549\,\Omega, X_i = 9.295\,\Omega, R_E = 50\,\Omega, X_E = 0\,\Omega, \quad \text{and} \quad Q = \pm\sqrt{0.383}.$$

$Q = +0.383$ provides $X_a = -8.980\ \Omega$ and $X_b = 19.15\ \Omega$, corresponding to $C_a = 0.0118\ \mu\text{F}$ and $L_b = 2.032\ \mu\text{H}$ (Fig. 4.7c). $Q = -0.383$ provides $X_a = -10.20\ \Omega$ and $X_b = -19.15\ \Omega$, corresponding to $C_a = 0.0104\ \mu\text{F}$ and $C_b = 0.0055\ \mu\text{F}$ (Fig. 4.7d).

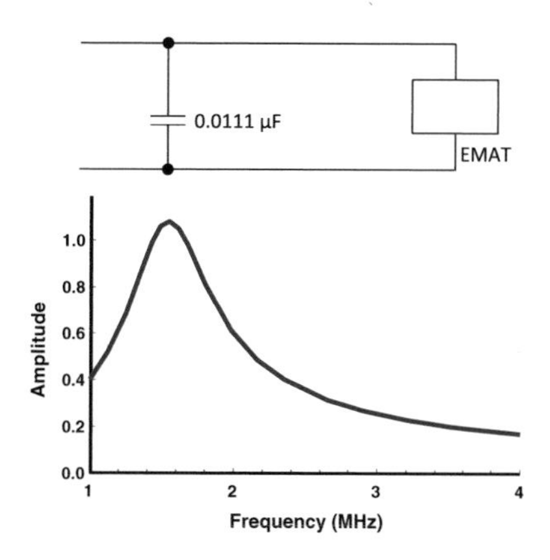

Fig. 4.8 Impedance matching with a single capacitor. Matching circuit (*upper*) and output response (*lower*)

The step-up ratio becomes

$$\frac{V_{\text{EMAT}}}{V_{\text{out}}} = \frac{|1.549 + j9.295|}{\sqrt{50 \times 1.549}} = 1.07.$$

The fact that the step-up ratio is so close to unity indicates that there is no significant voltage drop across the series matching element and it may be possible to eliminate one of the components. Equations (4.9) and (4.10) are used to calculate in parallel form the EMAT impedance:

$$X_P = 9.55\,\Omega \quad \text{and} \quad R_P = 57.3\,\Omega.$$

When 0.0111 μF is used to parallel resonate the inductance, it cancels X_P and the impedance becomes 57.3 Ω resistive. The difference between this and R_i (= 50 Ω) is completely trivial. Thus, this case is the obvious choice (Fig. 4.8).

Part II
Resonance Spectroscopy with EMATS-EMAR

Part II describes the principle of EMAR, which makes the most of contactless nature of EMATs and is the most successful amplification mechanism for precise velocity and attenuation measurements.

Chapter 5
Principles of EMAR for Spectral Response

Abstract The weak coupling efficiency with EMATs is sufficiently compensated for by the combination with resonance methods. This is called electromagnetic acoustic resonance, or EMAR. Tone burst input signals are normally used for this purpose, and an analog superheterodyne spectrometer is useful for detecting resonance frequencies. This chapter describes the principle and operation of EMAR measurement and shows electric circuit for the analog spectral analysis.

Keywords Band-pass filter · Gated amplifier · Intermediate frequency · Lorentzian function · Low-pass filter · Multiplier · Quadrature phase-sensitive detection

5.1 Through-Thickness Resonance

An easy way to understand the EMAR principle is to consider the through-thickness resonances of bulk waves (thickness oscillations). Consider that a transducer is placed on the surface of a plate of thickness d and excites an ultrasonic plane wave of single cycle propagating in the thickness direction, whose wavelength and period are λ and T, respectively. The pulse wave undergoes the repeated reflections at both free surfaces and is received by the same transducer each time it reaches the incident surface. In the case of $\lambda \ll d$, the received signals will be observed as shown in Fig. 5.1a. The horizontal axis in Fig. 5.1 is the time measured from the excitation, and the vertical axis is the wave amplitude. T_0 denotes the round-trip time through the thickness. The amplitude decays as the propagation path increases, depending on attenuation of the material.

Next, suppose an excitation with a radio-frequency (rf) burst signal composed of five cycles. In this case, before the first echo is completely received, the second echo reaches the incident surface; the end of the first echo and the head of the second echo are superimposed, causing interference between them. The same occurs between the second and third echoes, the third and fourth echoes, and so on. The signal amplitudes are canceled or enhanced in the overlapping area, depending

M. Hirao and H. Ogi, *Electromagnetic Acoustic Transducers*,
Springer Series in Measurement Science and Technology,
DOI 10.1007/978-4-431-56036-4_5

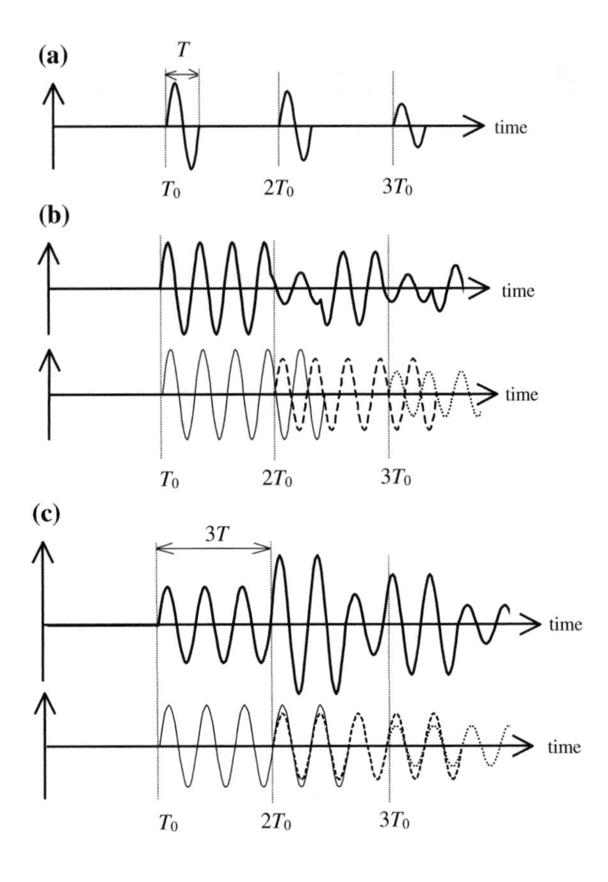

Fig. 5.1 Schematic explanation of through-thick resonance in a plate. **a** Single-cycle echoes, **b** destructive interference of five-cycle echoes, and **c** constructive interference of five-cycle echoes (ultrasonic resonance)

on the phase difference caused by the difference of propagation distances. In Figs. 5.1b and c, the destructive and constructive interferences are illustrated.

Then, suppose sweeping the carrier frequency to vary wavelength and watching the resultant amplitude of the overlapped signal. The resultant amplitude is a function of the frequency. The constructive interference occurs when the round-trip time T_0 is an integer multiple of the period T and all the echoes are overlapped in phase:

$$T_0 = nT. \tag{5.1}$$

Here, n is an integer, giving the number of cycles involved in the T_0 interval. The constructive interference gives rise to a large amplitude by simply summing up echo amplitudes. This is ultrasonic resonance. Figure 5.1c shows the case of n = 3. Expanding the number of cycles in the exciting rf bursts to say several hundreds makes the first echo overlap almost all other echoes coherently, and a much larger signal amplitude is observed. It is easily found from Eq. (5.1) that the resonance occurs at every frequency satisfying

$$f_n = nv/(2d). \tag{5.2}$$

Here, v is the phase velocity of the bulk plane wave, and f_n is the resonance frequency of the nth order. The resonance frequency therefore gives the velocity or the thickness, when the other is known.

The above explanation pertains to the thickness resonance. There are many ultrasonic resonances depending on the sample geometry and the wave mode, but their basic principle is the same as this case.

At a resonance, a large number of received signals are coherently overlapped, giving enhanced signal-to-noise ratio and a sharp spectral response. Such an ideal ultrasonic resonance is only available with noncontact transducers. Weak coupling with EMATs is advantageous in that they hardly disturb the elastic waves on reflections. Ultrasound lasts for a long time experiencing many reflections (see Fig. 5.1b) and contributing to a sharp resonance. On the other hand, contacting transducers absorb a large fraction of elastic wave energy and alter the phase angles of signals at every reflection. This aspect results in a poor ultrasonic resonance with inadequate frequency resolution.

Figure 5.2 illustrates the fundamental mode of the thickness resonance (n = 1) excited by a conventional piezoelectric transducer and by a bulk-wave EMAT. When a contacting transducer is in use, the ultrasonic wave propagates not only in the specimen but also in the coupling material and the transducer itself. This leads to a composite resonator encompassing all the participants. The resonance frequency is not given by Eq. (5.2), and the separation of the material's resonance frequency from the measured one needs tedious and unrealistic procedures (Mozurkewich 1989; Ogi et al. 2002). In particular, it requires the couplant thickness and elastic wave speeds, which are changeable with the pressure and temperature.

Moreover, the contacting technique poses an even more troublesome situation for attenuation measurement. Energy absorption at the transducers causes large extra losses and conceals the material's small attenuation. The separation is impossible within a necessary precision. The EMAT is, however, free from these problems. The weak coupling of the EMAT takes little energy of the wave, making it possible to measure the small amount of energy loss in the specimen. Low transfer efficiency of

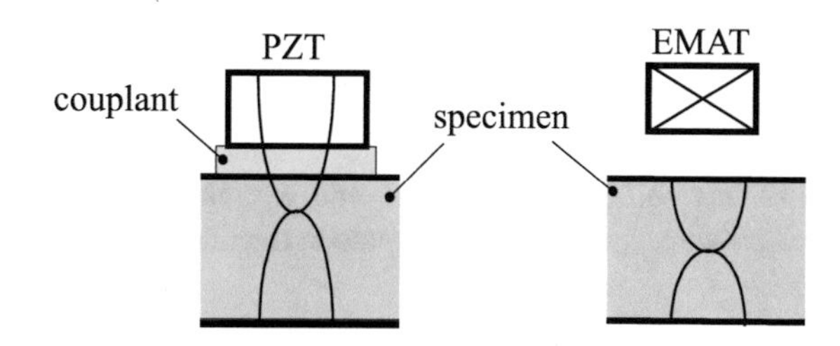

Fig. 5.2 Composite resonance caused by a contacting transducer (*left*) and isolated resonance with an EMAT (*right*)

EMATs brings a large benefit of isolating the specimen's ultrasonic resonance from the transducer (Fig. 5.2), leading to absolute measurements of phase velocity and attenuation. The EMAT and the resonance technique are thus a winning combination for materials characterization.

5.2 Spectroscopy with Analog Superheterodyne Processing

The instrumentation presented here has been developed by Petersen et al. (1992, 1994), Fortunko et al. (1992), and Chick (1999) originally for nuclear magnetic resonance (NMR) applications. For the first time, this instrumentation is incorporated with the EMATs to realize today's EMAR methodology (Fukuoka et al. 1993; Hirao et al. 1993). It is based on the sampled continuous-wave (CW) techniques of Bolef and Miller (1971). Sampled CW techniques typically involve the use of a low-level CW driving signal. The main advantage of using the long bursts and sampled CW technique is that the elastic wave displacements within the specimen can be gradually increased to a near-equilibrium value, which is generally determined by the loss within the specimen and external loading by transducers, without exceeding the breakdown limits of the transducers. When the electrical driving signals are abruptly interrupted, the elastic wave displacements begin to decay. The transient elastic wave signals, which follow the abrupt interruption of the driving signals, are then processed using an appropriate algorithm.

Although the measurement approach is conceptually similar to the sampled CW method, it takes advantage of several advances in the instrumentation and signal processing to improve the absolute accuracy of the measurement. Specifically, it uses a high-power gated rf amplifier to drive an EMAT and uses a phase-sensitive, gated analog integrator as a spectrometer to process the received transient ultrasonic signals. Clark (1964) has reported this method of using analog integration to obtain spectroscopic data. Avogadro et al. (1979) have investigated in considerable detail the application of Clark's method to NMR studies.

Analog spectroscopic methods, such as Clark's method, have been largely replaced by digital methods, such as those using the fast Fourier transform (FFT) algorithm. However, analog spectroscopic methods offer significant advantages in certain applications. For example, standard digital FFT signal processing requires that the bandwidth of the driving signal be larger than the line width of the resonance. This means that short driving pulses or rf bursts are needed, and the amount of energy available to build up a strong resonant standing wave is limited.

Figure 5.3 shows a block diagram of the apparatus developed by Petersen et al. (1992, 1994) and Chick (1999), including the experimental configuration with a bulk-wave EMAT of Fig. 3.1 and a plate specimen. The EMAT generates and detects shear waves that propagate in the thickness direction of the plate. The EMAT

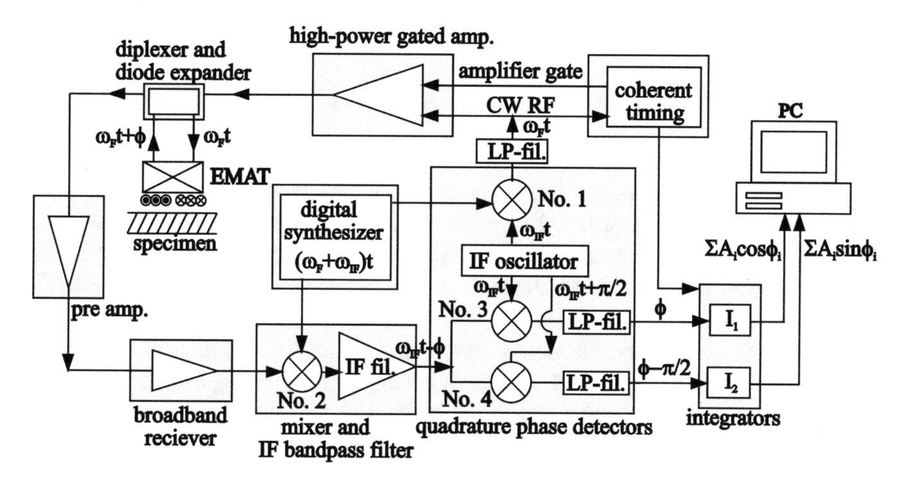

Fig. 5.3 Blockdiagram of a superheterodyne spectroscopy with EMAT

is driven with a long rf burst, and the same transducer detects the acoustic ringing in the specimen.

A narrow band rf burst is produced from a synthesized continuous-wave signal using a high-power gated amplifier capable of producing 0.1–5 kV peak-to-peak rf bursts into 50 Ω. The rf burst is applied to the EMAT. The rf burst is chosen with sufficient length to build up strong ultrasonic standing waves within the specimen. In practice, the rf burst length may not be long enough to establish steady-state, near-equilibrium resonance conditions. However, typical rf burst lengths are sufficient to establish favorable signal-to-noise conditions in most engineering materials. The amplitudes of the resonant ultrasonic fields are directly proportional to the voltage of the driving pulse and increase monotonically with rf burst length. In low-loss materials, the ultrasonic vibrations last long time to maintain the ultrasonic resonance. Thus, too long rf bursts are not needed. In lossy materials, too long rf bursts are also not required, because earlier parts of the bursts will not remain for a long time and fail to make the constructive interference with later parts of the bursts.

The CW driving signal is not generated directly by the synthesizer, but is instead obtained by using analog multiplier 1 to multiply the reference output R_{IF} of the intermediate frequency (IF) oscillator

$$R_{IF} = A_{IF}\cos(\omega_{IF}t + \theta_{IF}), \tag{5.3}$$

by the output R_S of a higher-frequency synthesizer

$$R_S = A_S\cos[(\omega_F + \omega_{IF})t + \theta_S]. \tag{5.4}$$

The multiplication produces their different and sum frequencies

$$M_1 = R_{\mathrm{IF}} \cdot R_{\mathrm{S}} = A_{\mathrm{F}}\{\cos(\omega_{\mathrm{F}} t + \theta_{\mathrm{S}} - \theta_{\mathrm{IF}}) + \cos[(\omega_{\mathrm{F}} + 2\omega_{\mathrm{IF}})t + \theta_{\mathrm{S}} + \theta_{\mathrm{IF}}]\}. \quad (5.5)$$

Here, the terms θ_{IF} and θ_{S} refer to the arbitrary phases of the IF (25 MHz) oscillator and the synthesizer, respectively. A_{F} is the amplitude of the output M_1. The sum frequency term, (ω_{F} + $2\omega_{\mathrm{IF}}$), is first removed using a low-pass filter at the output of the multiplier. The difference frequency term is then applied to the input of the gated amplifier, which produces the high-power, rf burst electrical signals needed to drive the EMAT.

The EMAT excites the ultrasonic wave traveling in the thickness direction. Just after the excitation, the same EMAT detects the highly overlapping echoes. Each echo has different amplitude A_i and phase ϕ_i, depending on the attenuation and propagated distance, and the *i*th echo is expressed by $A_i \cos(\omega_{\mathrm{IF}} t + \theta_{\mathrm{S}} - \theta_{\mathrm{IF}} + \phi_i)$. ϕ_i is the phase shift caused by the propagation with a velocity V over the distance $2id$:

$$\phi_i = 2i\,d\omega_{\mathrm{F}}/V, i = 1, 2, 3, \ldots \quad (5.6)$$

The signals are amplified by the variable gain (78 dB range) wideband rf amplifier after preamplification. The output of the wideband amplifier is then multiplied by the synthesizer output signal R_s at frequency (ω_{F} + ω_{IF}) using the analog multiplier 2. The resulting voltage is

$$M_2 = g_2 A_i \cos(\omega_{\mathrm{IF}} t + \theta_{\mathrm{IF}} - \phi_i) + \text{high-freq. terms.} \quad (5.7)$$

Here, the constant g_2 includes the rf gain, the amplitude of the reference R_s, and the conversion efficiency of the multiplier. The high-frequency terms of Eq. (5.7) involving the frequencies (ω_{IF} + $2\omega_{\mathrm{F}}$) are completely rejected by the IF band-pass filter, which has a fixed center frequency of 25 MHz and three computer-settable bandwidths: 0.4, 1, 4 MHz. The remaining signals, containing the phases of the ultrasonic signals, are amplified and multiplied by the IF oscillator outputs using analog multipliers 3 and 4. The output of multiplier 3 is

$$M_3 = g_{\mathrm{out}} A_i \cos(-\phi_i) + \text{high-freq. terms.} \quad (5.8)$$

Here, the total gain and conversion efficiencies are all included in the term g_{out}. The output of multiplier 4 is

$$M_4 = g_{\mathrm{out}} A_i \sin(-\phi_i) + \text{high-freq. terms.} \quad (5.9)$$

Equations (5.3)–(5.9) describe the quadrature phase-sensitive detection. The high-frequency terms are effectively removed by low-pass filters that follow analog multipliers 3 and 4. Only the zero-frequency (base band) terms survive. The phase terms θ_{IF} and θ_{S} are also removed.

After all of this, the remaining signals $I_1 = \Sigma g_{out} A_i \cos(-\phi_i)$ and $I_2 = \Sigma g_{out} A_i \sin(-\phi_i)$ are obtained. Both A_i and ϕ_i are functions of time. These outputs are integrated over the time region just after the excitation by the analog gated integrators, and their outputs are stored in the attached computer.

The *amplitude spectrum* is obtained from square root of the sum of squares of these integrator outputs as a function of the operating frequency ω_F. A frequency scan provides a resonance spectrum as shown in Fig. 5.4. In-phase and out-of-phase signals before the integration are also shown at three different frequencies around the resonance peak. When the frequency is equal to the nth resonant frequency, ϕ_i becomes $2ni\pi$, cos ϕ_i equals 1, and sin ϕ_i equals 0, regardless of i. The in-phase

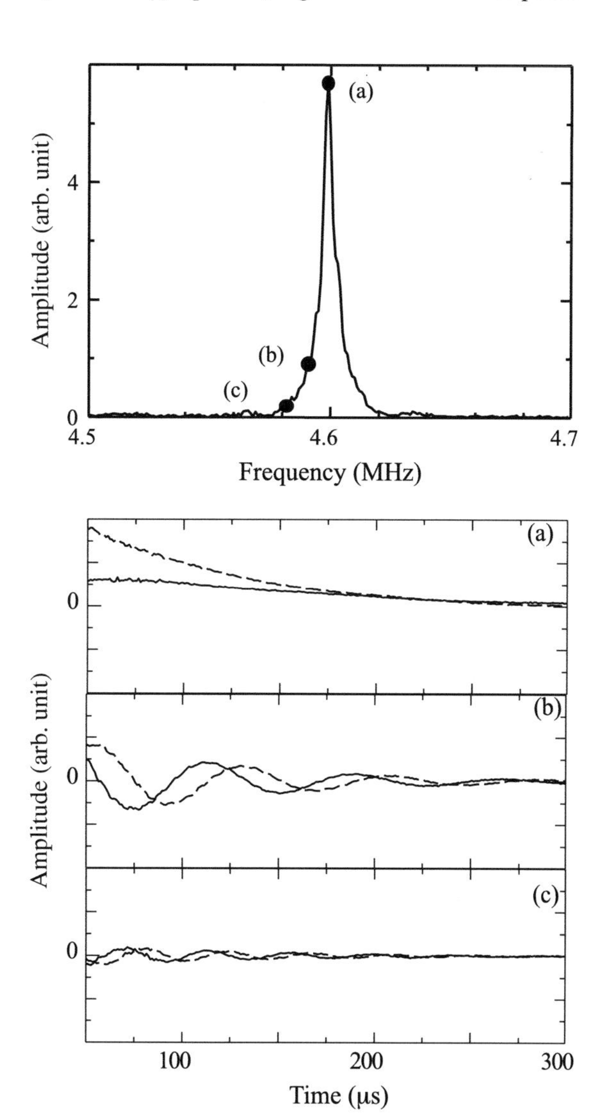

Fig. 5.4 Resonance spectrum and analog signals before the integration at three frequencies: **a** the resonance frequency, **b** an off-resonance frequency, and **c** a frequency far from resonance. *Solid lines* denote the out-of-phase signals, and *broken lines* denote the in-phase signals. Reprinted with permission from Hirao et al. 1993, Copyright [1993], AIP Publishing LLC

signal therefore remains positive and decays exponentially depending on the attenuation in the material. The amplitude spectrum becomes a large value (Fig. 5.4a). For a frequency of slightly off resonance, the signals show still high amplitudes, but decay with beats because ϕ_i changes with time. Their integrated values are not large any more because the beating signals take both positive and negative values along the time axis, which cancel each other by the integration (Fig. 5.4b). For a frequency far away from a resonance, the signal amplitudes are small because of destructive interference among reflections (Fig. 5.4c). Thus, two mechanisms, the cancelation within the integrator gate and the interference among overlapping echoes, work to build a sharp resonance peak.

5.3 Determination of Resonance Frequency and Phase Angle

Considering the case that the rf burst with a duration T_B is used to drive an EMAT to impinge a plane bulk wave into a plate of thickness d, the overlapping echo $A(t)$ received at the incident surface is expressed by

$$A(t) = A_1 e^{j\omega t} \sum_i \left[H(t - iT_0) e^{-i(\alpha T_0 + 2jkd)} \right],\ i = 1, 2, 3, \ldots \tag{5.10}$$

Here, A_1 is the amplitude of the first arriving echo, α the attenuation coefficient for time, and k the wavenumber (= ω/v). $H(t)$ is a function defined by

$$H(t) = \begin{cases} 0, & t \leq 0 \text{ or } t \geq T_B \\ 1, & 0 < t < T_B \end{cases}. \tag{5.11}$$

When the burst signal is long enough, that is, $T_B \gg T_0$, a continuous-wave approximation can be used and the factors $H(t - iT_0)$ all simultaneously equal unity. Equation (5.10) then reduces to

$$A(t) = \frac{A_1 e^{j\omega t}}{1 - e^{-(\alpha T_0 + j2kd)}}. \tag{5.12}$$

$|A(t)|$ is periodic in ω and shows peaks at the frequencies that satisfy Eq. (5.2). Focusing around one of the peaks, for example, $\omega = 2\pi/T_0$ and $kd \sim \pi$, and assuming a low-attenuation material ($\alpha T_0 \ll 1$), a Taylor series expansion simplifies $|A(t)|$ to

$$|A(t)|^2 = \frac{A_1}{T_0^2} \frac{1}{\alpha^2 + (\omega - 2\pi/T_0)^2}, \tag{5.13}$$

which is familiar Lorentzian function. It is, therefore, effective to obtain a resonance frequency by fitting a Lorentzian function to the measured spectrum around a peak and calculating the center axis.

The superheterodyne processing algorithm also provides the phase angle of the received signal. The in-phase and out-of-phase outputs of integrators, I_1 and I_2, for an isolated burst signal are proportional to $\cos(-\phi_i)$ and $\sin(-\phi_i)$ (see Eqs. (5.8) and (5.9)). From them, the phase angle ϕ_i is calculated by

$$\phi_i = \tan^{-1}(I_2/I_1). \tag{5.14}$$

References

Avogadro, A., Bonera, G., & Villa, M. (1979). The Clark method of recording lineshapes. *Journal of Magnetic Resonance, 35*, 387–407.

Bolef, D. I., & Miller, J. G. (1971). High-frequency continuous wave ultrasonics. In *Physical Acoustics* (Vol. 8, pp. 95–201). New York: Academic Press.

Chick, B. B. (1999). Research instruments and systems. In *Physical Acoustics* (Vol. 24, pp. 347–361). New York: Academic Press.

Clark, W. G. (1964). Pulsed nuclear resonance apparatus. *Review of Scientific Instruments, 35*, 316–333.

Fortunko, C. M., Petersen, G. L., Chick, B. B., Renken, M. C., & Preis, A. L. (1992). Absolute measurement of elastic-wave phase and group velocities in lossy materials. *Review of Scientific Instruments, 63*, 3477–3486.

Fukuoka, H., Hirao, M., Yamasaki, T., Ogi, H., Petersen, G. L., & Fortunko, C. M. (1993). Ultrasonic resonance method with EMAT for stress measurement in thin plate. In *Review of Progress in Quantitative Nondestructive Evaluation* (Vol. 12, pp. 2129–2136).

Hirao, M., Ogi, H., & Fukuoka, H. (1993). Resonance EMAT system for acoustoelastic stress evaluation in sheet metals. *Review of Scientific Instruments, 64*, 3198–3205.

Mozurkewich, G. (1989). Transducer and bond corrections for velocity measurements using continuous-wave ultrasound. *The Journal of the Acoustical Society of America, 86*, 885–890.

Ogi, H., Shimoike, G., Hirao, M., Takashima, K., & Higo, Y. (2002). Anisotropic elastic-stiffness coefficients of an amorphous Ni-P film. *Journal of Applied Physics, 91*, 4857–4862.

Petersen, G. L., Chick, B. B., & Fortunko, C. M. (1992). A versatile ultrasonic measurement system for flaw detection and material property characterization in composite materials. In *Nondestructive Characterization of Materials* (Vol. 5, pp. 847–856).

Petersen, G. L., Chick, B. B., Fortunko, C. M., & Hirao, M. (1994). Resonance techniques and apparatus for elastic-wave velocity determination in thin metal plates. *Review of Scientific Instruments, 65*, 192–198.

Chapter 6
Free-Decay Measurement for Attenuation and Internal Friction

Abstract Acoustically noncontacting measurement with an EMAT can be an important advantage for attenuation measurement, because it excludes coupling materials between transducer and specimen; contacting measurements allow wave propagation inside the coupling material and transducer as well as the specimen, causing significantly large extra energy losses. Even with EMAT measurement, diffraction loss appears. This chapter describes the attenuation measurement with electromagnetic acoustic resonance (EMAR) and a scheme for the diffraction loss correction for a bulk-wave EMAT.

Keywords Eddy current loss · Electromagnetic loss · EMAR · Internal friction · Seki parameter

6.1 Difficulty in Attenuation Measurement

Ultrasonic attenuation has a great utility in the wide variety of material characterization (Mason 1958; Truell et al. 1969; Beyer and Letcher 1969; Goebbels 1980; Ritchie and Fantozzi 1992). It is capable of sensing the dislocation mobility associated with deformation and heat treatment, and of determining the grain size of polycrystalline metals. The importance of attenuation for materials characterization is well recognized. Measurement difficulties are also well known and have prevented actual applications. This chapter discusses a principle for the attenuation measurement with plane bulk waves, relying on electromagnetic acoustic resonance (EMAR). Use of a noncontacting and weakly coupling EMAT has a distinct advantage of eliminating the extra energy losses, which otherwise occur in case of using a conventional contacting transducer.

In the attenuation measurement with a contacting transducer, on well-finished specimen surfaces, an ultrasonic beam loses its energy not only to attenuation in the specimen (absorption and scattering), but also to the following factors: (i) energy leakage into the transducer on reception, (ii) damping in the couplant and buffer, if any, and (iii) beam spreading (diffraction) loss. (This loss is absent when a free

M. Hirao and H. Ogi, *Electromagnetic Acoustic Transducers*,
Springer Series in Measurement Science and Technology,
DOI 10.1007/978-4-431-56036-4_6

vibration of a closed-shape specimen is concerned; Chap. 8). Being interested only in the attenuation in the sample, we must remove factors (i) to (iii) from the as-measured attenuation by proper correcting procedures. The diffraction loss can be estimated by a familiar formula (Seki et al. 1955; McSkimin 1960; Papadakis 1966), while corrections for the other factors are unrealistic because the acoustic parameters of all the components involved have to be known a priori.

Measurement with an EMAT is inherently free from losses associated with the interfaces because of the noncontact coupling, so that the energy loss arises only from attenuation through the specimen, the diffraction effect, and additionally the electromagnetic loss. The last factor occurs when the elastic wave travels through the magnetic field and induces eddy currents. But, the amount will be shown to be negligible compared with attenuation. The diffraction effect can be eliminated through a correcting algorithm at a resonance. It is then possible to evaluate an absolute value of ultrasonic attenuation by the EMAR measurement. Bulk-wave EMATs are used to measure the shear wave attenuation throughout this chapter, but the method is applicable to other configurations of EMAR measurements.

6.2 Isolation of Ultrasonic Attenuation

When an EMAT is used, the as-measured attenuation coefficient for time, α_m, consists of the material's attenuation (α), the diffraction loss (α_d), and the eddy current loss (α_e):

$$\alpha_m = \alpha + \alpha_d + \alpha_e. \tag{6.1}$$

(Note that the attenuation coefficient for distance equals that for time divided by velocity.) The eddy current loss occurs when the ultrasonic wave travels through the static magnetic field and the reversed Lorentz force mechanism gives rise to eddy currents in the material. A part of the eddy currents is finally picked up by the EMAT coil to give a received signal. For plane waves, the loss can be estimated by Ogi et al. (1995a)

$$\alpha_e = \frac{\eta B_0^2}{2\rho}. \tag{6.2}$$

Here, B_0 is the applied static magnetic flux density, and ρ and η denote the mass density and conductivity, respectively. Taking $B_0 = 0.5$ T, $\eta = 3.0 \times 10^6$ S/m and $\rho = 7850$ kg/m^3 for a standard steel, we obtain $\alpha_e = 4.7 \times 10^{-5}$ μs^{-1}, or $Q^{-1} = 1.5 \times 10^{-5}$ at 1 MHz. Q^{-1} represents internal friction and is related to the attenuation coefficient α with frequency f:

$$Q^{-1} = \frac{\alpha}{\pi f}. \tag{6.3}$$

The attenuation coefficients of plane bulk waves in common metals are of the order 10^{-3} μs^{-1} or larger at frequencies higher than 1 MHz. Thus, α_e can be neglected. But, α_e has to be taken into account for a material of a high conductivity and extremely low attenuation and/or for a strong magnetic field.

6.3 Measurement of Attenuation Coefficient

Resonance frequencies are measured by activating an EMAT with long, high-power rf bursts gated coherently, sweeping the operation frequency, and acquiring the amplitude spectrum (Chap. 5). We then generate a standing wave in the specimen by driving the EMAT with the measured resonance frequency. Only at a resonance, all the echoes become coherent with a phase, and the echo amplitudes can be summed to calculate the resultant amplitude. The in-phase components ($A_i \cos \phi_i$) decay exponentially with time, depending on the time coefficient α_m as illustrated in Fig. 6.1 (see also Fig. 5.4a). The out-of-phase components ($A_i \sin \phi_i$) also decay with α_m, but with much smaller amplitudes.

The ring-down curve, or the envelope of the decaying amplitude, is measurable by detecting the in-phase and out-of-phase outputs with a narrow gate, digitizing their amplitudes, and calculating the root of the sum of their squares. Figure 6.2 presents an example of the measured ring-down curve of a shear wave at the tenth resonance frequency around 2 MHz for a 6-mm-thick carbon steel. We obtain the relaxation time coefficient α_R at the resonance by fitting an exponential curve. The relation between α_R and α_m is investigated in details. The reverberation signal is

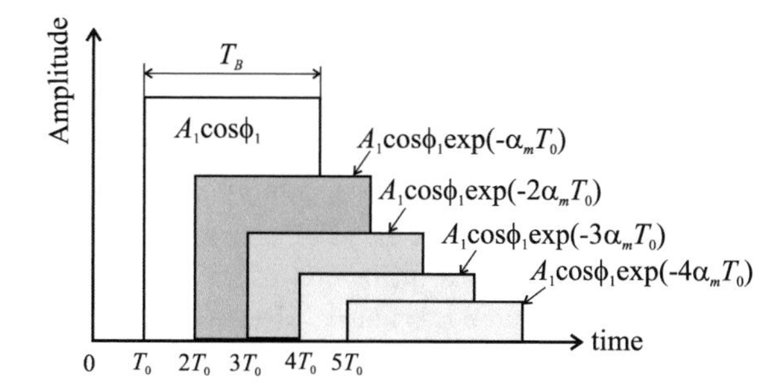

Fig. 6.1 Decay of the in-phase components of the received signal at a resonance. A_1 and ϕ_1 represent the amplitude and phase of the first echo, respectively. T_0 and T_B are the round-trip time through the specimen and the width of the driving rf bursts, respectively

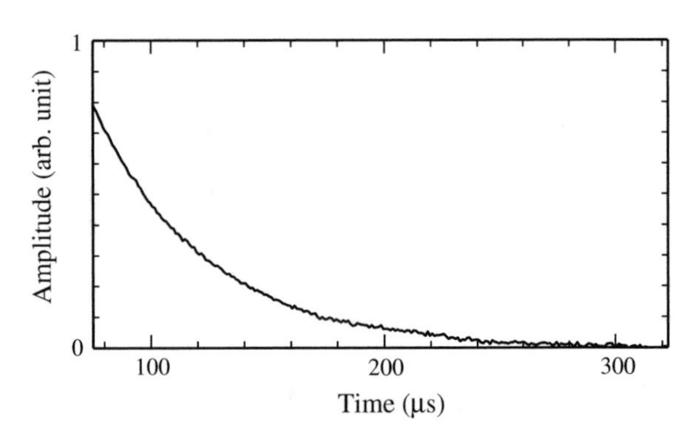

Fig. 6.2 Measured ring-down curve for a low-carbon steel at a shear wave resonance around 2 MHz. Reprinted with permission from Ogi et al. (1995b), Copyright (1995), Acoustical Society of America

numerically simulated using a damping coefficient α_m for individual reflection echoes in Fig. 6.1. The resultant ring-down signal reproduces α_R from the exponential decay constant, showing that $\alpha_R = \alpha_m$ (Ogi et al. 1994). This is true when the input burst width T_B is much longer than the round-trip time T_0.

The effects of liftoff and slightly off-resonance excitation are also studied, and both have shown to produce no significant influences on the attenuation measurement. The amplitude decreases with increasing liftoff, but the relaxation time coefficient (α_R) is independent of the amplitude, unless it is comparable with the noise intensity.

The resonance sharpness, or Q^{-1} value, also indicates α_m (Eq. (6.3)). However, the value for Q^{-1} is a derived quantity based on the geometry of the resonance measurement (EMAT size, sample thickness, liftoff, …). Moreover, the diffraction effect on the Q^{-1} value is unclear and cannot be removed.

6.4 Correction for Diffraction Loss

An ultrasonic beam radiated from a finite-size transducer spreads perpendicular to the propagation direction as it travels through the material. A part of the incident energy will never return to the transducer. This geometrical phenomenon, called *diffraction*, causes an apparent amplitude loss and a phase shift in the received echo signal. Following Seki et al. (1955), several authors (McSkimin 1960; Papadakis 1966) have studied the diffraction effect for a longitudinal wave radiated from a circular piston source transducer. However, the existing solutions are inapplicable to the EMAR measurements because of the noncircular transducer geometry, a nonuniform distribution of the body forces, and highly overlapped echoes at a resonance. Thus, a new approach should be developed.

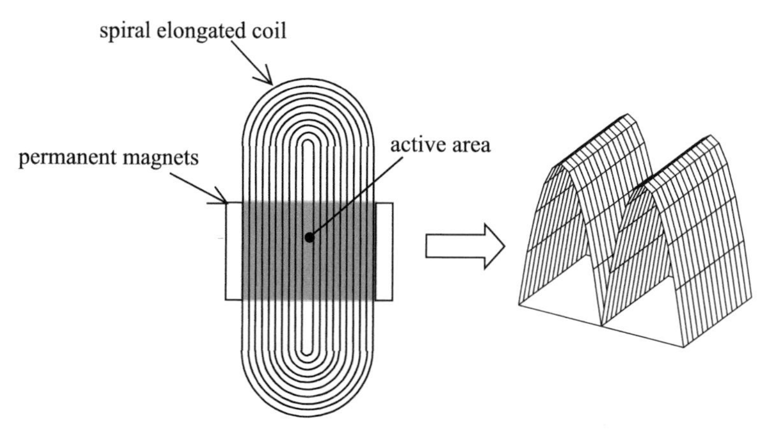

Fig. 6.3 A bulk-wave EMAT consisting of a spiral-elongated coil and a pair of permanent magnets causes a shearing force distribution at the specimen surface, which is approximated by twin parabolic curves. Reprinted with permission from Ogi et al. (1995c), Copyright (1995), Acoustical Society of America

6.4.1 *Diffraction Phenomena Radiated by a Bulk-Wave EMAT*

A three-dimensional distribution of the shearing body force is induced by a bulk-wave EMAT (Fig. 3.1), which consists of a spiral-elongated coil and two permanent magnets with the polarity normal to the surface. The active area of the EMAT is a superimposed section between the coil and magnets (Fig. 6.3). An FEM computation provides us with the detailed body-force distribution (Fig. 3.5), which can be simplified to be two-dimensional on the specimen surface and characterized by twin parabolic curves with the nodes at the edges and the center line as shown in Fig. 6.3. This simplification is allowable when the material has a good electrical conductivity and a high permeability, because such a material has a small electromagnetic skin depth, confining the body force in the thin surface region (see Eq. (2.19)).

Considering a shear wave propagating along the z-axis with the x polarization, the y component of the vector potential $\boldsymbol{\psi}$ for elastic displacement ($\mathbf{u} = \text{rot}\ \boldsymbol{\psi}$) governs the acoustic field. Once the shearing force distribution, or the distribution of $\boldsymbol{\psi}$, is given on the surface, one can calculate the acoustic field ψ_y at an arbitrary location $\mathbf{r}$ by integrating the radiation fields from all the source elements over the sending surface D_T as illustrated in Fig. 6.4:

$$\psi_y(\mathbf{r}) = \sum_i \frac{\Delta_i \psi_i}{2\pi} \left(\frac{z}{|\mathbf{r} - \mathbf{r}_i'|^3} + jk_s \frac{z}{|\mathbf{r} - \mathbf{r}_i'|^2} \right) e^{j(\omega t - k_s |\mathbf{r} - \mathbf{r}_i'|)}. \tag{6.4}$$

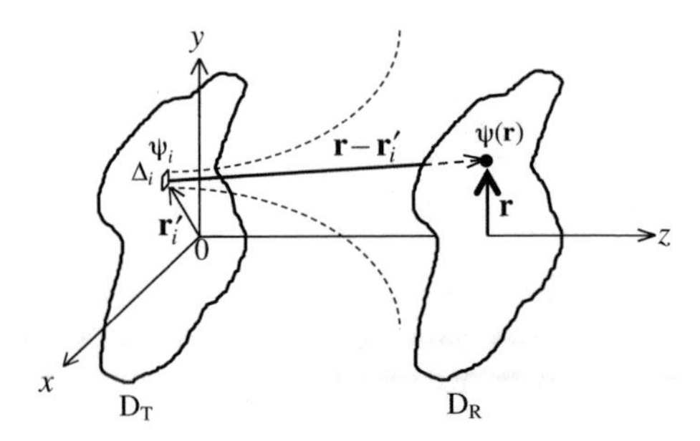

Fig. 6.4 Acoustic field $\psi(\mathbf{r})$ at $\mathbf{r}$ on the receiving area D_R radiated from all oscillating elements on the radiating region D_T. Δ_i and ψ_i represent the strength and area of a small element on the radiating region D_T. $\mathbf{r}_i'$ is the local coordinate on D_T. Reprinted with permission from Ogi et al. (1995c), Copyright (1995), Acoustical Society of America

Thus obtained acoustic field provides the amplitude and phase profiles on an arbitrary plane perpendicular to the propagation direction. The amplitude loss from diffraction is available by the ratio of the total power over the receiving area D_R to that on the radiating area D_T (Ogi et al. 1995c).

Figure 6.5 shows the calculated amplitude loss of the shear wave radiated from the bulk-wave EMAT, whose effective region is a square with 14 × 20 mm^2 area.

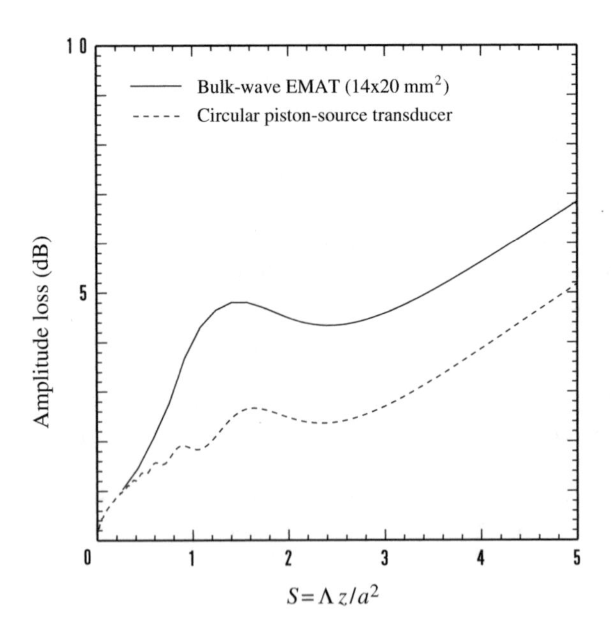

Fig. 6.5 Amplitude loss caused by diffraction of a shear wave launched by the bulk-wave EMAT with 14 × 20-mm^2 active area. The body-force distribution is approximated by the double-parabolic profile. Broken line shows the classical result for a circular piston source transducer of the same area given by Seki et al. (1955). Λ denotes the wavelength, z the propagation distance, and a the equivalent radius of the transducer. Reprinted with permission from Ogi et al. (1995b), Copyright (1995), Acoustical Society of America

For a comparison, the classical solution by Seki et al. (1955) is shown for a circular piston source transducer of the same aperture area. The phase shift from diffraction causes little influence on the attenuation measurement because the variation is limited to within the 0 to $\pi/2$ range and it is asymptote to the maximum $\pi/2$ as S ($=\Lambda z/a^2$) increases (Ogi et al. 1995c).

6.4.2 Correction at a Resonant State

Figure 6.6 sketches the correcting algorithm for the diffraction loss at a resonance. The ring-down signal at a resonance is composed of reflection echoes with a width T_B superimposed in phase. They are delayed by integer multiples of the round-trip time T_0 though the thickness. The amplitude of the echo decreases as the reflection proceeds due to attenuation and the diffraction loss. The diffraction loss of individual echo is determined from the calculated amplitude loss (Fig. 6.5), which

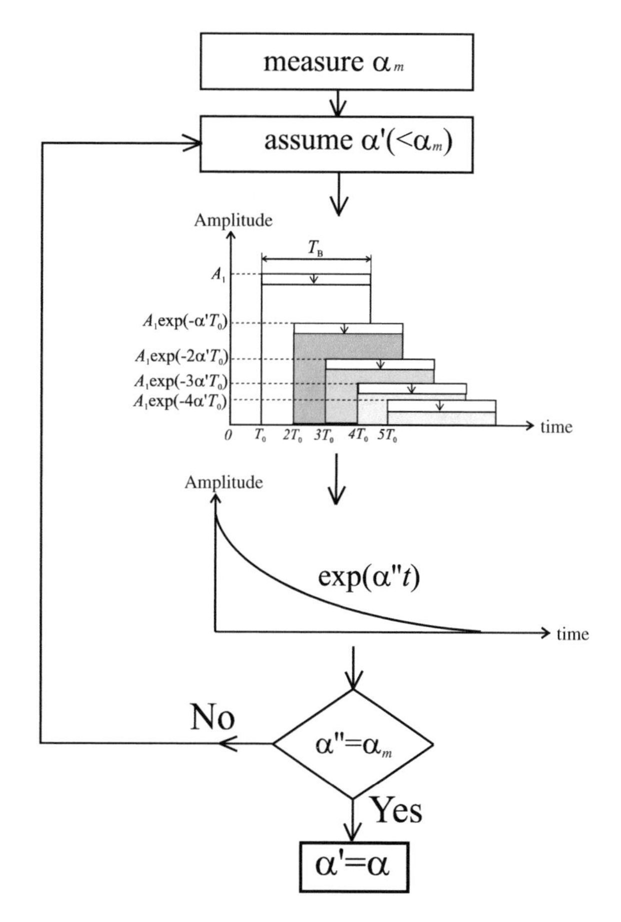

Fig. 6.6 Flowchart of correction algorithm to remove the diffraction loss in the resonance method. Reprinted with permission from Ogi et al. (1995b), Copyright (1995), Acoustical Society of America

depends on the EMAT size in use, frequency, and the propagation distance. The diffraction effect causes different losses to the individual echoes, because they propagate different distances. Since all the echoes are coherent, summing their amplitudes simply provides the ring-down curve. The correction proceeds in iteration calculation as follows:

(i) Measure the resonance frequency and the accompanying attenuation coefficient α_m.
(ii) Assume a diffraction-free attenuation coefficient α' ($\alpha' < \alpha_m$).
(iii) Calculate the amplitude of the echoes $A_i(t)$ by Eq. (5.5) with α'. Reduce the amplitudes $A_i(t)$ by introducing the diffraction loss from the calculation for the EMAT in use.
(iv) Construct the ring-down curve by summing up the echoes including the diffraction losses and calculate the time coefficient α''. Now, if the guessed α' is correct, α'' must equal the measured α_m, because α'' involves both the attenuation in the material (α) and the diffraction loss (α_d). This iteration calculation is repeated until $|\alpha_m - \alpha''|/\alpha_m < 10^{-4}$ is achieved.

6.5 Comparison with Conventional Technique

Figure 6.7 shows the frequency dependence of the attenuation coefficients measured by the EMAR method for a 25-mm-thick carbon steel with 100×100 mm^2 area. Two bulk-wave EMATs are used; their effective areas are 14×20 mm^2 and

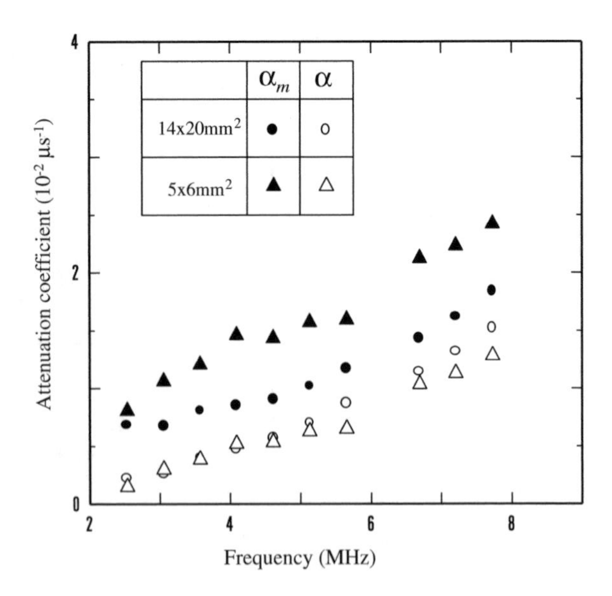

Fig. 6.7 Measurements of shear wave attenuation coefficients for a 25-mm-thick carbon steel with two bulk-wave EMATs with the active areas of 14×20 mm^2 and 5×6 mm^2. Solid marks are the as-measured coefficients and open marks the coefficients after correction for the diffraction loss

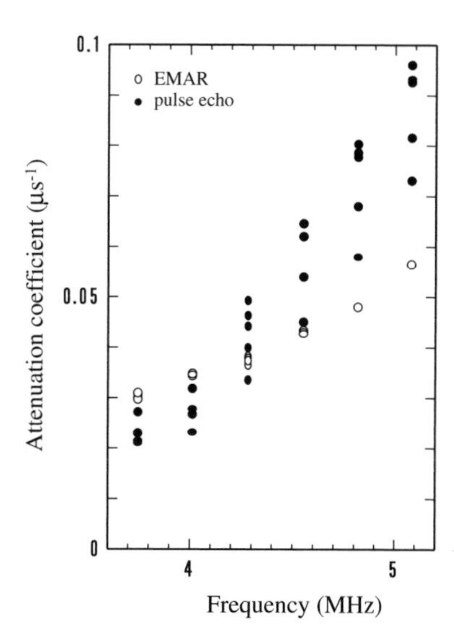

Fig. 6.8 Attenuation coefficients for a 6-mm-thick carbon steel measured by the EMAR and conventional pulse echo methods

5×6 mm^2. The burst signals of 120-μs duration drove the EMATs. The as-measured coefficients (α_m) with the two EMATs differ from each other, because the different transducer geometry causes different diffraction loss. But, the coefficients α corrected for the diffraction effect agree well with each other. A smaller EMAT induces more diffraction loss. The absolute value for the attenuation coefficients can then be obtained with the EMAR method by incorporating the calculated diffraction loss.

Figure 6.8 compares the EMAR measurements with the conventional pulse echo measurements. The specimen was a 6-mm-thick carbon steel plate of 100×100 mm^2. In the pulse echo measurements, a piezoelectric shear wave transducer was used, which had a circular area of 12 mm diameter and 5-MHz center frequency, and the transducer/buffer/specimen system proposed by Papadakis (1984) was used to minimize the reflection and transmission losses at the transducer-specimen interface. The received signals were digitized and the amplitudes at the resonance frequencies were determined by the conventional FFT method. The diffraction effect was corrected with Seki's solution (1955). The measurements were repeated five times at the same position for each result. It is obvious that the EMAR method is superior to the pulse echo method in reproducibility. This occurs not only because of using the noncontacting EMAT but also because a large number of echoes participate in the resonance, making the measurement stable and robust against noises.

References

Beyer, R. T., & Letcher, S. V. (1969). *Physical Ultrasonics*. New York: Academic Press.

Goebbels, K. (1980). Structure analysis by scattered ultrasonic radiation. In *Research Techniques in Nondestructive Testing* (Vol. 4, pp. 87–157). London: Academic Press.

Mason, W. P. (1958). *Physical Acoustics and Properties of Solids*. Princeton: Van Nostrand.

McSkimin, H. J. (1960). Empirical study of the effect of diffraction on velocity of propagation of high-frequency ultrasonic waves. *The Journal of the Acoustical Society of America, 32*, 1401–1404.

Ogi, H., Hirao, M., & Fukuoka, H. (1994). Grain size measurement in carbon steels with electromagnetic acoustic resonance. *Transactions of the Japan Society of Mechanical Engineers A, 60*, 258–263 (in Japanese).

Ogi, H., Hirao, M., Honda, T., & Fukuoka, H. (1995a). Absolute measurement of ultrasonic attenuation by electromagnetic acoustic resonance. In *In Review of Progress in Quantitative Nondestructive Evaluation* (Vol. 14, pp. 1601–1608).

Ogi, H., Hirao, M., & Honda, T. (1995b). Ultrasonic attenuation and grain size evaluation using electromagnetic acoustic resonance. *The Journal of the Acoustical Society of America, 98*, 458–464.

Ogi, H., Hirao, M., Honda, T., & Fukuoka, H. (1995c). Ultrasonic diffraction from a transducer with arbitrary geometry and strength distribution. *The Journal of the Acoustical Society of America, 98*, 1191–1198.

Papadakis, E. P. (1966). Ultrasonic diffraction loss and phase change in anisotropic materials. *The Journal of the Acoustical Society of America, 40*, 863–876.

Papadakis, E. P. (1984). Absolute measurements of ultrasonic attenuation using damped nondestructive testing transducers. *Journal of Testing and Evaluation, 12*, 273–279.

Ritchie, I.G. & Fantozzi, G. (1992). Internal friction due to the intrinsic properties of dislocations in metals. In *Dislocations in Solids* (Vol. 9, pp. 57–133). Amsterdam: Elsevier.

Seki, H., Granato, A., & Truell, R. (1955). Diffraction effects in the ultrasonic field of a piston source and their importance in the accurate measurement of attenuation. *The Journal of the Acoustical Society of America, 28*, 230–238.

Truell, R., Elbaum, C., & Chick, B. B. (1969). *Ultrasonic Methods in Solid State Physics*. New York: Academic Press.

Part III
Physical Acoustic Studies

In Part III, EMAR is applied to studying the physical acoustics. New measurements emerged on a number of subjects: in situ monitoring of dislocation behavior, determination of anisotropic elastic constants, antenna transmission technique for characterization of piezoelectric materials, resonance ultrasound microscopy, and acoustic nonlinearity evolution.

Chapter 7
In-Situ Monitoring of Dislocation Mobility

Abstract The noncontacting measurement with EMAR allows not only accurate evaluation of absorption loss inside materials but also such measurements at low and high temperatures and during deformation. This chapter shows many applications of EMAR for studying dislocation mobility and interactions between dislocations and point defects.

Keywords Acoustoelasticity · Activation energy · Activation volume · Dislocation density · Dislocation segment length · Granato and Lücke theory · Pinning and depinning · Recovery · Recrystallization · Second-/third-order elastic constants · Vacancy diffusion

7.1 Dislocation Damping Model for Low Frequencies

Dislocations are pinned by point defects such as vacancies and interstitials. They vibrate anelastically responding to the ultrasonic wave and absorb its energy, resulting in increase of attenuation α and decrease of modulus. Granato and Lücke (1956) established a dislocation-damping theory to relate the ultrasonic velocity and attenuation with dislocation characteristics such as the segment length L and density Λ. The detailed expressions appear in their original paper and in many monographs (for example, Mason 1958; Truell et al. 1969). For frequencies well below the resonance frequency of a single dislocation segment line, they can be reduced to

$$\alpha = \left(\frac{4GB|\mathbf{b}|^2 \omega^2}{\pi^6 C^2} \right) \Lambda L^4, \tag{7.1}$$

$$\frac{V - V_0}{V_0} = -\left(\frac{4G|\mathbf{b}|^2}{\pi^4 C} \right) \Lambda L^2. \tag{7.2}$$

Here, G denotes the shear modulus, B the specific damping constant, and $\mathbf{b}$ Burgers vector. V_0 is the purely elastic velocity (velocity without the dislocation effect).

M. Hirao and H. Ogi, *Electromagnetic Acoustic Transducers*,
Springer Series in Measurement Science and Technology,
DOI 10.1007/978-4-431-56036-4_7

C denotes the average tension of the dislocation line and is expressed in a wide range of metals by Granato and Lücke (1956)

$$C = AG|\mathbf{b}|^2. \tag{7.3}$$

Here, the constant A depends on the type of the dislocation (edge or screw), Poisson's ratio, and the shape of the bowing dislocation line (Hull and Bacon 1984). These simplifications apply in most EMAT studies, because EMATs usually use frequencies below 10 MHz, which is normally much lower than the resonance frequency of a single dislocation line (10^7–10^8 Hz in Cu).

Granato and Lücke's theory explained many measurements and is now well recognized as a key theory in this field. After their study, ultrasonics has been adopted for studying dislocation mobility in metallic materials. Few techniques are capable of evaluating the dynamic behavior of dislocations in bulk materials. Most previous studies (Hikata et al. 1956; Truell et al. 1969; Gremaud et al. 1987; Kuang and Zhu 1994; Zhu and Fei 1994) used the pulse echo technique with contacting transducers, in which two echoes were selected for the velocity and attenuation. The velocity was estimated from the difference of their transit times and the attenuation coefficient from the ratio of their amplitudes. But, as shown in Fig. 1.1 and discussed in Chaps. 5 and 6, the contact transducers absorb much energy from ultrasound and fatally disturb the attenuation measurement.

The EMAR measurements are truly free from these problems. Owing to their noncontacting nature, the resulting ultrasonic properties show high enough

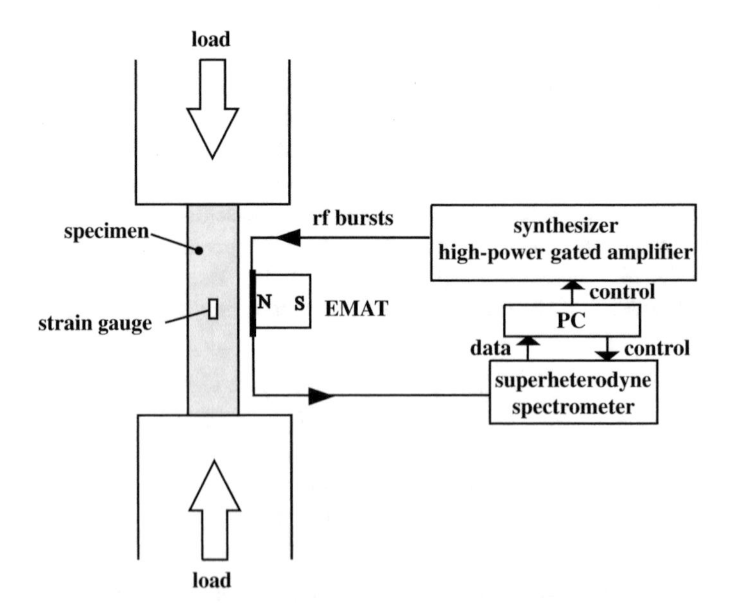

Fig. 7.1 Side view of the measurement setup with a bulk-wave EMAT for the continuous measurement of the attenuation coefficient and the phase velocity

sensitivity to study the minute dislocation characteristics. This chapter presents studies on the interaction between the ultrasonic properties and dislocations using the EMAR method.

7.2 Elasto-Plastic Deformation in Copper

The attenuation coefficients and the phase velocities of the polarized shear wave and the longitudinal wave were simultaneously measured in polycrystalline copper during compressive or tensile deformation process (Ogi et al. 1998). Plate specimens were machined from a commercial hot-rolled copper plate of 99.99 mass% purity. Grain size was determined to be about 35 μm with optical observation. The compression specimens were 7 mm thick, 35 mm wide, and 40 mm long. The tensile specimens were 7 mm thick, 35 mm wide, and 230 mm long. The uniaxial stress was applied in the specimen's longitudinal direction. They were annealed at 200 °C for 1 h before the measurements.

Figure 7.1 shows the measurement system. A bulk-wave EMAT was placed with a 0.2-mm liftoff from the specimen surface, adjusting the shear wave polarization parallel or perpendicular to the load. The EMAT consisted of a permanent-magnet block and a spiral coil (see Fig. 3.1). The active area was 10×10 mm^2. The Lorentz force mechanism generated and received simultaneously a shear wave polarized parallel to the surface and a longitudinal wave; both were propagated in the thickness direction.

Two strain gauges were attached on both sides of the specimen to monitor the transverse strains and thickness change. The thickness resonance frequencies and corresponding attenuations were continuously measured throughout the deformation process together with the load and strain. One set of measurements took a few seconds. The loading speed was kept constant at 0.3 MPa/s. For the other polarization, the same measurement was done with another specimen and the EMAT was rotated by 90°. Resonances around 2.3 MHz (n = 14 for the shear waves and n = 7 for the longitudinal wave) were used. Their spectra and the ring-down curve are shown in Fig. 7.2.

The evolutions of velocity (V) and attenuation (α) were monitored during the first loading process of the annealed specimens. Figure 7.3 shows typical responses of the transverse strain and the attenuation coefficients for the three bulk waves when the uniaxial compressive stress varied up to 170 MPa. For comparison, it includes the result for the shear wave with the parallel polarization from the tensile-stress tests. In every case, α increased with stress. Other general observations are as follows:

(i) At the beginning of the deformation, α of the shear waves increased with stress, while that of the longitudinal wave was nearly stable.

(ii) From the stress around 20–40 MPa, which was far below the macroscopic yield point, the increase rate of α decreased with the shear waves. After this, α of the longitudinal wave started to increase.

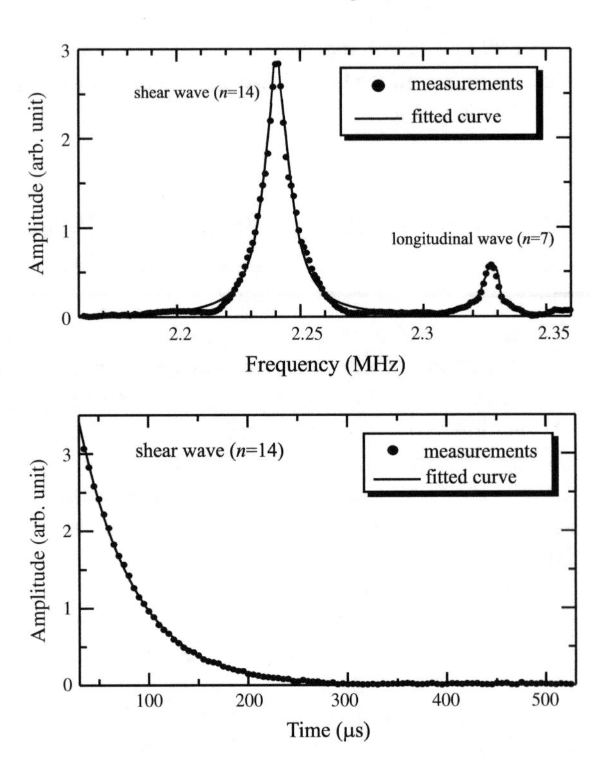

Fig. 7.2 Thickness resonance spectrum and the attenuation curve of the shear wave. (After Ogi et al. 1998.)

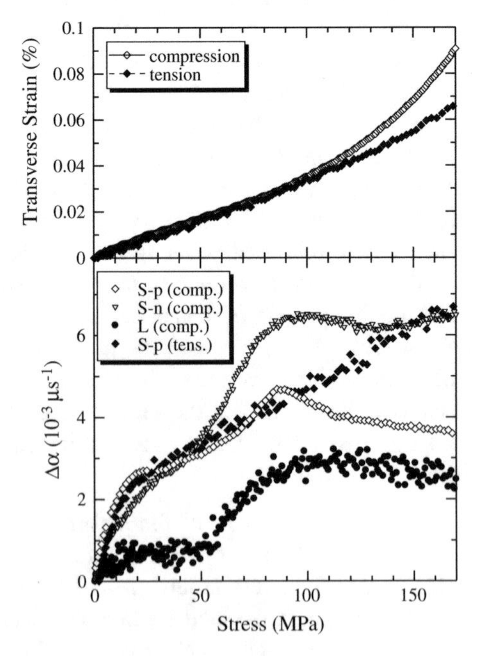

Fig. 7.3 Measured stress-strain curve and the evolution of the attenuation coefficient for the three bulk waves. S-p and S-n denote the results for the shear waves polarized parallel and normal to the stress, respectively. L denotes the measurements of the longitudinal wave. (After Ogi et al. 1998.)

(iii) α responded similarly for compressive and tensile tests below the yield point.
(iv) Around the yield stress apparently at 90 MPa, α became a maximum and then decreased with stress in the compressive tests, while it continued increasing in the tensile tests.
(v) The shear wave polarization made little difference in the evolution of α.
(vi) α changed more with the shear waves than with the longitudinal wave.

Contrary to the attenuation response, there was a remarkable difference in the velocity responses between the compression and the tension tests even in the low-stress region. Figure 7.4 plots the as-measured velocity change with solid

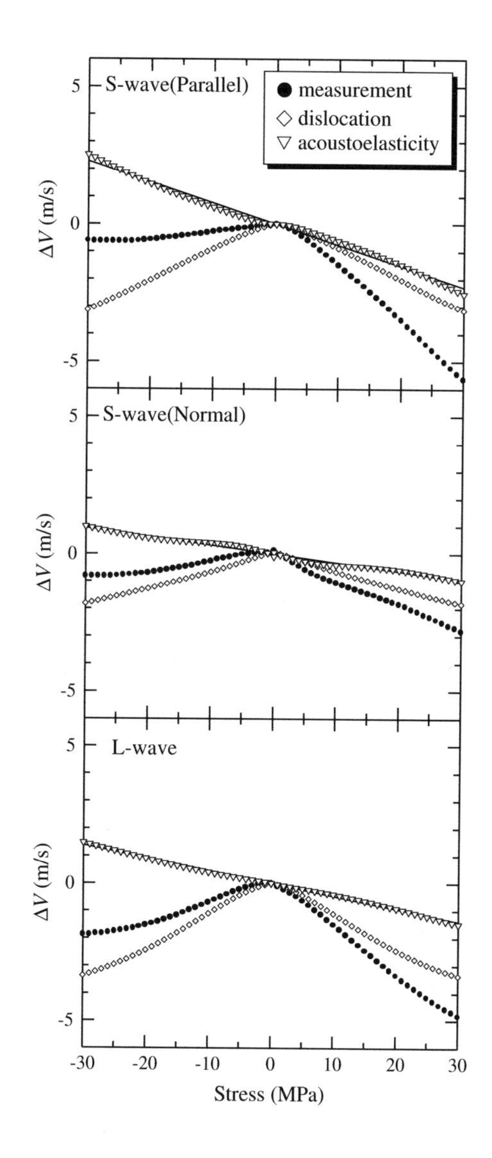

Fig. 7.4 Velocity changes of the bulk waves caused by the uniaxial compressive and tensile stresses. The contributions of the dislocation mobility and the acoustoelasticity are plotted separately. (After Ogi et al. 1998.)

circles. For all the cases, V decreased with the applied stress in the elastic region; the decrement rate was larger for the tensile stress cases.

The dislocation damping theory can explain the attenuation responses. α is proportional to the dislocation density Λ and the fourth power of the mean length of the vibrating dislocation segment L in Eq. (7.1). Nearly, the same change of α in the compression and tension tests below the yield point indicates that the dislocation movement in the first loading after annealing is independent of the stress direction. The anelastic vibration of dislocations also affects the phase velocity through Eq. (7.2). If the dislocation damping was the principal cause of the velocity change, then it should have shifted the velocity at the same rate for tension and compression tests as it did on the attenuation coefficient. However, there is an obvious dependence of the velocity on the stress direction in Fig. 7.4. This can be explained by considering the acoustoelastic effect (Chap. 12). Acoustoelasticity shifts the velocity in opposite directions for the compressive and tensile stresses. Ultrasonic velocity exhibits a linear dependence on the uniaxial stress σ; that is, $V(\sigma) = V_0(1 + C_U\sigma)$. C_U is the acoustoelastic constant, which can be expressed in terms of the second- and third-order elastic constants and is independent of the stress direction. This phenomenon is originated from the lattice's anharmonicity. The stress application changes the interatomic distance, which perturbs the binding force and the elastic constants.

The attenuation evolution is attributable to the dislocation damping, because it is independent of the stress direction. The velocity shift, however, consists of the contributions both of the dislocation movement and acoustoelasticity. Considering the acoustoelastic effect, Eq. (7.2) is modified as

$$\frac{V - V_0}{V_0} = -\left(\frac{4G|\mathbf{b}|^2}{\pi^4 C}\right)\Lambda L^2 + C_U\sigma. \tag{7.4}$$

This indicates that one can distinguish between the velocity changes caused by the dislocation motion and by the acoustoelastic effect by performing the compressive and tensile tests. Figure 7.4 plots separately the dislocation and acoustoelastic contributions to the velocity change. The two effects cause velocity changes in comparable magnitudes. By fitting a regression linear function to the velocity change for the acoustoelasticity, C_U's for the three bulk waves are obtained, which appear in Table 7.1 together with V_0's.

A number of efforts have been made to remove the effect of dislocation movement on the velocity to determine the higher-order elastic constants on the basis of acoustoelastic theory. Hiki and Granato (1966) deformed the specimen with the purpose of multiplying dislocations and making them immobile. Salama and Alers (1967) added a substitutional element to the specimen to pin the dislocations and applied neutron irradiation. Those efforts were made for deriving the third-order elastic constants of monocrystal copper. Fukuoka and Toda (1977) measured the velocity changes of the bulk waves with uniaxial stresses in polycrystalline copper and observed nonlinear acoustoelasticity that C_U is a function of stress and its

Table 7.1 Velocities at the initial state, the acoustoelastic constants for the three bulk waves, and the second-order and third-order elastic constants of polycrystalline copper determined by (a) the EMAR method by Ogi et al. (1998); (b) the prediction using data from Hiki and Granato (1966); and (c) the prediction using data from Salama and Alers (1967). The superscripts (L), (S1), and (S2) indicate the quantities for the longitudinal wave and the shear waves with polarizations parallel and perpendicular to the stress, respectively

	V_0 (m/s)			C_U (10^{-6} MPa^{-1})			Elastic constants (10^2 GPa)				
	$V_0^{(S1)}$	$V_0^{(S2)}$	$V_0^{(L)}$	$C_U^{(S1)}$	$C_U^{(S2)}$	$C_U^{(L)}$	λ	μ	ν_1	ν_2	ν_3
EMAR[a]	2243	2245	4670	−34	−14.1	−10	1.05	0.449	−42.1	−8.9	−1.3
Prediction[b]	—	—	—	−21	15	6.4	—	—	−2.1	−1.3	−1.9
Prediction[c]	—	—	—	−15	3.8	3.2	—	—	−2.6	−2.1	−1.2

direction. They ascribed the observation to the higher-order elasticity, ignoring the dislocation effect, and refrained from calculating the third-order elastic constants.

Assuming a quasi-isotropic material, one can derive the third-order elastic constants from the three acoustoelastic constants (Hughes and Kelly 1953; Fukuoka and Toda 1977; Pao et al. 1984; see Chap. 12) via

$$\left.\begin{aligned} \nu_1 &= \frac{2\mu(\lambda+2\mu)(3\lambda+2\mu)C_U^{(L)}+2\lambda(\lambda+2\mu)+2\nu_2(\lambda-\mu)+4\nu_3\lambda}{\mu} \\ \nu_2 &= \frac{4\mu^2(3\lambda+2\mu)C_S^{+}-(\mu+\nu_3)(2\mu-\lambda)}{2\mu} \\ \nu_3 &= 2\mu^2 C_S^{-}-\mu \end{aligned}\right\}, \tag{7.5}$$

$$\left.\begin{aligned} C_S^{-} &= \frac{V_0^{(S1)}C_U^{(S1)}-V_0^{(S2)}C_U^{(S2)}}{V_0^{(S)}} \\ C_S^{+} &= \frac{V_0^{(S1)}C_U^{(S1)}+V_0^{(S2)}C_U^{(S2)}}{2V_0^{(S)}} \end{aligned}\right\}. \tag{7.6}$$

The superscripts (L), (S1), and (S2) indicate the quantities for the longitudinal wave and the shear waves with polarizations parallel and perpendicular to the stress, respectively. $V_0^{(s)}$ denotes the shear wave velocity at the stress-free state. λ and μ are the second-order elastic (Lamé) constants and are determined from $V_0^{(S)}$ and $V_0^{(L)}$ and the mass density of 8930 kg/m^3.

Table 7.1 gives the *pure* third-order elastic constants of polycrystalline copper thus determined. The values are compared with those by Hiki and Granato (1966) and Salama and Alers (1967). For this, the Voigt-averaging scheme (Johnson 1982) was used to derive the acoustoelastic constants and the third-order elastic constants of isotropic polycrystalline copper from their monocrystal third-order elastic constants. The results are also shown in Table 7.1. There is a considerable difference except for ν_3. This suggests that the dislocation effect was not completely removed in the previous studies, where the measurements were done only for the compressive stress and the dislocation effect might have reversed the velocity increase with stress to decrease as demonstrated in Fig. 7.4. The predicted C_U from Salama and Alers' study is closer to those from the EMAR measurement, indicating that addition of a substitutional element and neutron irradiation are more effective than plastic deformation in suppressing the dislocation mobility. The values of ν_3 are close to each other. For weakly anisotropic metals, ν_3 is approximately proportional to $C_U^{(S1)} - C_U^{(S2)}$, that is, the difference of the acoustoelastic constants between the shear waves of the orthogonal polarizations. Figure 7.4 reveals that the dislocation effect tends to decrease the velocity irrespective of the shear wave polarization. Thus, taking the velocity difference can minimize the influence. This interpretation is supported by the similar attenuation responses of the two polarizations in Fig. 7.3.

Equations (7.1) and (7.2) were used to calculate Λ and L from the evolutions of α and V of the shear wave polarized in the stress direction (Fig. 7.5b). Figure 7.5c shows the results. This calculation neglects the dislocations that exist before the deformation, assuming that the variations of Λ and L arise from the dislocations that move with stress and vibrate with the shear wave. The parameters used were $G = 45$ GPa, $|\mathbf{b}| = 2.55 \times 10^{-10}$ m, $B = 10^{-2}$ Ns/m^2, and $C_U^{(S1)} = -34 \times 10^{-6}$ MPa^{-1}. The value for A in Eq. (7.3) ranges between 0.5 and 1.0, depending on the fraction of the edge and screw dislocations and their orientations (Hull and Bacon 1984). It was fixed to $A = 0.75$.

This calculation reveals the sequence of four regions in the dislocation evolution as shown in Fig. 7.5a. In region I, the dislocation segments are weekly pinned by the point detects and bow out responding to the applied stress, which increases Λ

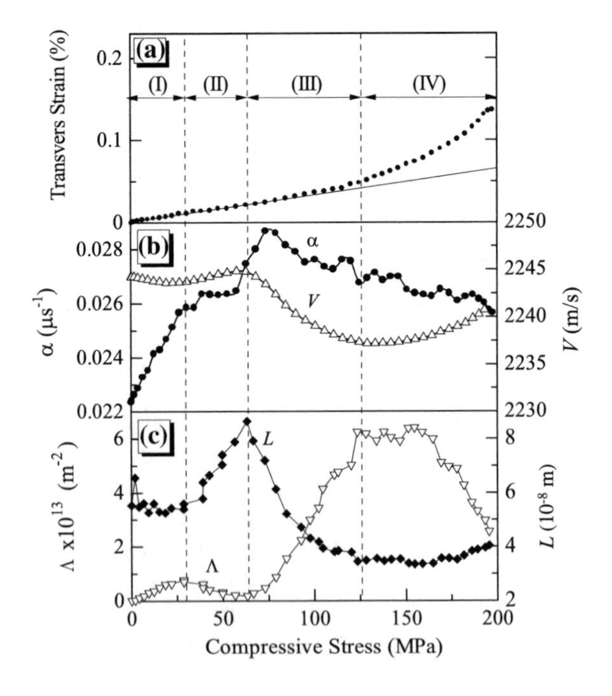

Fig. 7.5 **a** Transverse strain, **b** evolutions of attenuation coefficient α and the phase velocity V of the shear wave polarized in the compressive stress direction, and **c** calculated dislocation density Λ and dislocation segment length L. (After Ogi et al. 1998.)

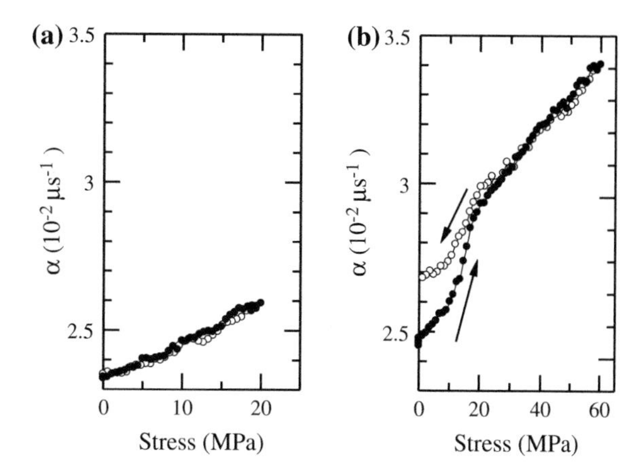

Fig. 7.6 Behavior of the attenuation coefficient α of the shear wave when the stress is released from **a** region I and **b** region II. Solid and open circles are for measurements during loading and unloading processes, respectively. The shear wave polarization is in the stress direction. (After Ogi et al. 1998.)

slightly and then α. At a larger stress, the dislocations break away from the pinning points and will stop at rigid obstacles such as precipitates. At this stage, the dislocation segment length becomes longer than the initial state and bowing out from the new pinning points will occur.

To support this interpretation, additional experiments were made for the shear wave, in which the compressive stress was released at different stress levels (Fig. 7.6). When the stress was released within region I, α showed a reversible behavior and returned completely to the initial value. However, when the stress was released from region II, it showed a reversible change in region II, but it followed a different path in region I. When the stress was completely released, α took a value larger than the original.

This observation suggests the following dislocation behavior: In region I, the dislocations do not break away from the pinning points and they trace the same path during the loading-unloading process, which results in a reversible change of α. When the stress reaches the depinning level, the discontinuous change occurs in the segment length L because of the repinning at the rigid pinning points, which causes the discontinuous incremental rate of α. Bowing out from the new equilibrium state corresponds to reversible deformation during further loading and subsequent unloading processes, so that a reversible attenuation change is observed in region II. The different trace of α in region I in Fig. 7.6b indicates that the dislocations did not return to the initial equilibrium state and the longer segment length remained after unloading. The larger incremental rate of L in region II in Fig. 7.5c is therefore understood by the bowing out of the dislocations from the new, sparsely scattered, pinning points.

At the beginning of region III, there is a steep increase in Λ and a decrease in L, which can be interpreted as dislocation multiplication. Dislocations multiply from the rigid pinning points (Frank-Read sources) with a sufficiently large stress, which is followed by piling up against the obstacles such as grain boundaries and other dislocations. Highly tangled dislocations make their segments shorter and reduce the attenuation, then causing the attenuation peak in region III. It should be noted that the stress at which Λ starts increasing (beginning of region III) corresponds to the point at which the stress-strain curve departs from the linear relation (Fig. 7.5a). The attenuation peak thus tells us about the yield stress. Further stress causes dislocation tangling and finally reduces the effective dislocations, which can vibrate responding to the shear wave and then leads to the decrease of Λ in region IV.

The attenuation peak was absent for the tensile loading in Fig. 7.3. This is because of longer specimen geometry and work hardening. The parallel parts of the specimens for the tension test were five times longer than those for the compression test. The work hardening occurs accompanying the dislocation multiplication and tangling. The process is not necessarily uniform in space. The EMAT can well detect the evolution in the case of compression test because of the limited space, while it may fail to detect it for the larger space of the specimens for the tension test. The smaller transverse strain of the tensile test after yielding in Fig. 7.3 indicates that the dislocations were less active in the particular area of measurement and did not multiply enough to produce the attenuation peak.

The most important observation is the decreased slope of the attenuation increase at the end of region I (Fig. 7.5), which always appears in the shear wave measurements, being independent of the polarization direction and the stress directions. This observation points out the first depinning point, or the microscopic yielding stress, which must be related to mechanical properties such as the fatigue limit. The longitudinal wave is incapable of sensing this event in the attenuation response. The shear waves are more sensitive to the dislocation mobility than the longitudinal wave. After all, the dislocations alter the ultrasonic properties through their slip on the lattice planes, being activated by the ultrasonic shear stress.

7.3 Point-Defect Diffusion toward Dislocations in Aluminum

7.3.1 Diffusion under Stress

In deformed metals, ultrasonic attenuation and sound velocity tend to return to their values before deformation. Recovery of the ultrasonic properties closely relates to the kinetics of point-defect migration toward dislocations. Thus, such experiments help us understand the diffusion process of the point defects.

Because this kind of recovery occurs below the recrystallization temperature, it is caused by decrease either of dislocation mobility or density. Possible mechanisms are pinning by point defects, annihilation, and interactions among dislocations (rearrangement). Granato et al. (1958) showed that only the pinning mechanism could explain the attenuation evolution during the loading-holding-unloading process of aluminum. They proposed a theory (GHL theory) that combined the vibrating-string model (Granato and Lücke 1956) and the migration model for impurity atoms given by Cottrell and Bilby (1949). GHL theory was consistent with measurements at early stages of the recovery, and several researchers used it to study the diffusion of point defects toward dislocations (Phillips and Pratt 1970; Anderson and Pollard 1979). It, however, fails to explain long-time recovery, because the Cottrell-Bilby model neglects saturation of the pinning points on a dislocation. Further efforts were made to improve the model and several modifications were suggested (Harper 1951; Ham 1959; Bullough and Newman 1962a, b, 1970; Hartley and Wilson 1963; Bratina 1966).

The main purpose in this section is to consider the stress dependence of point-defect diffusion to dislocations in polycrystalline aluminum using shear wave thickness resonance. Faster recovery of the ultrasonic properties in aluminum, reflecting lower migration energy of point defects, permits us to apply GHL theory. The theory includes a recovery parameter β that governs the attenuation decrease rate (see Eq. (7.10)). β depends on the diffusion coefficient of the mobile point defects. The dependence of β on external stress provides us with the activation volume for point-defect diffusion (Ogi et al. 1999).

The specimens were 6-mm-thick plate of polycrystalline aluminum with 99.99 mass% purity, measuring 50 mm by 40 mm. They were annealed at 200 °C for 1 h before loading. The yield strength was 15 MPa, which was determined from the stress beyond which the macroscopic stress-strain curve departs from linearity. Uniaxial compressive stress was applied in the longitudinal direction.

The measurement setup was the same as that in Fig. 7.1. The bulk-wave EMAT (Fig. 3.1) with 10×10 mm^2 active area was placed close to the specimen surface by adjusting the shear wave polarization parallel to the stress direction. It was driven by rf bursts of 500 V_{PP} and 50-μs duration. The resonance frequency of the shear wave around 6.25 MHz (24th resonance) was used. The system automatically recorded the resonance frequency f and attenuation coefficient α together with the load and transverse strain. One set of measurements took two seconds. The shear wave velocity V was obtained from the resonance frequency and transverse strain. At first, the uniaxial compressive stress was applied up to 15 MPa at a constant loading speed of 0.06 MPa/s. Then, the unloading-holding (for 20 min) stress sequence was made. The first loading, up to 15 MPa (yield strength), would change the dislocation structure and introduce additional vacancies and self-interstitials. But, during the unloading-holding sequence, no essential change would occur in the dislocation structure and the number of point defects. Also, focusing on the unloading process minimizes the artifacts caused by the friction between the specimen and the loading plates.

Figure 7.7 shows typical evolutions of α and V observed during the unloading-holding stress sequence. During holding stress, α decreased and V increased with time. As soon as the stress was decreased after a holding, α increased, and V decreased to the values before the holding. Almost the same magnitude of attenuation was recovered at each holding. The recovery phenomena observed here can be understood by point-defects migration toward dislocations to pin them. Other mechanisms for recovery (dislocation annihilation and rearrangement) cannot explain the jump of α and drop of V at the beginning of an unloading process. As shown in Eqs. (7.1) and (7.2), α is proportional to the fourth power of the average dislocation segment length L, and velocity change is proportional to the

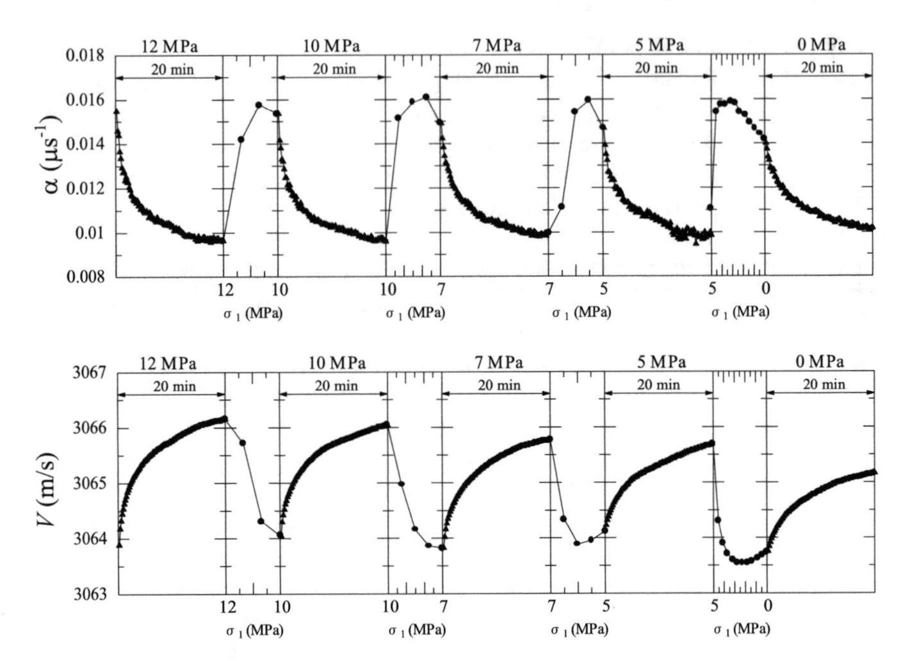

Fig. 7.7 Evolutions of the attenuation coefficient α (*upper*) and the phase velocity V (*bottom*) during the unloading-holding stress sequence. σ_1 denotes the uniaxial compressive stress. Reprinted from Ogi et al. 1999, Copyright (1999), with permission from Elsevier

second power of L. They also depend on the dislocation density Λ, but it remains unchanged during the holding. When the stress is held, the point defects migrate to the dislocation lines and pin them, decreasing L and α, and increasing V. Because these pinnings are so weak that they break upon releasing the stress, causing a temporary increase of α and decrease of V.

Figure 7.8 shows the recovery behaviors of α and V at various holding stresses, showing obvious stress dependence of the recovery rate. Recovery occurred in approximately 30 min. GHL theory gives the recovering attenuation caused by the pinning of dislocations by the mobile point defects as

$$\alpha = \frac{A_1}{(n_1 + n_2)^4} \cdot \frac{1}{(1 + \beta t^{2/3})^4} + \alpha_b. \tag{7.7}$$

Here, n_1 and n_2 denote concentrations of the mobile and immobile point defects, respectively, t the time measured after recovery starts, and β the recovery parameter. A_1 is a constant, which is independent of stress and time. The background attenuation, α_b, which was introduced by Ogi et al. (1999), includes the time-independent dislocation damping (i.e., dislocation damping after complete recovery), grain-scattering loss (Sect. 15.2), and diffraction loss (Sect. 6.4). The first one occurs because the number of point defects on a dislocation line will saturate

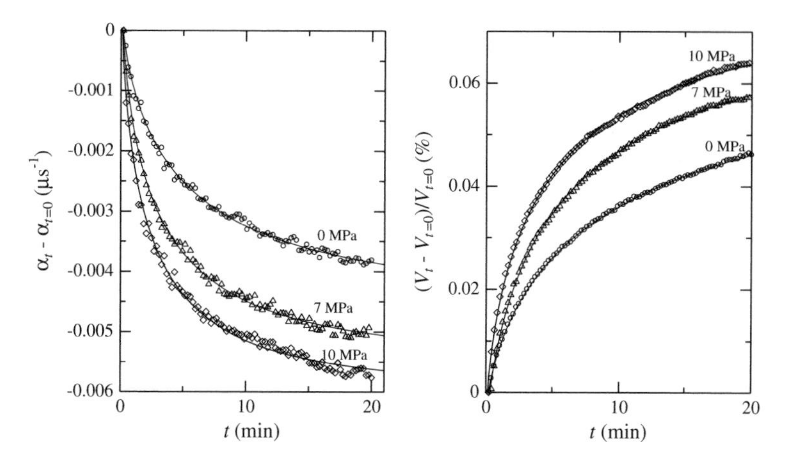

Fig. 7.8 Recovery of attenuation *left* and velocity *right* at various holding stresses. Solid lines are the modified GHL theory. Reprinted from Ogi et al. 1999, Copyright (1999), with permission from Elsevier

and the average dislocation segment length will converge to a finite value at infinite time. The latter two change very little during recovery. Hence, α_b is independent of stress, if the number of the mobile point defects and the segment length after recovery are unchanged. Concerning the velocity change, GHL theory also provides a similar equation. In the present case, however, one has to consider the acoustoelastic effect (Chap. 12) as discussed in deriving Eq. (7.4). Then, the change of V during the recovery will be of the form

$$\frac{V - V_0}{V_0} = \frac{\Delta V}{V_0} = \frac{A_2}{(n_1 + n_2)^2} \cdot \frac{1}{(1 + \beta t^{2/3})^2} + \left(\frac{\Delta V}{V_0}\right)_0, \tag{7.8}$$

$$\left(\frac{\Delta V}{V_0}\right)_0 = C_{\mathrm{U}}^{(\mathrm{S1})}\sigma + \left(\frac{\Delta V}{V_0}\right)_{\mathrm{b}}. \tag{7.9}$$

Here, A_2 is another stress-independent and time-independent constant. σ is the uniaxial stress (positive for tension). $C_{\mathrm{U}}^{(\mathrm{S1})}$ denotes the acoustoelastic constant for the shear wave polarized in the stress direction and $(\Delta V/V_0)_{\mathrm{b}}$ is the background caused by the dislocation vibration after recovery. V_0 is the velocity in the stress-free state. Thus, $(\Delta V/V_0)_0$ is independent of time, but depends on the stress. Figure 7.8 demonstrates faster recovery of α than that of V, which is understood from their different time dependences.

The GHL theory provides an expression for the recovery parameter β:

$$\beta = A_3 \frac{n_1}{n_1 + n_2} \left(\frac{D}{T}\right)^{2/3}. \tag{7.10}$$

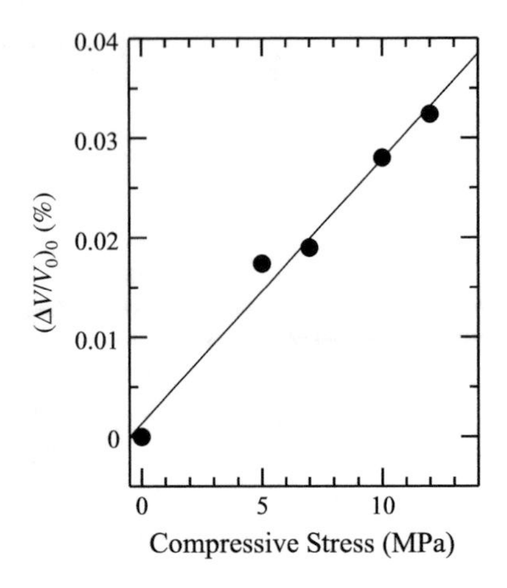

Fig. 7.9 Stress dependence of the time-independent velocity change. Reprinted from Ogi et al. 1999, Copyright (1999), with permission from Elsevier

Here, T denotes temperature and D the diffusion coefficient of the mobile point defects. A_3 is a constant, depending on the lattice parameter. By fitting Eqs. (7.7) and (7.8) to the measurements, one can determine the recovery parameter β and the time-independent terms α_b and $(\Delta V/V_0)_0$. These regression curves are drawn with solid lines in Fig. 7.8. α_b is almost independent of stress, being $9.34 \pm 0.10 \times 10^{-3}$ μs^{-1}. This indicates that the stress holding did not influence the dislocation structure and the average segment length L after recovery; the stress only affected the point-defect kinetics. On the other hand, $(\Delta V/V_0)_0$ shows a linear dependence on stress as shown in Fig. 7.9, supporting Eq. (7.9). The slope is -2.6×10^{-5} MPa^{-1}, being close to the acoustoelastic constant of a 0.99-polycrystalline aluminum for a shear wave propagating normal to and polarized parallel to the uniaxial stress, $C_U^{(S1)} = -2.7 \times 10^{-5}$ MPa^{-1} (Fukuoka and Toda 1977). This agreement manifests that the background velocity change $(\Delta V/V_0)_b$ is independent of the biasing (holding) stress just like α_b. Thus, introducing the stress-independent terms and the acoustoelastic effect into GHL theory results in better agreement between the measurements and the theory as shown in Fig. 7.10, which plots logarithmically the recovering attenuation and velocity versus $(1 + \beta t^{2/3})$ for the biasing stress at 10 and 0 MPa. The attenuation plot yields a line of slope −4 and the velocity changes a slope of −2, supporting Eqs. (7.7) and (7.8), respectively.

The recovery parameter β monotonically increases with the biasing compressive stress as shown in Fig. 7.11 and ranges from 0.08 to 0.2 min$^{-2/3}$. In companion measurements for a 99.6 mass% polycrystalline aluminum, β ranges between 0.05 and 0.1 min$^{-2/3}$ in the macroscopic elastic region (<60 MPa) (Ogi et al. 1999). This suggests that the mobile point defects are vacancies or self-interstitials, not impurities, because Eq. (7.10) shows that an increase of the concentration of immobile point defects reduces β.

Many previous studies regarded vacancies as the dominant pinning factor at and just below room temperature. The measurement of the recovery parameter with various temperatures in copper yielded an activation energy that showed good

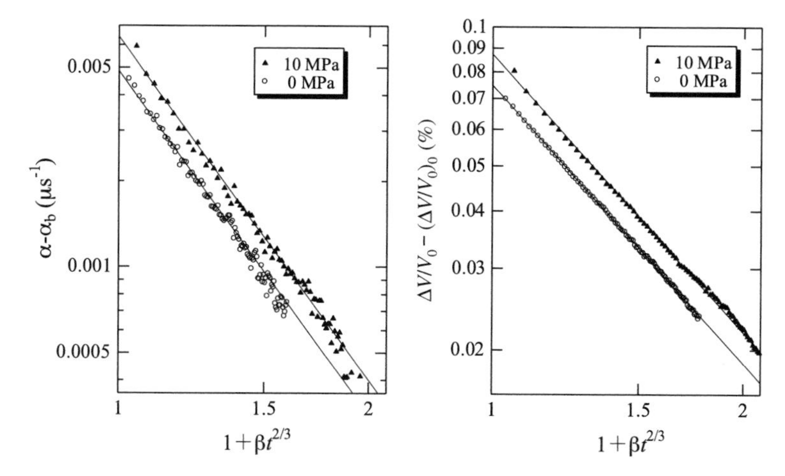

Fig. 7.10 Comparison of measurements and modified GHL theory. The slopes of *solid lines* are –4 for the attenuation change *left* and –2 for the velocity change *right*. Reprinted from Ogi et al. 1999, Copyright (1999), with permission from Elsevier

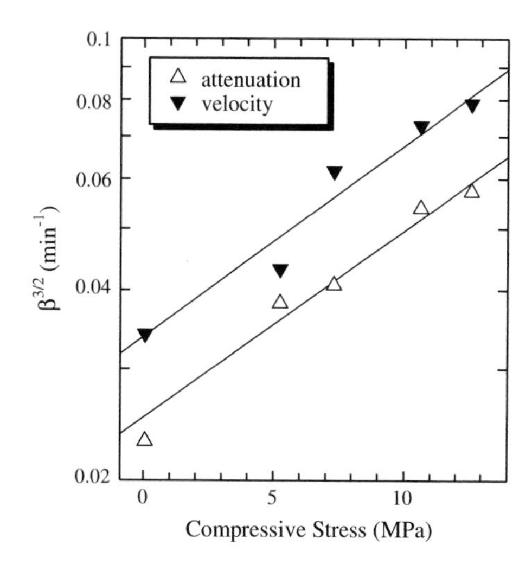

Fig. 7.11 Dependence of the recovery parameter β on stress. Slope gives activation volume $V_a = 13b^3$, where b (=2.863 × 10^{-10} m) denotes the magnitude of the Burgers vector. Reprinted from Ogi et al. 1999, Copyright (1999), with permission from Elsevier

agreement with the vacancy-migration energy (Smith 1953). Lentz et al. (1971) measured attenuation evolution in a copper monocrystal during temperature increase from 220 K, at which vacancies were introduced by γ-rays. They found a remarkable attenuation decrease at the temperature beyond which vacancies can move. For aluminum, similar results were reported by Garr and Sosin (1967) with resistivity changes. On the other hand, Lauzier et al. (1989 and 1990) studied the dislocation-point-defect interaction in an ultra-high-purity aluminum by measuring the temperature dependence of elastic modulus and internal friction. They asserted that vacancies act as the lubrication agents of dislocation motion, not as pinning points; the pinning agents are self-interstitials.

β's determined by the EMAR method are much smaller than those measured previously on a 0.99 polycrystalline aluminum by a factor three or more (Granato et al. 1958). This is attributed to the contacting method used in the past: A contacting method might involve additional apparent recovery associated with the changeable coupling condition. To explain the stress dependence of β, consider a mechanism that the diffusion is activated not only by temperature, but also by the biasing uniaxial stress σ_1:

$$D = D_0 \exp\left(-\frac{E_0 - V_a|\sigma_1|}{kT}\right). \tag{7.11}$$

Here, D_0 denotes the diffusion coefficient at the high-temperature limit, E_0 the activation energy at the stress-free state, and k the Boltzmann's constant. V_a denotes the activation volume of point-defect diffusion under uniaxial stress, which refers to the partial derivative of the activation energy with respect to the uniaxial stress. Concerning vacancies from grain boundaries, for example, compressive stress makes the vacancy-formation energy higher at a grain boundary normal to the stress and lower at a boundary parallel to the stress, inducing diffusional vacancy flow (Herring 1950). Thus, applied stress can activate point-defect migration, which causes creep. The idea of stress-dependent diffusion was pointed out many times. In particular, diffusion under hydrostatic stress and shear stress was studied extensively (Flynn 1972).

Combining Eqs. (7.10) and (7.11), one find that plotting $\log(\beta^{3/2})$ versus stress should yield straight lines whose slopes yield V_a. Figure 7.11 shows such plots for the 0.9999 polycrystalline aluminum. Values for β determined from the attenuation recovery were smaller than from the velocity recovery, which remains unclear. However, both slopes yielded the same activation volume $V_a = 13b^3$, where b (=2.863 $\times$ 10^{-10} m) denotes the magnitude of the Burgers vector. This is considerably larger than the self-diffusion activation volume ($\sim 1b^3$) in aluminum under hydrostatic stress (Flynn 1972).

7.3.2 *Activation Energy for Diffusion*

Further study on the point-defect diffusion was performed with various temperatures using the measurement setup shown in Fig. 7.12 (Ogi et al. 2005): The plate-shape specimen was polycrystalline aluminum with 99.995 mass% purity, measuring 50 $\times$ 6 $\times$ 1 mm^3. A tensile load was applied in the longitudinal direction of the specimen through fixtures and bearings to avoid the bending stress. A solenoid coil was set near the specimen with a 0.5-mm liftoff from the specimen surface. The specimen and coil were located in a furnace to control temperature, and a pair of permanent magnets was located outside the furnace to apply a static magnetic field needed for the Lorentz force coupling.

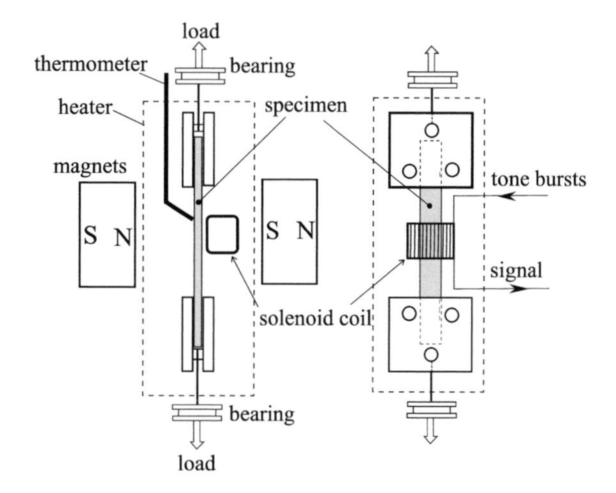

Fig. 7.12 Measurement setup for shear wave attenuation by electromagnetic acoustic resonance technique. Reprinted from Ogi et al. 2005, Copyright (2005), with permission from Elsevier

The β value after deformation was measured by fitting the theory (Eq. 7.7) to the time attenuation recovery curve as shown in Figs. 7.13 and 7.14, which increased with temperature but decreased beyond 333 K. This was attributed to dislocation movements, because the stress remained unchanged, dislocations continuously moved, dragging the point defects and keeping their effective segment length long.

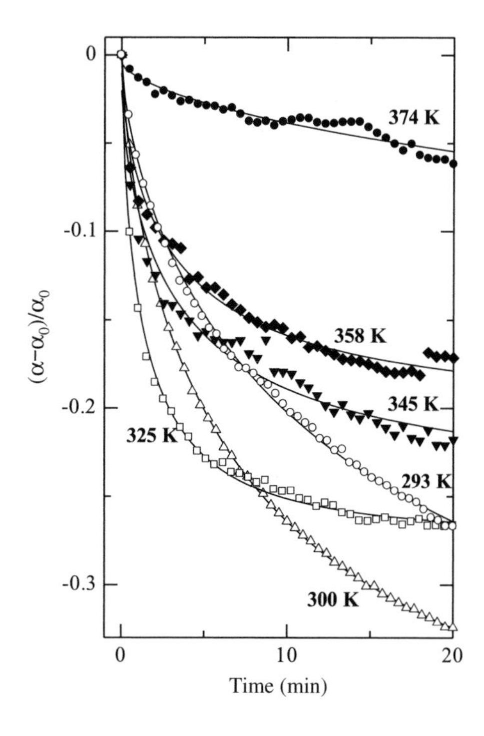

Fig. 7.13 Attenuation recovery curves under 4.9 MPa at various temperatures. Solid lines denote the fitted theory. Reprinted from Ogi et al. 2005, Copyright (2005), with permission from Elsevier

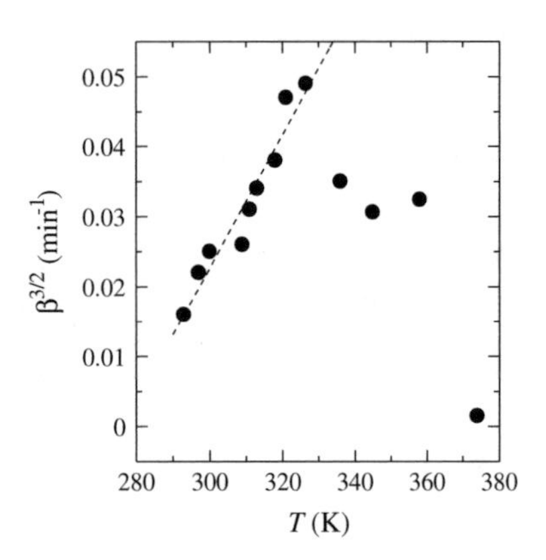

Fig. 7.14 Temperature dependence of the β value of 99.995 mass% polycrystalline aluminum. Reprinted from Ogi et al. 2005, Copyright (2005), with permission from Elsevier

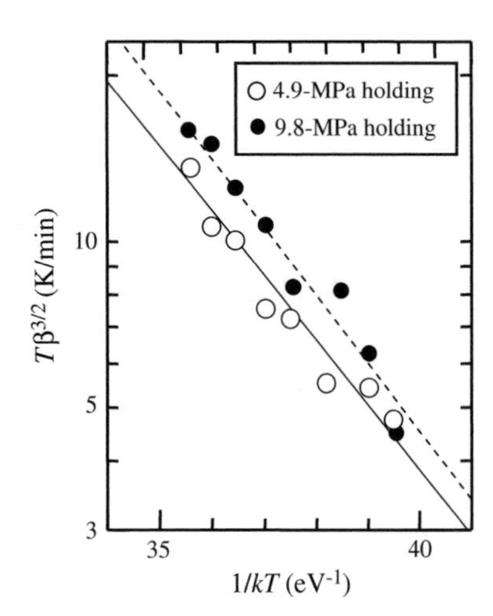

Fig. 7.15 Logarithmic plot of $T\beta^{3/2}$ versus $1/kT$ for 99.995 mass% polycrystalline aluminum. Slopes give the activation energy. Reprinted from Ogi et al. 2005, Copyright (2005), with permission from Elsevier

Bremnes et al. (1997) suggested that such a dislocation movement occurred at temperatures higher than 335 K. The activation energy of the point-defect movement was then determined by using β values below 333 K with Eqs. (7.10) and (7.11). The slopes in Fig. 7.15 yield the activation energy for point-defect movement to be $E_m = 0.28 \pm 0.03$ eV for the biasing stresses of 4.9 and 9.8 MPa. This value is considerably smaller than the activation energy of vacancy's bulk diffusion in aluminum (0.61–0.65 eV (DeSorbo and Turnbull 1959)) and larger than that of diffusion of self-interstitials in aluminum (0.1 eV (Johnson 2001)). The attenuation-monitoring measurement after quenching the specimen revealed that this activation energy corresponded to that of fast diffusion of vacancies along dislocations (Ogi et al. 2005).

7.4 Dislocation Damping after Elastic Deformation in Al-Zn Alloy

A similar approach to that in the previous section was pursued by Johnson (2001) to explore the interaction of dislocations with point defects in an Al-0.2 at.% Zn binary alloy. The applied stress was kept below the plastic region, and no irreversible dislocation motion and associated permanent dimensional changes occurred; however, reversible dislocation displacements did occur. This limiting of the stress has the advantages of providing information on the evolution of dislocation/point-defect interactions without the complication of simultaneous changes in dislocation densities and providing information on the temperature dependence through repeatable measurements on a single specimen.

Polycrystalline Al (0.2 at.% Zn), produced from 99.999 % pure starting materials, was swaged into a rod, machined, and annealed at 285 °C for 5.2 h to induce recovery of the dislocation structure without recrystallization. The specimen had a cylindrical trapped-mode geometry similar to a conventional dumbbell-shaped tensile specimen but with a central section of slightly larger diameter (Fig. 7.16). The diameter at the center was 10.42 mm and the step in diameter was 1.6 %. The length of the larger-diameter central section was 41.7 mm. The length between the threaded sections was 152 mm. The specimen was mounted in a mechanical testing machine and subjected to tensile stress.

The EMAR measurements were performed on trapped axial shear modes of the specimen using the technique described in Sect. 3.11. The trapping of resonance vibrations in the central section reduces acoustic losses through the ends of the specimen, which are attached to the machine grips, and thus makes measurements more accurate and reproducible. Electromagnetic-acoustic coupling was implemented with a solenoid coil surrounding the central section of the specimen and permanent magnets that produced a transverse magnetic field. The resonance frequency f and internal friction Q^{-1} were measured.

The specimen was subjected to a three-step loading sequence that involved (i) holding at 0 MPa, (ii) applying 10.2 MPa tensile stress for 2000 s, and

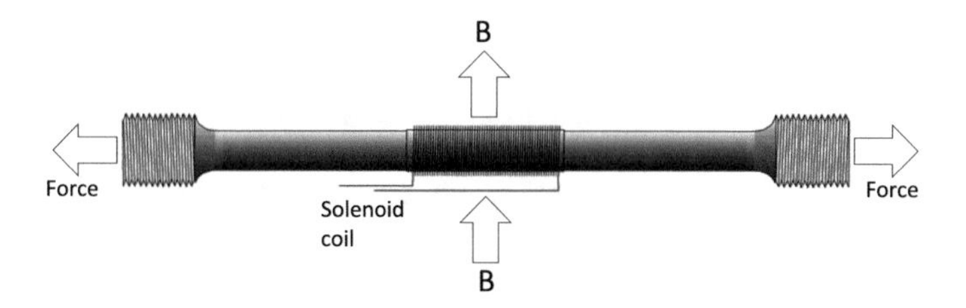

Fig. 7.16 Geometry of the specimen and ultrasonic transducer. The slightly larger diameter of the central section localizes resonance vibration. A solenoid coil and a static magnetic field are used to excite and detect resonance vibrations with Lorentz force coupling. Tensile forces are applied with a mechanical testing machine. Reprinted from Johnson 2001, Copyright (2001), with permission from Elsevier

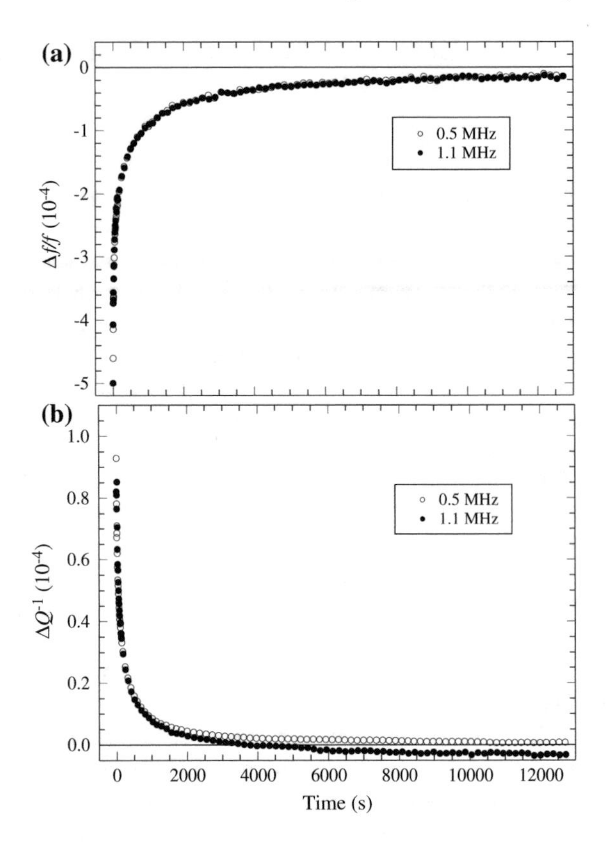

Fig. 7.17 **a** Changes in 0.5-MHz and 1.1-MHz resonance frequencies and **b** internal friction vs. time following application of a 10.2 MPa tensile stress for 2000 s at 308.2 K. The time is defined to be 0 s when the stress was returned to 0 MPa. Reprinted from Johnson 2001, Copyright (2001), with permission from Elsevier

(iii) holding at 0 MPa. Figure 7.17 shows changes in frequency and internal friction for 0.5 and 1.1-MHz resonance modes versus time t at 308.2 K during the third stage of the loading sequence, where $\Delta f/f = (f - f_0)/f_0$ and $\Delta Q^{-1} = Q^{-1} - Q_0^{-1}$. f_0 and Q_0^{-1} are measured at 0 MPa before loading. The measurements show no significant dependence of either $\Delta f/f$ or ΔQ^{-1} on the mode frequency. The fact that $\Delta f/f$ approaches 0 at long times reflects the lack of permanent deformation, since f is highly sensitive to dimensional changes. The deviation of the 1.1-MHz ΔQ^{-1} below 0 after 4000 s in Fig. 7.17b is attributed to experimental error; measurements of Q^{-1} were generally found to be less repeatable than f, occasionally showing small long-term variations of unknown origin in the absence of applied loads.

The 0.5-MHz data from Fig. 7.17 were plotted on a semilogarithmic scale. The slopes of the regression lines were $\mathrm{d}(\Delta f/f)/\mathrm{d}(\log t) = 1.48 \times 10^{-4}$ and $\mathrm{d}(\Delta Q^{-1})/\mathrm{d}(\log t) = -0.33 \times 10^{-4}$ (Johnson 2001). Therefore, the rate of change of $\Delta f/f$ is -4.5 times that of ΔQ^{-1} over this time interval, which is inconsistent with the Granato-Hikata-Lücke (GHL) theory (1958); the theory predicts that the slope of $\Delta f/f$ is only -0.3 times that of ΔQ^{-1} (an order of magnitude less than the factor of -4.5 observed here) if ΔQ^{-1} is weakly dependent on frequency (near the maximum of the peak). The theory predicts also that the change of $\Delta f/f$ is strongly dependent

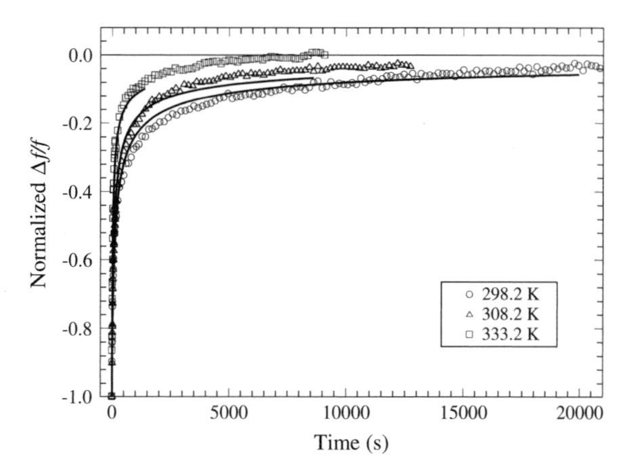

Fig. 7.18 Change in 0.5-MHz resonance frequency versus time following application of a 10.2-MPa tensile stress for 2000 s at 298.2, 308.2, and 333.2 K. The solid lines are fits of the form $A[(1 - N(t))/N(\infty)]$, where A is a fitting parameter. Reprinted from Johnson 2001, Copyright (2001), with permission from Elsevier

on frequency if ΔQ^{-1} is independent of frequency. But, there is no frequency dependence in Fig. 7.17a. This means that $\Delta f/f$ and ΔQ^{-1} are dominated by different physical mechanisms.

The changes in f may be attributed to stiffening of the material caused by diffusion of point defects toward new and more stable positions, which interact with dislocations more strongly. When the stress is applied, a dislocation breaks away from point defects, and the point defects then diffuse to the new dislocation position while the stress is held constant. When the stress is released, the same type of process occurs, eventually returning the system to its original equilibrium configuration. Thus, the recovery after loading is an evolution to a more tightly bound configuration with higher elastic stiffness (and higher resonance frequencies). This view is supported by measurements of the temperature dependence of the frequency change. Figure 7.18 shows such a result. A number of empirical and physically based functions have been proposed to describe the time-dependent pinning of dislocations by point defects (Bullough and Newman 1970). The results from one physical model presented by Bullough and Newman (1962a) are considered here. In this model, the number of diffusing point defects, $N(t)$, that are trapped at a dislocation is given by

$$\frac{N(t)}{N(\infty)} = 1 - \frac{4}{\pi}\int_0^{\infty} \frac{\exp(-x^2 t/\tau)}{x} \cdot \frac{H_2(x) - Y_2(x) - 2(1+x^2/3)/(\pi x)}{J_2^2(x) + Y_2^2(x)} \mathrm{d}x, \quad (7.12)$$

where

$$\tau = \frac{r_c^2}{D}. \quad (7.13)$$

Here, H_2 is the Struve function, J_2 and Y_2 are the Bessel functions, D is the diffusion coefficient of the point defects, and r_c is the core radius, inside which defects are trapped. If the observed recovery of $\Delta f/f$ is a stiffening of the crystal arising from point defects migrating to lower-energy sites, then since $\Delta f/f$ is referenced to $t \to \infty$, it will be proportional to $(1 - N(t))/N(\infty)$. Numerical fits to this function for the measurements, as shown in Fig. 7.18, provided values of τ as a function of temperature, which obeys an Arrhenius expression

$$\tau = \tau_0 \exp\left(\frac{E}{kT}\right). \tag{7.14}$$

This yielded an activation energy of 0.690 ± 0.02 eV for the diffusion of the point defects. In high-purity Al (Zn), there are few simple candidates for the diffusing species. These include the vacancy, self-interstitial (Al_I), substitutional Zn (Zn_S), interstitial Zn (Zn_I), and the interstitial aluminum-zinc complex (Al_I-Zn_I). The activation energy for vacancy migration is approximately 0.66 eV near room temperature and increases at higher temperatures (Mondolfo 1976). Al_I and Al_I-Zn_I have activation energies of 0.11 and 0.36 eV, respectively (Wallace et al. 1985). Reported values for Zn_S are higher than 1.1 eV (Mondolfo 1976). The activation energy for Zn_I is not known but is expected to be low (comparable to that of Al_I). Therefore, the measured activation energy agrees with values reported for the vacancy and is significantly different from values reported for the other candidate defects.

7.5 Recovery and Recrystallization in Aluminum

Monitoring of ultrasonic damping during heat treatment of a deformed metal provides valuable information about changes in the dislocation network arising from recovery and recrystallization. Johnson (1998) adopted the EMAR method for studying recovery and recrystallization of high-purity aluminum. The spherical specimens were 99.999 % pure polycrystalline aluminum. The original material was annealed in air at 400 ± 5 °C for 10 min to induce recrystallization and slowly cooled. Then, it was cold worked under uniaxial stress in a mechanical testing machine. Spherical specimens with a diameter of $\sim$6 mm were fabricated from these materials.

Figure 7.19 shows the measurement setup. The specimen was supported by a stainless steel tube and surrounded by a solenoid coil between two water-cooled permanent magnets. The magnets provided a static field that was transverse to the coil axis in the region of the specimen. A direct-current resistive heater surrounded the specimen and coil, and reflective shielding surrounded the heater. The temperature was varied in the 92–1000 °C range and was measured by an optical pyrometer. The Lorentz force mechanism works for generation and detection of vibration with this configuration (Johnson et al. 1992). A torsional mode was used.

Figure 7.20 shows the change of decrement δ ($=\pi Q^{-1}$) of a deformed specimen after a plastic strain of $\sim$0.1 during a series of temperature ramps. During the first heating (Ih), δ increased up to $\sim$210 °C, dropped irreversibly from 210 to 255 °C,

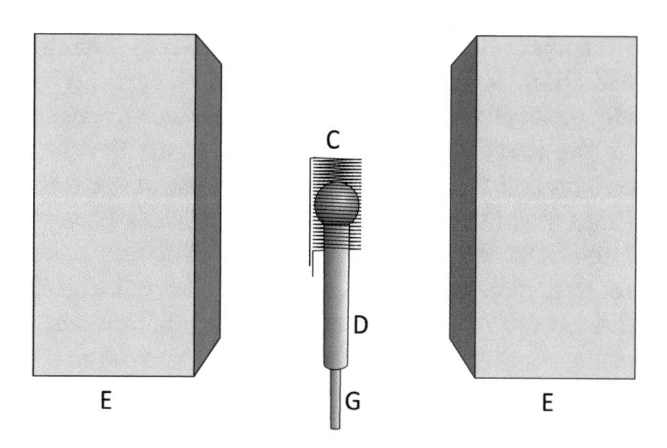

Fig. 7.19 Schematic diagram of transducer and specimen. C: solenoid coil. D: tube supporting the spherical specimen. E: permanent magnets. G: sapphire rod that transmits black-body radiation to a pyrometer. Reprinted with permission from Johnson 1998, Copyright (1998), AIP Publishing LLC

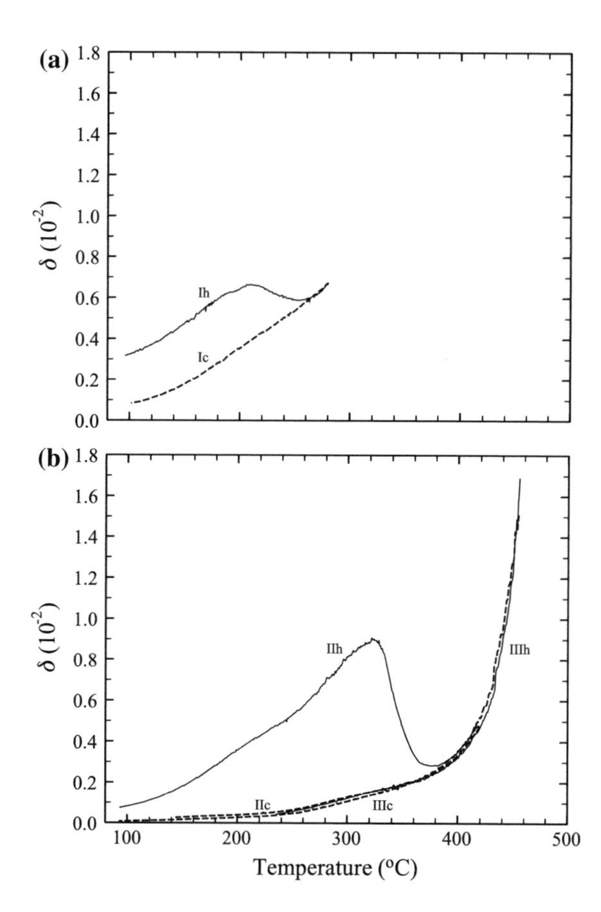

Fig. 7.20 Change of decrement δ of a deformed aluminum specimen versus temperature. **a** During heating (Ih) and cooling (Ic) in the first thermal cycle. **b** During heating (IIh) and cooling (IIc) in the second thermal cycle, and during heating (IIIh) and cooling (IIIc) in the third thermal cycle. Solid lines: heating. Dashed lines denote cooling. Reprinted with permission from Johnson 1998, Copyright (1998), AIP Publishing LLC

and then increased with further heating to 280 °C. On cooling (Ic), δ decreased monotonically with decreasing temperature. During the second heating (IIh), δ followed the previous cooling curve and continued increasing with further heating up to 320 °C. Then, a second irreversible drop occurred between 320 and 380 °C.

On cooling (IIc), δ decreased monotonically. During the third thermal cycle, (IIIh) and (IIIc), δ followed curve IIc and showed no irreversible changes, increasing monotonically up to the highest measured temperature (455 °C).

The irreversible change in δ during the first thermal cycle is attributed to recovery, and that during the second thermal cycle is attributed to recrystallization. (The definitions of recovery and recrystallization are discussed by Cahn and Haasen (1996)). Since any anelastic response of dislocations causes damping (see Eq. (7.2), for example), the change in δ is consistent with the decrease in dislocation densities that occurs during recovery and recrystallization. The temperature range of the change during the second heating (IIh) is consistent with the temperature range for recrystallization determined by Anderson and Mehl (1945) for 99.97 % aluminum and Bay and Hansen (1984) for 99.4 % aluminum, considering that recrystallization is more rapid for higher purity or more heavily deformed material (Cahn and Haasen 1996).

The association of recovery with the irreversible change during the first heating was supported by results of isothermal annealing of the specimen. A specimen with a plastic strain of 0.095 was heated to 186.3 °C in 75 min and then held at this temperature for 480 min. As shown in Fig. 7.21, δ decreased monotonically with time in a nonexponential way. The form of time dependence of the decay shown in Fig. 7.21 is typical of that observed for mechanical properties such as hardness and yield stress during recovery (Humphreys and Hatherly 1995). Properties change relatively rapidly in the initial stage of recovery as dislocations are annihilated and then change more slowly as more stable structures are formed from the surviving dislocations. Several studies support the idea that recovery kinetics in polycrystalline metals is dominated by thermally activated glide or cross-slip of dislocations and that the internal stress σ_I relates to the time t

$$\sigma_I = \sigma_0 - A\ln\left(1 + \frac{t}{t_0}\right). \tag{7.15}$$

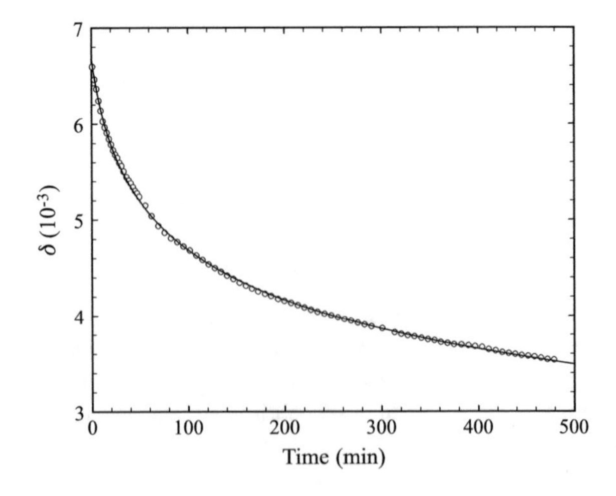

Fig. 7.21 Decay of decrement δ of the deformed specimen during isothermal annealing at 186.3 °C. The solid line is a least-square fit of the form given by Eq. (7.16). Reprinted with permission from Johnson 1998, Copyright (1998), AIP Publishing LLC

Here, σ_0, A, and t_0 are functions of temperature (Cottrell and Aytekin 1950). Because the dislocation density is approximately proportional to σ_I^2 in a wide range of materials (Humphreys and Hatherly 1995), one expects the time dependence of δ during recovery annealing to be given by

$$\delta = \left[\delta_0 - B\ln\left(1 + \frac{t}{t_0}\right)\right]^2, \tag{7.16}$$

if the high-temperature damping was caused by a dislocation interaction with a magnitude proportional to the dislocation density. The solid line in Fig. 7.21 is a least-square fit of this form, which closely follows the measurements.

References

Anderson, A. R., & Pollard, H. F. (1979). Changes in internal friction and dislocation charge in sodium chloride crystals following plastic deformation. *Journal of Applied Physics, 50*, 5262–5265.

Anderson, W. A., & Mehl, R. F. (1945). Recrystallization of aluminum in terms of the rate of nucleation and the rate of growth. *Transactions of the AIME, 161*, 140–172.

Bay, B., & Hansen, N. (1984). Recrystallization in commercially pure aluminum. *Metallurgical Transactions A, 15*, 287–297.

Bratina, W. (1966). Internal friction and basic fatigue mechanisms in body-centered cubic metals, mainly iron and carbon steels. In *Physical Acoustics* (Vol. 3A, pp. 223–291). New York: Academic Press.

Bremnes, Ø., Progin, O., Gremaud, G., & Benoit, W. (1997). Complex interaction mechanisms-between dislocations and point defects studied in pure aluminium by a two-wave acoustic coupling technique. *Physical Status Solidi (a), 160*, 395–402.

Bullough, R., & Newman, R. (1962a). The growth of impurity atmospheres around dislocations. *Proceedings of the Royal Society of London A: Mathematical, Physical and Engineering Sciences, 266*, 198–208.

Bullough, R., & Newman, R. (1962b). Impurity precipitation on dislocations—A theory of strain aging. *Proceedings of the Royal Society of London A: Mathematical, Physical and Engineering Sciences, 266*, 209–221.

Bullough, R., & Newman, R. (1970). The kinetics of migration of point defects to dislocations. *Reports on Progress in Physics, 33*, 101–148.

Cahn, R. W., & Haasen, P. (1996). *Physical Metallurgy* (Vol. 3). Amsterdam: Elsevier.

Cottrell, A. H., & Aytekin, V. (1950). The flow of zinc under constant stress. *Journal of the Institute of Metals, 77*, 389–422.

Cottrell, A. H., & Bilby, B. A. (1949). Dislocation theory of yielding and strain aging of iron. *Proceedings of the Physical Society. Section A, 62*, 49–62.

DeSorbo, W., & Turnbull, D. (1959). Kinetics of vacancy motion in high-purity aluminum. *Physical Review, 115*, 560–563.

Flynn, C. (1972). *Point Defects and Diffusion*. New York: Oxford.

Fukuoka, H., & Toda, H. (1977). Preliminary experiment on acoustoelasticity. *Archives of Mechanics, 29*, 673–686.

Garr, K., & Sosin, A. (1967). Recovery of electron-irradiated aluminum and aluminum alloys II. Stage II. *Physical Review, 162*, 669–681.

Granato, A., Hikata, A., & Lücke, K. (1958). Recovery of damping and modulus changes following plastic deformation. *Acta Metallurgica, 6*, 470–480.
Granato, A., & Lücke, K. (1956). Theory of mechanical damping due to dislocations. *Journal of Applied Physics, 27*, 583–593.
Gremaud, G., Bujard, M., & Benoit, W. (1987). The coupling technique: A two-wave acoustic method for the study of dislocation dynamics. *Journal of Applied Physics, 61*, 1795–1805.
Ham, F. (1959). Stress-assisted precipitation on dislocations. *Journal of Applied Physics, 30*, 915–926.
Harper, S. (1951). Precipitation of carbon and nitrogen in cold-worked alpha-iron. *Physical Review, 83*, 709–712.
Hartley, C., & Wilson, R. (1963). Dislocation pinning effects in unalloyed molybdenum. *Acta Metallurgica, 11*, 835–845.
Herring, C. (1950). Diffusional viscosity of a polycrystalline solid. *Journal of Applied Physics, 21*, 437–445.
Hikata, A., Truell, R., Granato, A., Chick, B., & Lücke, K. (1956). Sensitivity of ultrasonic attenuation and velocity changes to plastic deformation and recovery in aluminum. *Journal of Applied Physics, 27*, 396–404.
Hiki, Y., & Granato, A. (1966). Anharmonicity in noble metals; higher order elastic constants. *Physical Review, 144*, 411–419.
Hughes, D. S., & Kelly, J. L. (1953). Second-order elastic deformation of solids. *Physical Review, 92*, 1145–1149.
Hull, D., & Bacon, D. J. (1984). *Introduction to Dislocations*. New York: Pergamon Press.
Humphreys, F. J., & Hatherly, M. (1995). *Recrystallization and Related Annealing Phenomena*. New York: Elsevier.
Johnson, G. C. (1982). Acoustoelastic response of polycrystalline aggregates exhibiting transverse isotropy. *Journal of Nondestructive Evaluation, 3*, 1–8.
Johnson, W. (1998). Ultrasonic damping in pure aluminum at elevated temperatures. *Journal of Applied Physics, 83*, 2462–2468.
Johnson, W. (2001). Ultrasonic dislocation dynamics in Al (0.2 at % Zn) after elastic loading. *Materials Science and Engineering, A, 309–310*, 69–73.
Johnson, W., Norton, S., Bendec, F., & Pless, R. (1992). Ultrasonic spectroscopy of metallic spheres using electromagnetic-acoustic transduction. *The Journal of the Acoustical Society of America, 91*, 2637–2642.
Kuang, G., & Zhu, Z. (1994). A study of dislocation movement during push-pull fatigue by ultrasonic attenuation. *Physical Status Solidi (a), 142*, 357–363.
Lauzier, J., Hillairet, J., Gremaud, G., & Benoit, W. (1990). Lubrication agents of dislocation motion at very low temperature in cold-worked aluminum. *Journal of Physics: Condensed Matter, 2*, 9247–9256.
Lauzier, J., Hillairet, J., Vieux-Champagne, A., Benoit, W., & Gremaud, G. (1989). The vacancies, lubrication agents of dislocation motion in aluminum. *Journal of Physics: Condensed Matter, 1*, 9273–9282.
Lentz, D., Edenhofer, B., & Lücke, K. (1971). On a dislocation-pinning memory-effect in γ-irradiated and stress-annealed copper. *Scripta Metallurgica, 5*, 387–393.
Mason, W. P. (1958). *Physical Acoustics and Properties of Solids*. Princeton: Van Nostrand.
Mondolfo, L. F. (1976). *Aluminum Alloys, Structure and Properties*. Boston: Butterworths.
Ogi, H., Suzuki, N., & Hirao, M. (1998). Noncontact ultrasonic spectroscopy on deforming polycrystalline copper dislocation damping and acoustoelasticity. *Metallurgical and Materials Transactions A, 29*, 2987–2993.
Ogi, H., Tsujimoto, A., Hirao, M., & Ledbetter, H. (1999). Stress-dependent recovery of point defects in deformed aluminum: An acoustic-damping study. *Acta Materialia, 47*, 3745–3751.
Ogi, H., Tsujimoto, A., Nishimura, S., & Hirao, M. (2005). Acoustic study of kinetics of vacancy diffusion toward dislocations in aluminum. *Acta Materialia, 53*, 513–517.
Pao, Y.-H., Sachse, W., & Fukuoka, H. (1984). Acoustoelasticity and ultrasonic measurements of residual stresses, In *Physical Acoustics* (Vol. 17, pp. 61–143). New York: Academic Press.

Phillips, D. C., & Pratt, P. L. (1970). The recovery of internal friction in sodium chloride. *Philosophical Magazine, 21*, 217–243.
Salama, K., & Alers, G. (1967). Third-order elastic constants of copper at low temperature. *Physical Review, 161*, 673–680.
Smith, A. D. (1953). The effect of small amounts of cold-work on young's modulus. *Philosophical Magazine, 44*, 453–466.
Truell, R., Elbaum, C., & Chick, B. B. (1969). *Ultrasonic Methods in Solid State Physics*. New York: Academic Press.
Wallace, P. W., Hultman, K. L., Holder, J., & Granato, A. V. (1985). Migration of the interstitial-impurity mixed (100) dumbbell configuration in Al-Zn. *Journal de Physique Colloques: Tous les numéros, 46*, C10-59-C10-61.
Zhu, Z., & Fei, G. (1994). Variation in internal friction and ultrasonic attenuation in aluminum during the early stage of fatigue loading. *Journal of Alloys and Compounds,* 211/212, 93–95.

Chapter 8
Elastic Constants and Internal Friction of Advanced Materials

Abstract Elastic constants are important material parameters not only for designing structures but also for studying thermodynamics of materials. They are determined inversely by measuring free-vibration resonance frequencies of the material, and the EMAR method has been applied for this purpose. By controlling the magnetic force symmetry on the material, a specific vibrational group can be excited and detected, contributing to mode identification. The noncontacting nature of EMAR allows accurate internal friction measurement as well. This chapter shows the mode-selective principle of the EMAR and its application for measuring elastic anelastic coefficients of various materials including metals, composites, ceramics, porous materials, thin films, and piezoelectric materials.

Keywords Cross-ply composite · Eigenstrain · Eshelby tensor · Gallium nitride (GaN) · High-temperature measurement · Internal friction tensor · Inverse calculation · Legendre polynomials · Micromechanics · Ni_3Al · Phase transformation · Rafted structure · Silicon carbide fiber · Vibration group

8.1 Mode Control in Resonance Ultrasound Spectroscopy by EMAR

8.1.1 Difficulty of Mode Identification

Resonant ultrasound spectroscopy (RUS) (Demarest 1971; Ohno 1976; Maynard 1996; Migliori and Sarrao 1997; Leisure and Willis 1997) is recognized as a useful method to determine elastic constants of solids, including those with a lower crystallographic symmetry. The usual RUS configuration consists of a well-shaped specimen such as sphere, cylinder, or rectangular parallelepiped and two piezoelectric transducers which sandwich the specimen at opposite corners. The specimen dimensions are typically 1–10 mm. One transducer feeds a continuous-wave (CW) sinusoidal vibration, and the other detects ultrasonic oscillation. On sweeping the excitation frequency, the received amplitude shows peaks at the free-vibration

M. Hirao and H. Ogi, *Electromagnetic Acoustic Transducers*,
Springer Series in Measurement Science and Technology,
DOI 10.1007/978-4-431-56036-4_8

resonance frequencies of the specimen. The frequency response of the amplitude, or *resonance spectrum*, thus includes many resonance peaks. The resonance frequencies are then used in an inverse calculation to find the set of elastic constants that give the measured resonance frequencies.

The most attractive feature of RUS is that one can basically obtain all independent elastic constants with a single frequency scan on a single specimen. To make full use of it, many efforts have been made both for measurement development and for numerical calculation of the resonance frequencies. Concerning instrumentation and applications, Migliori et al. (1993) made major development for acquiring the resonance spectrum of very small specimens (~ 0.001 cm^3) as a function of temperature (20–400 K). Kuokkala and Schwarz (1992) deposited a nickel film to realize a noncontacting ultrasonic transduction and measured internal friction of an amorphous $Ni_{80}P_{20}$ alloy up to 520 K. Ledbetter et al. (1995a) applied RUS to measure elastic constants of a boron-aluminum fiber-reinforced composite. Tanaka and Koiwa (1996), Tanaka et al. (1996a, b) measured elastic constants of monocrystals of intermetallic compounds, including γ-TiAl. Isaak et al. (1998) measured the shear modulus pressure derivative of fused silica and discussed the effect of mass density of pressurizing gases. Concerning the numerical calculation, Holland (1968) used a Fourier series to approximate deformation in a cubic specimen. Demarest (1971) adopted Legendre polynomials and obtained a satisfactory solution with a smaller number of harmonic functions for cubic specimens. Ohno (1976) also used Legendre functions and established an effective method for obtaining free-vibration resonance frequencies of rectangular parallelepiped specimen with orthorhombic symmetry.

Successful determination of elastic constants by the RUS method requires finding correct correspondence between the observed and calculated resonance modes in the inverse calculation, that is, mode identification. An oriented rectangular parallelepiped-shaped solid of orthorhombic symmetry has eight vibration groups depending on the deformation symmetry about the three principal axes x_1, x_2, and x_3 (Ohno 1976). They are breathing vibration, torsional vibration, and shearing vibrations about the x_1-, x_2-, and x_3-axes, and flexural vibrations about the x_1-, x_2-, and x_3-axes; they are labeled as A_g, A_u, B_{3g}, B_{2g}, B_{1g}, B_{3u}, B_{2u}, and B_{1u}, respectively (Mochizuki 1987), as shown in Table 8.1. We must know the vibration groups and the harmonic orders of individual resonance peaks. However, this has never been straightforward, because the resonance spectrum contains a large number of resonance peaks and there is no a priori information as to the modes that cause individual resonance peaks. On the other hand, the calculation exactly tells the resonance modes. If the correspondence between the calculations and measurements was wrong, the elastic constants would converge to a false minimum or fail to converge. For this reason, one has to know beforehand a set of elastic constants close to the true values to correctly compare measurements with calculations. For a material with less information on elastic constants, the mode identification procedure easily goes to erroneous results, which could be avoided by supplementing RUS with other methods such as a pulse echo or rod resonance method. To overcome this inherent difficulty, several efforts have been made for

Table 8.1 Deformation symmetry of eight vibration groups of a rectangular parallelepiped specimen with orthorhombic symmetry. u_i denotes the displacement component along the x_i-axis. E and O mean even and odd functions of the axis. The origin is located at the center of the rectangular parallelepiped. Vibration group notation follows Mochizuki (1987)

Group	Displacement	x_1	x_2	x_3	Group	Displacement	x_1	x_2	x_3
A_g	u_1	O	E	E	A_u	u_1	E	O	O
	u_2	E	O	E		u_2	O	E	O
	u_3	E	E	O		u_3	O	O	E
B_{3g}	u_1	O	O	O	B_{3u}	u_1	E	E	E
	u_2	E	E	O		u_2	O	O	E
	u_3	E	O	E		u_3	O	E	O
B_{2g}	u_1	E	E	O	B_{2u}	u_1	O	O	E
	u_2	O	O	O		u_2	E	E	E
	u_3	O	E	E		u_3	E	O	O
B_{1g}	u_1	E	O	E	B_{1u}	u_1	O	E	O
	u_2	O	E	E		u_2	E	O	O
	u_3	O	O	O		u_3	E	E	E

resonance mode identification. Ohno (1976) varied the specimen size and used different rates of changes of resonance frequencies. Maynard (1992) switched assignments of frequencies during the iteration calculation for finding the best fit. Migliori et al. (1993) changed the sample orientation relative to the transducers and monitored the signal amplitude. These are not general methods. They are inapplicable, for example, with a solid having large internal friction, where broad resonance peaks overlap with one another.

8.1.2 Mode Selection Principle

A simple way to correctly identify the modes is to separately excite and detect only one group of vibrations, filtering all others out. This is made possible with the advantage of EMATs that they can control the body-force directions through the static field direction relative to the coil arrangement. Figure 8.1 shows the typical measurement setup to mechanize mode selection with EMAR (Ogi et al. 1999a). The specimen is inserted into a solenoid coil located between two permanent magnet blocks. The solenoid coil is loose, and the specimen is unconstrained. Mechanical contact between the specimen and coil's inner surface is minimum because no external force is applied except the specimen's own weight, and therefore, an ideal *free* vibration occurs. The permanent magnets provide the static magnetic field for Lorentz force coupling. Driving the coil with rf bursts causes eddy currents on the specimen surfaces, which interact with the magnetic field and generate the Lorentz forces. These forces are the sources of the vibration. After

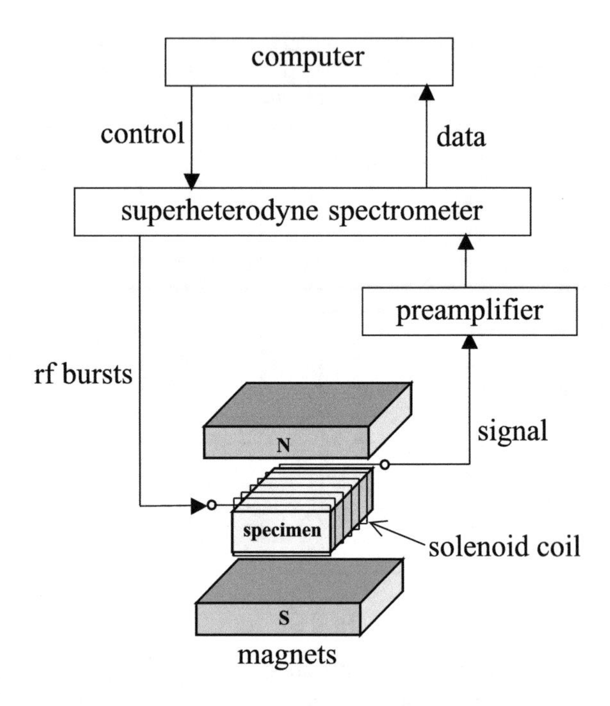

Fig. 8.1 Measurement setup of the EMAR method for measuring free-vibration resonance frequencies of a rectangular parallelepiped specimen, consisting of a solenoid coil, within which the specimen is located, and a pair of permanent magnets

driving, the same coil works to detect the vibration through the reversed mechanism of the generation (Chap. 2).

Symmetry of the Lorentz forces about the three axes can be controlled through the geometrical configuration between the specimen and magnetic field. For example, three configurations shown in Fig. 8.2 are considered. When the magnetic field is applied along the x_1-axis, which is the axial direction of the solenoid coil, the resultant Lorentz forces occur normal to the x_1-axis (Fig. 8.2a). In this case, u_3 on the x_1-x_2 faces can be detected by the reversed-Lorentz force mechanism only when it is an odd function about x_3 and an even function about x_1 and x_2. Also, u_2 on x_1-x_3 faces can be detected only when it is an odd function about x_2 and an even function of x_1 and x_3. Among the eight vibration groups in Table 8.1, only the A_g group (breathing vibration) satisfies this deformation symmetry. Thus, only A_g vibration modes are excited and detected with the configuration of Fig. 8.2a. In the setup shown in Fig. 8.2b, where the magnetic field is applied along the x_3-axis and the Lorentz forces arise parallel to the x_2-axis on the x_1-x_2 faces, the deformation u_2 must be an odd function of x_3 and an even function of x_1 and x_2. Only the B_{3g} vibration group can satisfy this requirement, and the resonance modes belonging to B_{3g} group are chosen for excitation and detection in this measurement setup. Similarly, the configuration of Fig. 8.2c measures only the B_{2g} vibration modes. Thus, one can easily select the vibration group by changing the relationship among the specimen orientation, coil's axial direction, and the field direction.

Some contrivances enable us to measure other vibration groups of different deformation symmetry. One of the options is to design a solenoid coil where the

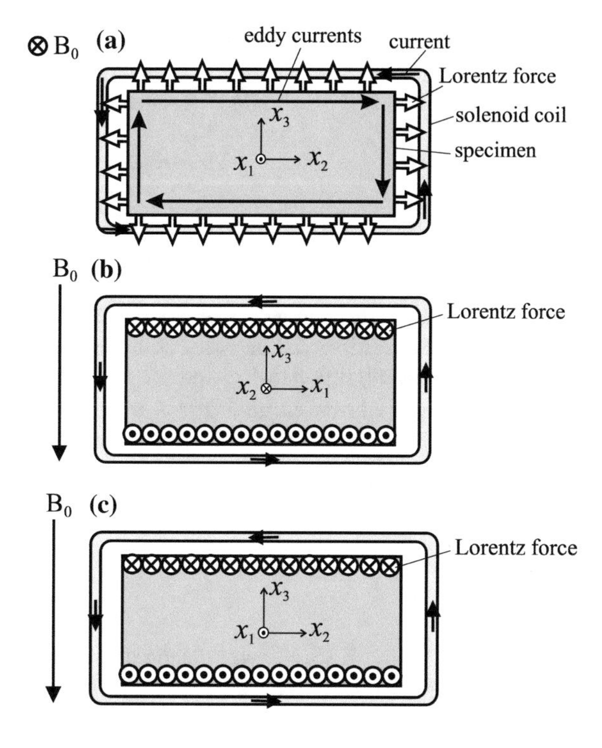

Fig. 8.2 Relationships among the specimen orientation, solenoid coil axis, and static field direction for independently generating and detecting **a** A_g modes, **b** B_{3g} modes, and **c** B_{2g} modes

direction of the driving current is reversed at the center of the specimen. Considering that the axial direction of such a coil is along the x_1-axis and the static field is applied along the x_1-axis (Fig. 8.2a), the displacement u_3 on the x_1-x_2 faces can be detected only if it is an odd function of x_3, an even function of x_2, and an odd function of x_1. This requirement applies to the B_{3u} group (symmetric flexure vibration about the x_1-axis). In principle, one can select one of the eight groups of vibration modes with the coil magnet arrangements.

8.2 Inverse Calculation for C_{ij} and Q_{ij}^{-1}

Measured resonance frequencies are compared with calculations, and a least-square procedure determines the best-fitting elastic constants C_{ij} after iterations. A calculation scheme for the resonance frequencies of free vibrations of rectangular parallelepipeds has been established by Demarest (1971), Ohno (1976, 1990), and Migliori et al. (1993). Here, we review Ohno's work (1990) for a piezoelectric material, which of course applies to nonpiezoelectric materials by neglecting the piezoelectric coefficients.

The constitutive equations relating the elastic and electric properties are

$$\sigma_{ij} = C_{ijkl}S_{kl} - e_{kij}E_k, \quad (8.1)$$

$$D_i = e_{ikl}S_{kl} + \varepsilon_{ij}E_j, \quad (8.2)$$

$$S_{kl} = \frac{1}{2}\left(\frac{\partial u_k}{\partial x_l} + \frac{\partial u_l}{\partial x_k}\right). \quad (8.3)$$

Here, σ_{ij} and S_{kl} are components of the stress and strain tensors, respectively. e_{kij} and ε_{ij} are piezoelectric and dielectric coefficients, respectively. E_i, D_i, and u_i denote electric field, electric flux density, and displacement along the x_i-axis, respectively. The electric field can be divided into a rotational component and an irrotational component (or quasistatic field). In the megahertz frequency region, the rotation component is negligible and the quasistatic electric field dominates, which is expressed by the electric potential ϕ (Auld 1973),

$$E_k = -\frac{\partial \phi}{\partial x_k}. \quad (8.4)$$

(As is shown in Fig. 8.35, this quasistatic approximation actually causes negligible errors in the resonance frequency calculation.) Substituting Eqs. (8.1)–(8.4) into the equation of motion

$$-\rho\omega^2 u_i = \frac{\partial \sigma_{ij}}{\partial x_j}, \quad (8.5)$$

with the boundary conditions for the elastic and electric fields leads to the free-vibration resonance frequencies, ω, of the system.

Exact solutions for the displacements and electric potential are unavailable for rectangular parallelepiped piezoelectric solids. Then, calculus of variations with a Rayleigh-Ritz method has been used. Eer Nisse (1967) derived the Lagrangian of a vibrating piezoelectric material as

$$L = \frac{1}{2}\int_V \left(S_{ij}C_{ijkl}S_{kl} - \frac{\partial \phi}{\partial x_m}\varepsilon_{mn}\frac{\partial \phi}{\partial x_n} + 2\frac{\partial \phi}{\partial x_m}e_{mkl}S_{kl} - \rho\omega^2 u_i u_i\right)\mathrm{d}V. \quad (8.6)$$

The stationary point of the Lagrangian ($\delta L = 0$) gives the resonance modes and the accompanying resonance frequencies of the system. Ohno (1990) approximated the displacements and electric potential in terms of linear combinations of the basis functions Ψ consisting of the normalized Legendre polynomials:

$$u_i(x_1, x_2, x_3) = \sum_k a_k^i \Psi_k^i(x_1, x_2, x_3), \tag{8.7}$$

$$\phi(x_1, x_2, x_3) = \sum_k a_k^\phi \Psi_k^\phi(x_1, x_2, x_3). \tag{8.8}$$

Here,

$$\Psi_k(x_1, x_2, x_3) = \sqrt{\frac{8}{L_1 L_2 L_3}} \bar{P}_l(2x_1/L_1)\bar{P}_m(2x_2/L_2)\bar{P}_n(2x_3/L_3). \tag{8.9}$$

$\overline{P}_\lambda$ denotes the normalized Legendre polynomial of degree λ and L_i denotes the edge length along the x_i-axis of the rectangular parallelepiped. Substituting Eqs. (8.7)–(8.9) into Eq. (8.6) and finding a stationary point of the Lagrangian determine the resonance frequencies ω together with the associated sets of expansion coefficients, a_k^i and a_k^ϕ.

Just as the C_{ij} are obtained from measurements of many resonance frequencies, all the independent internal friction can be determined by measuring internal friction at many resonance modes. The internal friction tensor Q_{ij}^{-1} is defined as the ratio of imaginary to real part of the complex elastic stiffness $\widetilde{C}_{ij}$ (Ledbetter et al. 1995b):

$$\widetilde{C}_{ij} = C_{ij}(1 + jQ_{ij}^{-1}). \tag{8.10}$$

(Summation convention is not implied.) Note that the same symmetry of internal friction holds as for the elastic constants. Determination of all independent C_{ij} and their companion internal friction Q_{ij}^{-1} permits one to predict unmeasurable mechanical loss of any ultrasonic modes and then to select a less-lossy mode, propagation direction, and surface orientation. This is an important procedure in designing acoustic devices including surface-acoustic-wave (SAW) and film-bulk-acoustic-resonator (FBAR) filters. Because the EMAR method provides internal friction Q_p^{-1} of individual resonance modes through the resonance ring-down measurements (Eq. 6.3), their measurements over many modes can deduce all the independent internal friction components Q_{ij}^{-1} through an inverse calculation. This procedure is based on a reasonable assumption that the square of the complex frequency can be expressed by a linear combination of complex elastic stiffness coefficients:

$$\tilde{f}_p^2 = \sum_q \frac{\partial f_p^2}{\partial C_q} \widetilde{C}_q = \sum_q 2 f_p b_{pq} C_q \left(1 + jQ_q^{-1}\right), \tag{8.11}$$

Here, $\tilde{f}_p$ denotes the complex frequency of the pth resonance and q denotes the matrix notation subscripts (i.e., q = 11, 12, 13, …). b_{pq} $(=\partial f_p/\partial C_q)$ is the sensitivity of the resonance frequency f_p to the elastic stiffness C_q, which is obtainable in the inverse calculation for C_{ij} (Migliori et al. 1993). On the other hand, one can define a particular modulus C_p and particular internal friction Q_p^{-1} constructing the complex resonance frequency as

$$\tilde{f}_p^2 = \text{const. } \tilde{C}_p = \text{const. } C_p\left(1 + jQ_p^{-1}\right) = f_p^2\left(1 + jQ_p^{-1}\right). \tag{8.12}$$

Comparison of Eq. (8.11) with Eq. (8.12) leads to

$$Q_p^{-1} = \frac{2\sum_q b_{pq} C_q Q_q^{-1}}{f_p} \text{ and } \frac{2\sum_q b_{pq} C_q}{f_p} = 1. \tag{8.13}$$

The quantities f_p and Q_p^{-1} correspond to the measured resonance frequencies and internal friction, respectively. Then, a standard least-square procedure yields the Q_q^{-1} (=Q_{ij}^{-1}).

8.3 Monocrystal Copper

The mode-selective EMAR technique was first applied to a monocrystal copper (Ogi et al. 1999a). Figure 8.3 shows the dimensions and three orthogonal axes of the specimen. The dimension fluctuation was within 0.3 %. Archimedes method mass density was 8940 kg/m^3. Measurements were taken at ambient temperature.

Figure 8.4 shows the EMAR resonance spectra for the three configurations illustrated in Fig. 8.2 together with the resonance spectrum measured by the usual RUS method with a broken line for comparison. While the RUS spectrum consists of many resonance peaks from all vibration groups, an EMAR spectrum contains only one vibration group as expected. EMAR fails to detect all the possible vibration modes for a given configuration. For example, A_g-11 and A_g-12 peaks are missing in Fig. 8.4. Such an absence occurs because the solenoid coil detects the integral displacement over the specimen faces. If this value is too small, the spectrum will not contain that mode. Although EMAR detects fewer resonance peaks than RUS, this will not obstruct determining a set of elastic constants. Mode

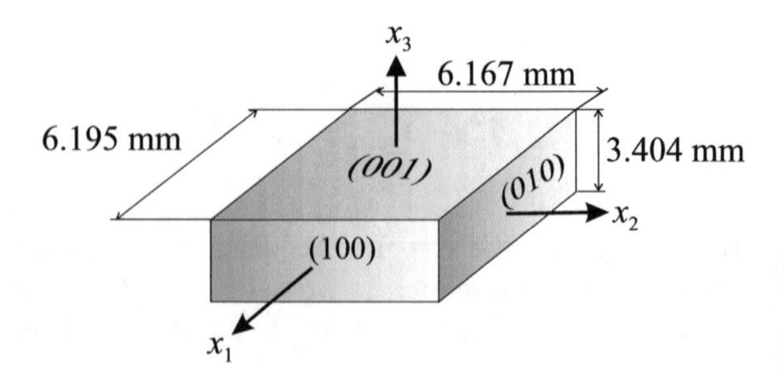

Fig. 8.3 Rectangular parallelepiped specimen of copper monocrystal. Reprinted with permission from Ogi et al. (1999a). Copyright (1999), Acoustical Society of America

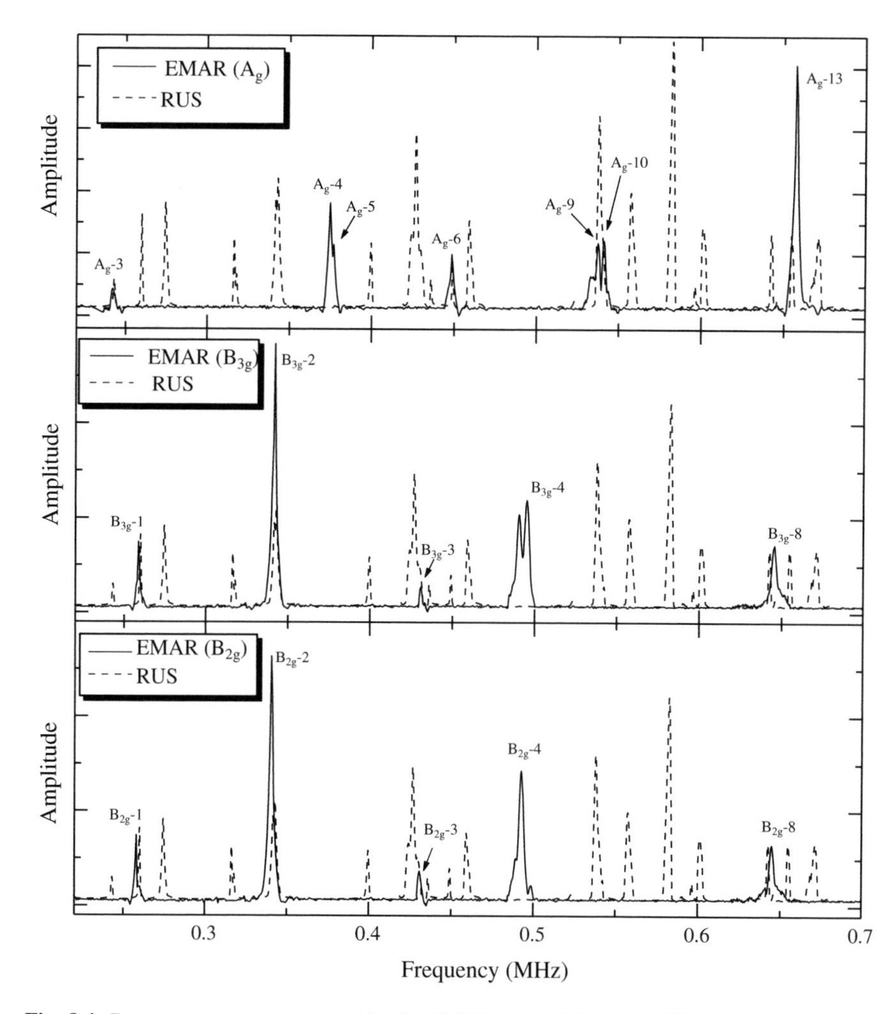

Fig. 8.4 Resonance spectra measured by the EMAR method for three different configurations in Fig. 8.2. The RUS spectrum is shown with a *broken line*. Reprinted with permission from Ogi et al. (1999a). Copyright (1999), Acoustic Society of America

identification is straightforward in EMAR. For example, the B_{3g}-2 and B_{2g}-2 resonances occur at close frequencies and they overlap each other in the RUS spectrum as shown in Fig. 8.5. However, the EMAR can select either of them, and then, the mode identification and separation can be made easily, giving no ambiguity.

The EMAR measurements entered into the inverse calculation to derive the three cubic symmetry elastic constants (C_{11}, C_{12}, and C_{44}). Measured and calculated resonance frequencies after the iteration calculation are shown in Table 8.2. Twenty-five modes were used. They agreed with each other within 0.4 %, which is comparable with the fractional error of the specimen dimensions.

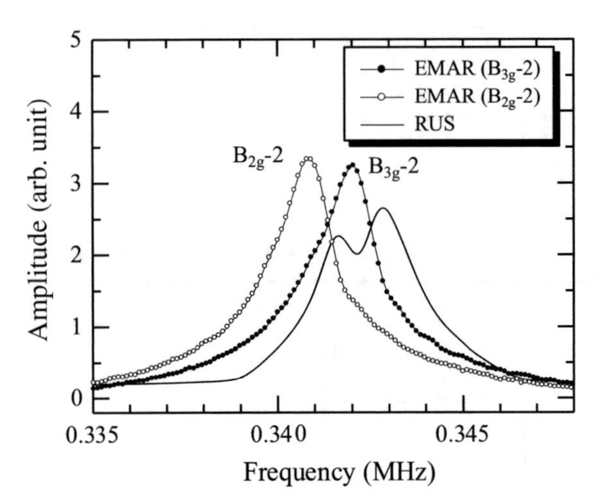

Fig. 8.5 Close-up of Fig. 8.4 and comparison between resonance spectra measured by EMAR and RUSReprinted with permission from Ogi et al. (1999a). Copyright (1999), Acoustic Society of America

Table 8.2 Resonance frequencies measured by EMAR (f_{meas}) and those calculated (f_{calc}) for the monocrystal copper specimen. Reprinted with permission from Ogi et al. (1999a). Copyright (1999), Acoustic Society of America

Mode	f_{meas} (MHz)	f_{calc} (MHz)	Difference (%)
A_g-1	–	0.183706	–
A_g-2	–	0.21136	–
A_g-3	0.242323	0.242549	−0.09
A_g-4	0.374916	0.373805	0.30
A_g-5	0.375139	0.374833	0.08
A_g-6	0.448678	0.450268	−0.35
A_g-7	–	0.472787	–
A_g-8	–	0.494272	–
A_g-9	0.535587	0.53339	0.41
A_g-10	0.542134	0.543198	−0.20
A_g-11	–	0.567346	–
A_g-12	–	0.64316	–
A_g-13	0.657429	0.65774	−0.05
B_{3g}-1	0.258849	0.258944	−0.04
B_{3g}-2	0.341657	0.341765	−0.03
B_{3g}-3	0.429726	0.428519	0.28
B_{3g}-4	0.492967	0.494452	−0.30
B_{2g}-1	0.257887	0.258572	−0.26
B_{2g}-2	0.340418	0.340449	−0.01
B_{2g}-3	0.430686	0.42946	0.29
B_{2g}-4	0.49249	0.49367	−0.24

Table 8.3 Elastic constants and internal friction of monocrystal copper determined by EMAR and RUS. Reprinted with permission from Ogi et al. (1999a). Copyright (1999), Acoustic Society of America

	C_{ij} (GPa)			Q_{ij}^{-1}	
	EMAR	RUS	Hearmon (1966)	EMAR	RUS
C_{11}	167.0	168.7	168.4	1.14	1.74
C_{12}	120.9	121.6	121.4	0.30	1.34
C_{44}	74.64	75.46	75.4	2.59	2.46
$B = (C_{11} + 2C_{12})/3$	136.2	137.3	137.1	0.64	1.50
$C' = (C_{11} - C_{12})/2$	23.05	23.52	23.5	3.34	2.80
$C_{110,110} = (C_{11} + C_{12} + 2C_{44})/2$	218.6	220.6	220.3	1.40	1.87
$C_{111,111} = (C_{11} + 2C_{12} + 4C_{44})/3$	235.7	237.9	237.7	1.47	1.90
$C_{111,arb} = (C_{11} - C_{12} + C_{44})/3$	40.25	40.84	40.8	2.88	2.59
C_L	199.6	201.6	–	1.34	1.84
G	47.54	48.22	–	2.83	2.57
E	127.8	129.5	–	2.60	2.46
ν	0.3437	0.3428	–	–	–

This determination requires the initial guess of elastic constants. The iterative calculation started with several sets of elastic constants covering a wide range (C_{11}: 100–230 GPa, C_{12}: 70–150 GPa, and C_{44}: 50–150 GPa), but the inverse calculation reached the same answer for C_{ij}. In the RUS measurement, on the other hand, much closer initial guesses (C_{11} = 168 GPa, C_{12} = 120 GPa, and C_{44} = 74 GPa) were necessary to achieve the correct convergence, although the measured and calculated frequencies again agreed within a discrepancy of 0.4 %.

Table 8.3 presents the elastic constants deduced from the two methods and those obtained previously, including other useful elastic constant combinations obtained from C_{11}, C_{12}, and C_{44}. Kröner's method (1978) was applied to calculate the usual isotropic material averaged-over-direction elastic constants: shear modulus G, longitudinal modulus C_L, Young's modulus E, and Poisson's ratio ν. EMAR provides slightly smaller stiffnesses than those measured by RUS. This is caused by the mechanical contacts between the piezoelectric transducers and the specimen in RUS; RUS does not achieve ideal free vibration. Applied load partially constrains the specimen deformation and increases the resonance frequencies (see Fig. 8.5), leading to larger apparent elastic constants (Sumino et al. 1976).

The EMAR's attractive feature that no energy loss occurs from mechanical contact is most emphasized in measuring internal friction. Table 8.4 compares internal friction measured by EMAR and RUS. In RUS measurements, internal friction for individual peak is obtained from the resonance peak width: $Q_p^{-1} = \Delta f/f_p$, where Δf denotes the peak width at half-power amplitude. In EMAR, internal friction is measured from the resonance ring-down as described in Chap. 6.

Table 8.4 Internal friction of copper monocrystal measured by EMAR (Q_{EMAR}^{-1}) and RUS (Q_{RUS}^{-1}). Reprinted with permission from Ogi et al. (1999a). Copyright (1999), Acoustic Society of America

Mode	Q_{EMAR}^{-1} (10^{-3})	Q_{RUS}^{-1} (10^{-3})	Difference (%)
A_g-1	–	2.63	
A_g-2	–	1.36	
A_g-3	2.73	3.52	−29
A_g-4	4.46	2.68	40
A_g-5	4.33	2.8	35
A_g-6	2.1	3	-43
A_g-7	–	–	–
A_g-8	–	–	–
A_g-9	3.01	3.04	−1
A_g-10	1.44	1.51	−5
A_g-11	–	3.38	–
A_g-12	–	1.71	–
A_g-13	1.55	1.74	−12
B_{3g}-1	2.2	2.38	−8
B_{3g}-2	3.12	3.78	−21
B_{3g}-3	1.94	1.83	6
B_{3g}-4	4.01	–	–
B_{2g}-1	2.42	2.82	−17
B_{2g}-2	3.16	3.43	−9
B_{2g}-3	2.03	2.06	−1
B_{2g}-4	3.87	–	–

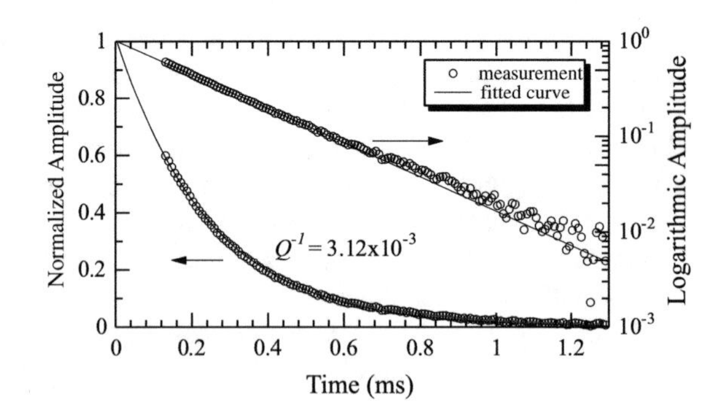

Fig. 8.6 Ring-down curve of B_{3g}-2 mode measured by EMAR. Solid line is a fitted exponential function. Logarithmic-scale amplitude for the same measurements is also shown. Reprinted with permission from Ogi et al. (1999a). Copyright (1999), Acoustic Society of America

Figure 8.6 shows an example of a measured ring-down curve and the fitted exponential function. Departure from the regression curve at low amplitudes is caused by the background noise. For most modes, Q^{-1} from RUS exceeds that from EMAR. Measurements in Table 8.4 were used to calculate the internal friction tensor Q_{ij}^{-1} with Eq. (8.13). The results are given in Table 8.3. Internal friction for the EMAR's bulk modulus B is considerably smaller. This observation is reasonable because a large amount of mechanical loss is caused by the dislocation damping in monocrystal copper and dislocations will not move during hydrostatic loading. The internal friction for shear waves (internal friction for C_{44}, C', $C_{111,\mathrm{arb}}$, and G) is larger than those of longitudinal waves (C_{11}, $C_{110,110}$, $C_{111,111}$, and C_L), which indicates that shear waves more effectively cause dislocation vibration about their pinning points than longitudinal waves do. This is supported by a simultaneous measurement of shear wave and longitudinal wave attenuation on deformed polycrystalline copper (see Fig. 6.3).

Comparing between EMAR and RUS, we find larger RUS Q_{ij}^{-1} for longitudinal waves and, especially, for the bulk modulus, while the Q_{ij}^{-1} for shear waves are comparable. This indicates that the vibration modes accompanying volume change produce energy losses into the sandwiching transducers in an RUS measurement, even when the applied force is minimized.

8.4 Metal Matrix Composites (SiC_f/Ti-6Al-4V)

8.4.1 *Difficulty of Measuring C_{ij} of Anisotropic Composites*

Fiber-reinforced metal matrix composites show outstanding mechanical properties and they are promising for structural materials for advanced engineering applications. Their elastic stiffnesses C_{ij} are indispensable for designing such a component. These composites, however, pose several severe obstacles for measuring their elastic constants and internal friction with conventional pulse echo and rod resonance methods. The difficulty arises from the following: (i) low elastic symmetry and many independent C_{ij}, (ii) heterogeneous phase distribution, and (iii) interfaces that scatter ultrasonic waves and cause longitudinal-shear wave-mode conversions. Whereas measurement on a high-quality monocrystal or polycrystal yields the elastic constants C_{ij} within one part in 10^3–10^4, or better, a composite sometimes shows errors of several percents. An attempt to reduce this error was described by Lei et al. (1994), who emphasized combining C_{ij} and S_{ij} measurements made in the principal directions, avoiding measurements in nonprincipal directions and the impure-mode problems.

The difficulties enumerated above hardly affect the measurements of the free-vibration resonance frequencies, for which the acoustic energy is confined within a small space of specimen. The EMAR method can, therefore, be useful for determining all the elastic constants and internal friction of composite materials. This section introduces the EMAR measurements for silicon carbide (SiC) fiber-reinforced titanium alloy composites (Ogi et al. 1999b, 2000).

8.4.2 Cross-Ply and Unidirectional Composites

The composites studied were polycrystalline Ti-6Al-4V matrix reinforced by continuous SCS-6 SiC fibers. They are candidate materials for components of jet engine and aerospace structures because they show high strength, stiffness, and toughness at elevated temperatures (Wadsworth and Froes 1989). Two types of the composite were measured. One was a cross-ply composite as illustrated in Fig. 8.7; the titanium alloy matrix is alternately reinforced by the fibers by $(90°/0°)_{2S}$. Figure 8.7 shows microstructure seen along the x_1-axis. The other was a unidirectional composite as illustrated in Fig. 8.8; the matrix was reinforced with the SiC fibers unidirectionally in the x_3 direction. Figure 8.8 shows microstructure on the x_3 surface. Both composites were fabricated by the foil-fiber-foil stacking technique (8 plies) with 65-MPa hydrostatic compression at 900 °C. The SiC fiber itself is a composite consisting of three layers: carbon core at the center (7.5 vol%), SiC surrounding the core, and thin carbon layer coating the SiC (see also Figs. 8.17 and 9.8). The fiber outer diameter was 140 μm. For both materials, the volume fraction of the SiC fiber was 0.35 and average grain size of the matrix was 9.9 μm. Archimedes method mass densities were 3.930 and 3.886 g/cm^3 for the cross-ply and unidirectional composites, respectively.

Three rectangular parallelepiped specimens were machined from each material. Their dimensions were a = 4.976–5.031 mm, b = 3.990–4.015 mm, and c = 1.890–1.905 mm for the cross-ply specimens, and a = 3.986–4.943 mm, b = 1.742–1.815 mm, and c = 3.023–3.979 mm for the unidirectional specimens (see Figs. 8.7 and 8.8). Because of the reinforcement geometry, the cross-ply composite shows tetragonal symmetry with six independent C_{ij}:

$$[C_{ij}] = \begin{bmatrix} C_{11} & C_{12} & C_{13} & 0 & 0 & 0 \\ C_{12} & C_{11} & C_{13} & 0 & 0 & 0 \\ C_{13} & C_{13} & C_{33} & 0 & 0 & 0 \\ 0 & 0 & 0 & C_{44} & 0 & 0 \\ 0 & 0 & 0 & 0 & C_{44} & 0 \\ 0 & 0 & 0 & 0 & 0 & C_{66} \end{bmatrix}, \tag{8.14}$$

and the unidirectional composite shows orthorhombic symmetry with nine C_{ij}:

$$[C_{ij}] = \begin{bmatrix} C_{11} & C_{12} & C_{13} & 0 & 0 & 0 \\ C_{12} & C_{22} & C_{23} & 0 & 0 & 0 \\ C_{13} & C_{23} & C_{33} & 0 & 0 & 0 \\ 0 & 0 & 0 & C_{44} & 0 & 0 \\ 0 & 0 & 0 & 0 & C_{55} & 0 \\ 0 & 0 & 0 & 0 & 0 & C_{66} \end{bmatrix}. \tag{8.15}$$

Also, they possess corresponding internal friction components Q_{ij}^{-1} defined by Eq. (8.10).

8.4.3 C_{ij} and Q_{ij}^{-1} at Room Temperature

The EMAR measurement setup was the same as that in Fig. 8.1. The solenoid coil was driven by high-power rf bursts with 300–1000 V_{PP} and 100-μs duration for generating the Lorentz forces. After the excitation, the same coil received the vibration through the reversed-Lorentz force mechanism. A frequency scan detected resonance peaks, and the resonance frequencies were determined by a Lorentzian function fitting. The free vibrations of a rectangular parallelepiped with orthorhombic or higher symmetry fall into eight groups shown in Table 8.1. A_g, B_{2g}, and B_{3g} vibration modes were independently measured for the cross-ply specimens, and A_g, B_{1g}, and B_{3g} modes were measured for the unidirectional specimens. About forty modes were used for each case in the inverse calculation.

Along with the EMAR measurements, the conventional RUS method and pulse echo method were applied. In the RUS method, two piezoelectric transducers sandwiched a specimen at the opposite corners. In the pulse echo measurements, a 19-MHz longitudinal wave and 10-MHz shear wave were used for measuring some diagonal elastic constants. They propagated in the thickness direction. The pulse echo method was inapplicable for other propagation directions because of poor echo waveforms caused mainly by small specimen cross sections relative to the transducer aperture.

Figure 8.9 compares between the resonance spectra measured by the RUS and EMAR methods. The RUS spectrum contains many resonance peaks from all eight groups of vibration, while different EMAR configurations led to individual subgroups, resulting in a smaller number of available peaks. But, the determination of a set of elastic constants was unambiguous in EMAR because the vibration groups of the observed modes are known. Figure 8.10 compares the measured and calculated resonance frequencies after the inverse calculation. For almost all modes, disagreement was less than 1 %. Some modes showed larger differences (up to 1.5 %). Such differences are expected for composite materials. Imperfect fiber alignment (see Figs. 8.7 and 8.8) may provide the main error.

The inverse iterative calculation requires a first guess of elastic constants. A 20 % variety of initial guesses reached the same results in the case of EMAR. In RUS, the inverse calculation failed to converge or it led to physically wrong solutions with this range of initial guesses. This emphasizes the importance of complementary C_{ij} measurements for the RUS method, which is vulnerable to the initial guess. When the C_{ij} obtained by EMAR are used as the initial guess for RUS calculation, the first 45 modes can be fully identified and the C_{ij} are determined with the similar accuracy with EMAR. Table 8.5 shows the resulting elastic constants. Fluctuations among the three specimens are also given. Elastic constants determined by EMAR, RUS, and pulse echo methods agree within a few percent.

Concerning the unidirectional composite, the results of $C_{11} \cong C_{22}$, $C_{44} \cong C_{55}$, and $C_{13} \cong C_{23}$ suggest that this material has nearly tetragonal symmetry, which is understandable from the manufacturing procedure (foil-fiber-foil technique). Furthermore, transverse isotropy (hexagonal symmetry) applies with the additional

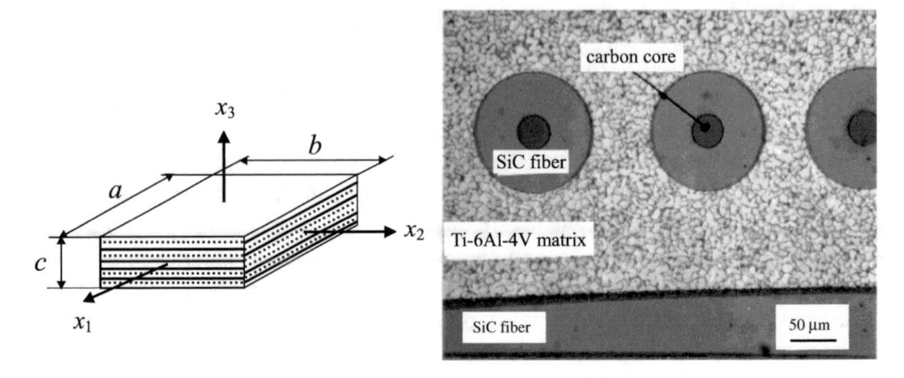

Fig. 8.7 Illustration of the cross-ply SiC_f/Ti-6AL-4V composite specimen (*left*) and microstructure seen along the x_1-axis (*right*)

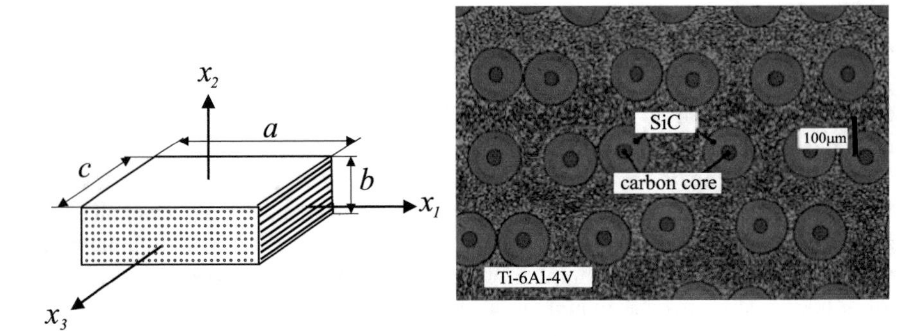

Fig. 8.8 Illustration of the unidirectional SiC_f/Ti-6AL-4V composite specimen (*left*) and microstructure seen along the x_3-axis (*right*). Reprinted with permission from Ogi et al. (2000). Copyright (2000), AIP Publishing LLC

requirement that $C_{66} = (C_{11} - C_{12})/2$. This is true within 5 % error, which is permissible for such a composite material of heterogeneous phases. Thus, the unidirectional composite can be treated as a transversely isotropic material showing five independent elastic constants:

$$[C_{ij}] = \begin{bmatrix} C_{11} & C_{12} & C_{13} & 0 & 0 & 0 \\ C_{12} & C_{11} & C_{13} & 0 & 0 & 0 \\ C_{13} & C_{13} & C_{33} & 0 & 0 & 0 \\ 0 & 0 & 0 & C_{44} & 0 & 0 \\ 0 & 0 & 0 & 0 & C_{44} & 0 \\ 0 & 0 & 0 & 0 & 0 & (C_{11} - C_{12})/2 \end{bmatrix}, \tag{8.16}$$

where the fiber longitudinal direction is along the x_3-axis.

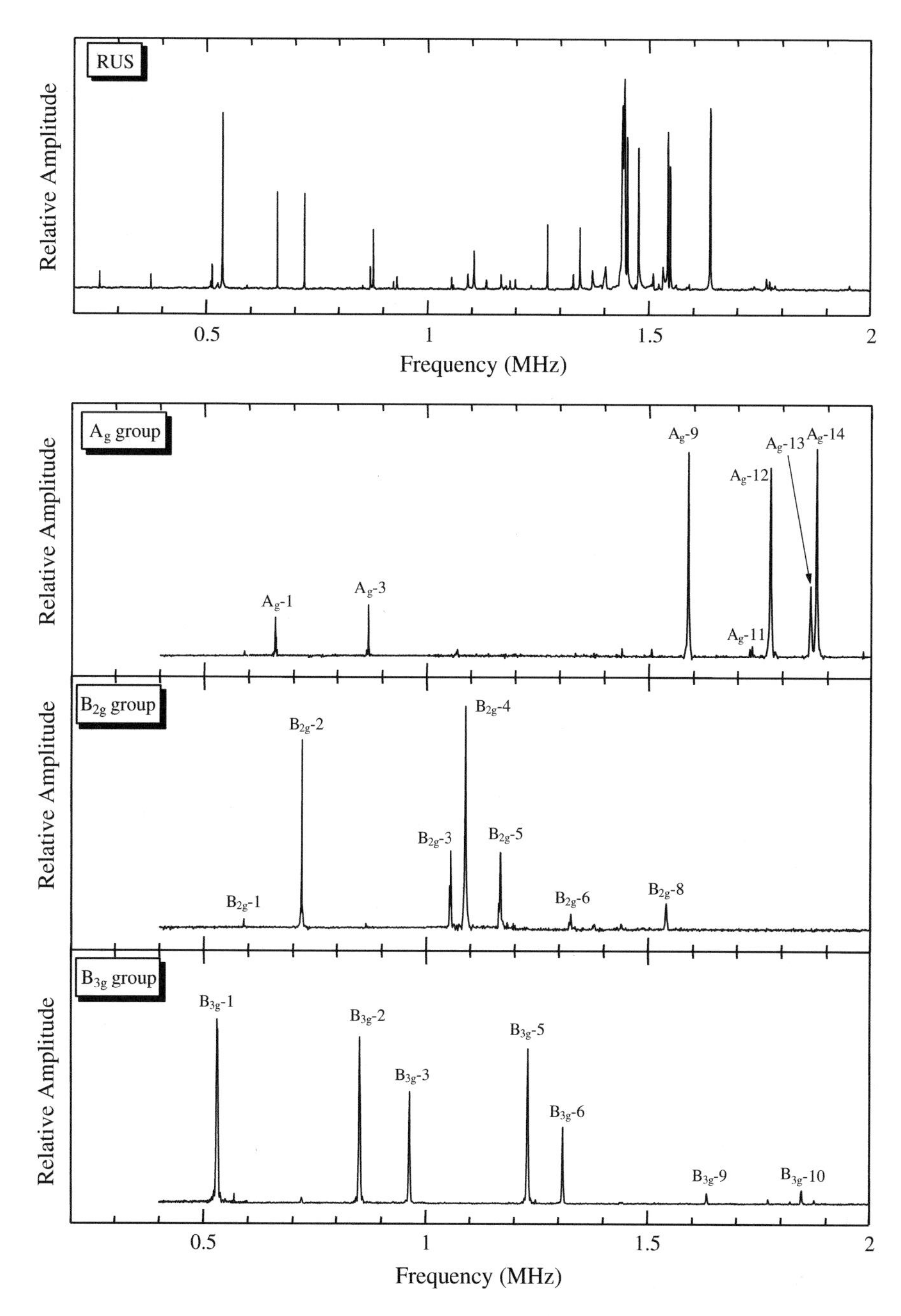

Fig. 8.9 Resonance spectra measured by RUS (*upper*) and EMAR (*bottom*) for a cross-ply SiC_f/Ti-6Al-4V composite specimen measuring 4.975 × 4.015 × 1.905 mm^3. Reprinted from Ogi et al. (1999b). Copyright (1999), with permission from Elsevier

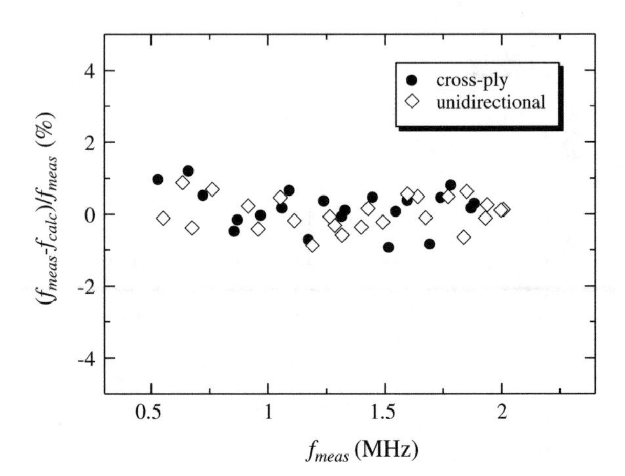

Fig. 8.10 Comparison between measurements and calculations of resonance frequency after convergence of inverse calculation

Table 8.5 Elastic constants of cross-ply and unidirectional SiC_f/Ti-6Al-4V composites at room temperature measured by EMAR, RUS, and pulse echo (PE) methods and internal friction of the composites measured by EMAR

	Cross-ply (tetragonal)				Unidirectional (orthorhombic)			
	C_{ij} (GPa)			Q_{ij}^{-1} (10^{-4})	C_{ij} (GPa)			Q_{ij}^{-1} (10^{-4})
	EMAR	RUS	PE		EMAR	RUS	PE	
C_{11}	224.6 ± 2.7	229.1 ± 1.8	–	7.8	190.5 ± 0.5	190.0 ± 0.6	–	3.1
C_{22}	224.6 ± 2.7	229.1 ± 1.8	–	7.8	191.3 ± 1.7	190.6 ± 0.2	192	6.9
C_{33}	187.8 ± 4.7	190.0 ± 5.8	188	15.6	249.5 ± 6.8	250.0 ± 7.2	–	3.5
C_{44}	54.59 ± 0.03	54.95 ± 0.05	53.7	9.3	56.40 ± 0.37	55.5 ± 1.3	56.1	10.8
C_{55}	54.59 ± 0.03	54.95 ± 0.05	–	9.3	51.7 ± 1.6	56.4 ± 1.2	–	11.7
C_{66}	51.7 ± 1.1	52.5 ± 1.1	–	6.3	54.3 ± 0.8	54.7 ± 0.7	53.8	6.8
C_{12}	79.3 ± 4.6	83.7 ± 2.0	–	16	74.2 ± 3.7	77.0 ± 2.5	–	–
C_{13}	76.9 ± 4.9	80.5 ± 2.9	–	24	71.9 ± 1.5	69.4 ± 1.0	–	–
C_{23}	76.9 ± 4.9	80.5 ± 2.9	–	24	66.8 ± 6.0	68.3 ± 6.9	–	–
E_1	181.2 ± 1.3	182.3 ± 1.5	–	3.5	152.2 ± 1.8	151.0 ± 0.3	–	5.1
E_2	181.2 ± 1.3	182.3 ± 1.5	–	3.5	155.5 ± 2.8	152.1 ± 3.8	–	13.2
E_3	147.5 ± 1.9	146.7 ± 2.5	–	10	212.9 ± 5.0	214.4 ± 3.9	–	4.3
B	122.6 ± 4.2	126.4 ± 2.2	–	15.2	117.5 ± 2.6	117.7 ± 3.1	–	1.9

For internal friction, the amplitude decay was measured after an excitation. The free-decay amplitude curve provides the internal friction at the resonance frequency. Figure 8.11 shows a typical measurement at B_{1g}-4 mode. Internal frictions for all observed modes were used to deduce the internal friction tensor Q_{ij}^{-1} (see Sect. 8.2), which appear in Table 8.5. In the frequency range used here (0.5–2.0 MHz), the main contributions to internal friction are dislocation damping and energy loss at the imperfectly bonded fiber matrix and layer boundaries. No diffraction loss occurs because the wave energy is trapped within the specimen. Concerning the dislocation

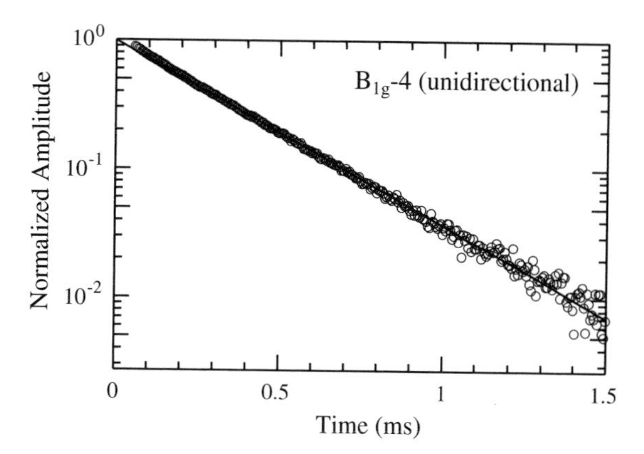

Fig. 8.11 Decay of resonating amplitude at B_{1g}-4 mode (~1.2 MHz) after excitation for a unidirectional composite material. *Open marks* denote measurements, and *solid line* is the fitted exponential function. Logarithmic scale is used. Reprinted with permission from Ogi et al. (2000). Copyright (2000), AIP Publishing LLC

damping, internal friction for shear waves exceeds that for longitudinal waves, typically by a factor of 2, being attributed to the effective shearing force to oscillate dislocations about their pinning points (see Sect. 8.3). Also, internal friction for the bulk modulus B is much smaller because of lower mobility of dislocations with breathing vibration. However, the results for the cross-ply composite oppose expectation: Internal friction for longitudinal waves (Q_{11}^{-1} and Q_{33}^{-1}) is comparable with or larger than that for shear waves (Q_{44}^{-1} and Q_{66}^{-1}), and the bulk modulus internal friction Q_B^{-1} is larger. An interesting observation is the larger internal friction for the wave modes propagating perpendicular to the lamination layer interfaces (i.e., in the x_3 direction) than others. Namely, $Q_{33}^{-1} > Q_{11}^{-1}$, $Q_{44}^{-1} > Q_{66}^{-1}$, and $Q_{E3}^{-1} > Q_{E1}^{-1}$. This trend also applies for the unidirectional composite, for which the x_2 direction is perpendicular to the lamination layer and we see that $Q_{22}^{-1} > Q_{11}^{-1} \cong Q_{33}^{-1}$, $Q_{44}^{-1} \cong Q_{55}^{-1} > Q_{66}^{-1}$, and $Q_{E2}^{-1} > Q_{E1}^{-1} \cong Q_{E3}^{-1}$. Thus, the principal loss arises from the presence of imperfect bond boundaries between ply layers. When an ultrasonic wave passes across a partially delaminated or weak-bonded boundary, the wave energy is lowered by friction at the boundary and by higher-harmonic generation from a nonlinear stress-strain response (see Chap. 10). Measuring Q_{ij}^{-1} provides a method for probing the quality of lamellar interfaces (Ledbetter 2003).

8.4.4 Micromechanics Modeling

When one obtains a complete set of elastic constants of a composite material and the matrix elastic constants, the elastic constants of the reinforcing inclusions can be predicted with micromechanics calculations. It is well worth a try, because a direct measurement of the elastic constants of a single inclusion is normally unavailable, while the inclusion's elastic constants are required for predicting complete sets of elastic constants of any other composites containing the inclusions.

Such a modeling scheme has been established by many researchers after Eshelby (1957) and Mori and Tanaka (1973), as reviewed by Ledbetter et al. (1995c).

Confirmation of the calculated moduli has received intensive study by Ledbetter and Dunn, and their coresearchers (Ledbetter et al. 1995a; Dunn and Ledbetter 1995a, b, 1997; Ogi et al. 1999b, 2000). Here, a brief overview of the micromechanics calculation is provided.

Consider a composite consisting of two phases: a homogeneous matrix and inclusions embedded in it. When a stress $\boldsymbol{\sigma}_\mathbf{A}$ is applied to the composite, the stress and strain in the matrix and inclusions are distributed in a complicated way. However, their volume averages over individual regions can be related to the applied stress and the average strain over the composite $\langle \boldsymbol{\varepsilon} \rangle$ as

$$\boldsymbol{\sigma}_\mathbf{A} = c_1 \langle \boldsymbol{\sigma}_\mathbf{M} \rangle + c_2 \langle \boldsymbol{\sigma}_\mathbf{I} \rangle, \tag{8.17}$$

$$\langle \boldsymbol{\varepsilon} \rangle = c_1 \langle \boldsymbol{\varepsilon}_\mathbf{M} \rangle + c_2 \langle \boldsymbol{\varepsilon}_\mathbf{I} \rangle, \tag{8.18}$$

where c_1 and c_2 denote volume fractions of the matrix and inclusion, respectively. $\boldsymbol{\sigma}_\mathbf{M}$ and $\boldsymbol{\sigma}_\mathbf{I}$ denote stress, and $\boldsymbol{\varepsilon}_\mathbf{M}$ and $\boldsymbol{\varepsilon}_\mathbf{I}$ denote strain in the matrix and inclusion, respectively. Quantities with the brackets $\langle \rangle$ indicate the volume average. Hooke's law is satisfied in each phase:

$$\langle \boldsymbol{\sigma}_\mathbf{M} \rangle = \mathbf{C}^\mathbf{M} \langle \boldsymbol{\varepsilon}_\mathbf{M} \rangle, \tag{8.19}$$

$$\langle \boldsymbol{\sigma}_\mathbf{I} \rangle = \mathbf{C}^\mathbf{I} \langle \boldsymbol{\varepsilon}_\mathbf{I} \rangle. \tag{8.20}$$

Hence, $\mathbf{C}^\mathbf{M}$ and $\mathbf{C}^\mathbf{I}$ denote elastic stiffness matrices of the matrix and inclusion, respectively. Thus, Eq. (8.21) defines the overall composite elastic stiffnesses $\mathbf{C}^\mathbf{C}$:

$$\boldsymbol{\sigma}_\mathbf{A} = \mathbf{C}^\mathbf{C} \langle \boldsymbol{\varepsilon} \rangle. \tag{8.21}$$

Equations (8.17)–(8.21) provide

$$c_1 \mathbf{C}^\mathbf{M} \langle \boldsymbol{\varepsilon}_\mathbf{M} \rangle + c_2 \mathbf{C}^\mathbf{I} \langle \boldsymbol{\varepsilon}_\mathbf{I} \rangle = \mathbf{C}^\mathbf{C} (c_1 \langle \boldsymbol{\varepsilon}_\mathbf{M} \rangle + c_2 \langle \boldsymbol{\varepsilon}_\mathbf{I} \rangle). \tag{8.22}$$

One can calculate $\mathbf{C}^\mathbf{C}$ through Eq. (8.22) if the relationship between $\langle \boldsymbol{\varepsilon}_\mathbf{M} \rangle$ and $\langle \boldsymbol{\varepsilon}_\mathbf{I} \rangle$ is known. This relationship can be predicted using the Mori and Tanaka's mean field approximation (1973) as follows.

Consider the case that the same type of inclusion is newly embedded in the matrix region, where the average stress and strain are $\langle \boldsymbol{\sigma}_\mathbf{M} \rangle$ and $\langle \boldsymbol{\varepsilon}_\mathbf{M} \rangle$, respectively (see Fig. 8.12). The average strain in that inclusion will be slightly different from $\langle \boldsymbol{\varepsilon}_\mathbf{M} \rangle$ by $\boldsymbol{\varepsilon}'$. Thus, Hooke's law in that inclusion is of the form

$$\langle \boldsymbol{\sigma}'_\mathbf{I} \rangle = \mathbf{C}^\mathbf{I} (\langle \boldsymbol{\varepsilon}_\mathbf{M} \rangle + \boldsymbol{\varepsilon}'). \tag{8.23}$$

Here, $\langle \boldsymbol{\sigma}'_\mathbf{I} \rangle$ is the average stress in that inclusion. According to Eshelby's equivalent inclusion method (Eshelby 1957; Mura 1987), the stress inside the inclusion can be expressed by a summation of the matrix stress $\langle \boldsymbol{\sigma}_\mathbf{M} \rangle$ and the internal stress of the

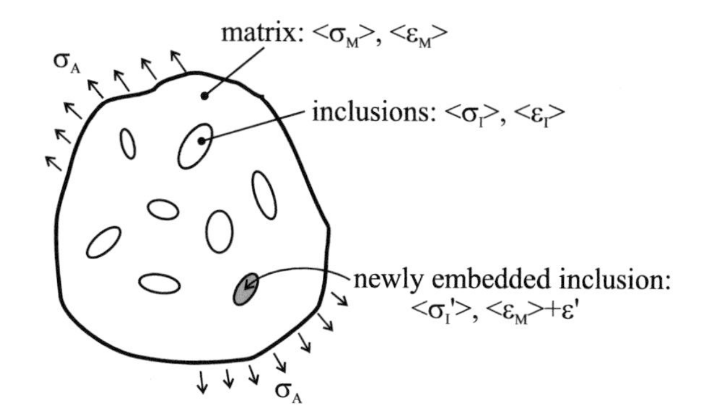

Fig. 8.12 Composite consisting of a homogeneous matrix and inclusions. When an inclusion is newly embedded, the stress and strain in that inclusion become $\langle \boldsymbol{\sigma}_{\mathbf{I}}' \rangle$ and $\langle \boldsymbol{\varepsilon}_{\mathbf{M}} \rangle + \boldsymbol{\varepsilon}'$, respectively

equivalent inclusion when it would be embedded in the matrix phase with an eigenstrain $\boldsymbol{\varepsilon}^*$:

$$\langle \boldsymbol{\sigma}_{\mathbf{I}}' \rangle = \mathbf{C}^{\mathbf{M}} \langle \boldsymbol{\varepsilon}_{\mathbf{M}} \rangle + \mathbf{C}^{\mathbf{M}} (\mathbf{S}\boldsymbol{\varepsilon}^* - \boldsymbol{\varepsilon}^*), \tag{8.24}$$

$$\boldsymbol{\varepsilon}' = \mathbf{S}\boldsymbol{\varepsilon}^*. \tag{8.25}$$

S denotes Eshelby's tensor, which is a function of inclusion's shape and elastic constants of matrix. When an ellipsoidal inclusion and an isotropic matrix are considered, the nonzero components of **S** become remarkably simple as tabulated in Mura's monograph (1987). For example, for the inclusion of continuous circular cylinder, they are

$$\left.\begin{aligned} &S_{11} = S_{22} = \frac{5 - 4\nu}{8(1 - \nu)}, \\ &S_{12} = S_{21} = \frac{4\nu - 1}{8(1 - \nu)}, \quad S_{13} = S_{23} = \frac{\nu}{2(1 - \nu)}, \\ &S_{44} = S_{55} = \frac{1}{2}, \quad S_{66} = \frac{3 - 4\nu}{4(1 - \nu)}, \\ &S_{33} = S_{32} = S_{31} = 0. \end{aligned}\right\} \tag{8.26}$$

in contracted notation. Comparing Eqs. (8.23) with (8.24) yields

$$\langle \boldsymbol{\varepsilon}_{\mathbf{M}} \rangle + \boldsymbol{\varepsilon}' = \left[\mathbf{S} \left\{ \mathbf{C}^{\mathbf{M}} \right\}^{-1} \left(\mathbf{C}^{\mathbf{I}} - \mathbf{C}^{\mathbf{M}} \right) + \mathbf{I} \right]^{-1} \langle \boldsymbol{\varepsilon}_{\mathbf{M}} \rangle \equiv \mathbf{A}^{\mathbf{d}} \langle \boldsymbol{\varepsilon}_{\mathbf{M}} \rangle. \tag{8.27}$$

Here,

$$\mathbf{A}^{\mathbf{d}} = \left[\mathbf{S} \left\{ \mathbf{C}^{\mathbf{M}} \right\}^{-1} \left(\mathbf{C}^{\mathbf{I}} - \mathbf{C}^{\mathbf{M}} \right) + \mathbf{I} \right]^{-1}, \tag{8.28}$$

is called the dilute strain concentration factor, indicating the strain concentration factor for a single inclusion embedded in an infinite matrix subjected to uniform far-field-stress or far-field-strain boundary condition (Dunn and Ledbetter 1995a). $\mathbf{I}$ is the unit matrix. When the number of inclusions is large enough, the embedding of such a single inclusion will hardly affect the average stress in the matrix and inclusion. Thus,

$$\langle \boldsymbol{\sigma}'_{\mathbf{I}} \rangle = \langle \boldsymbol{\sigma}_{\mathbf{I}} \rangle. \tag{8.29}$$

Equation (8.29) is the Mori and Tanaka's mean-field approximation and usually applies for composite materials reinforced by *many* inclusions. $\langle \boldsymbol{\varepsilon}_{\mathbf{M}} \rangle$ and $\langle \boldsymbol{\varepsilon}_{\mathbf{I}} \rangle$ are then related to each other using Eqs. (8.20), (8.23) and (8.29):

$$\langle \boldsymbol{\varepsilon}_{\mathbf{I}} \rangle = \mathbf{A}^{\mathbf{d}} \langle \boldsymbol{\varepsilon}_{\mathbf{M}} \rangle. \tag{8.30}$$

Substituting into Eq. (8.22) yields $\mathbf{C}^{\mathbf{C}}$:

$$\begin{aligned} \mathbf{C}^{\mathbf{C}} &= \mathbf{C}^{\mathbf{M}} + c_2 \left(\mathbf{C}^{\mathbf{I}} - \mathbf{C}^{\mathbf{M}} \right) \mathbf{A}^{\mathbf{d}} \left(c_1 \mathbf{I} + c_2 \mathbf{A}^{\mathbf{d}} \right)^{-1} \\ &\equiv \mathbf{C}^{\mathbf{M}} + c_2 \left(\mathbf{C}^{\mathbf{I}} - \mathbf{C}^{\mathbf{M}} \right) \mathbf{A}^{\mathbf{I}}. \end{aligned} \tag{8.31}$$

Here,

$$\mathbf{A}^{\mathbf{I}} = \mathbf{A}^{\mathbf{d}} \left(c_1 \mathbf{I} + c_2 \mathbf{A}^{\mathbf{d}} \right)^{-1}, \tag{8.32}$$

is the strain concentration factor for the inclusion and clearly satisfies

$$\langle \boldsymbol{\varepsilon}_{\mathbf{I}} \rangle = \mathbf{A}^{\mathbf{I}} \langle \boldsymbol{\varepsilon} \rangle, \tag{8.33}$$

with Eqs. (8.17)–(8.21) and (8.31). Thus, $\mathbf{A}^{\mathbf{I}}$ implies the fraction of the strain in the inclusions to the total strain. From Eq. (8.31), the elastic stiffnesses of the inclusion are then of the form

$$\mathbf{C}^{\mathbf{I}} = \left[\left(\mathbf{C}^{\mathbf{C}} - \mathbf{C}^{\mathbf{M}} \right) \mathbf{S} \left\{ \mathbf{C}^{\mathbf{M}} \right\}^{-1} - \frac{c_2}{c_1} \mathbf{I} \right]^{-1} \left[\left(\mathbf{C}^{\mathbf{C}} - \mathbf{C}^{\mathbf{M}} \right) \mathbf{S} + \mathbf{C}^{\mathbf{M}} - \frac{1}{c_1} \mathbf{C}^{\mathbf{C}} \right]. \tag{8.34}$$

Thus, one can calculate the elastic constants of inclusion if one knows the composite and matrix moduli, and volume fraction of the inclusion.

For a composite consisting of three phases (phase 1: matrix, phase 2: inclusion, and phase 3: the other inclusion), the overall stiffnesses are

$$\begin{aligned} \mathbf{C}^{\mathbf{C}} &= c_1 \mathbf{C}_1 \mathbf{A}_1 + c_2 \mathbf{C}_2 \mathbf{A}_2 + c_3 \mathbf{C}_3 \mathbf{A}_3 \\ &= \mathbf{C}_1 + c_2 (\mathbf{C}_2 - \mathbf{C}_1) \mathbf{A}_2 + c_3 (\mathbf{C}_3 - \mathbf{C}_1) \mathbf{A}_3. \end{aligned} \tag{8.35}$$

Here, c_i, $\mathbf{C_i}$, and $\mathbf{A_i}$ denote volume fraction, the elastic stiffnesses, and the strain concentration factor of the ith phase, respectively. $\mathbf{A_i}$ is given by

$$\mathbf{A_i} = \mathbf{A_i^d}\left(c_1\mathbf{I} + c_2\mathbf{A_1^d} + c_3\mathbf{A_3^d}\right)^{-1}, \tag{8.36}$$

$$\mathbf{A_i^d} = \left[\mathbf{I} + \mathbf{S}\mathbf{C_1^{-1}}(\mathbf{C_i} - \mathbf{C_1})\right]^{-1}. \tag{8.37}$$

8.4.5 *Elastic Constants of a Single* SiC *Fiber*

The SiC fiber consists of a carbon core surrounded by SiC as shown in Figs. 8.7 and 8.8. Such an annular structure macroscopically exhibits transverse isotropy with the five independent C_{ij} of Eq. (8.16). The unidirectional composite in Fig. 8.7 is approximately a transverse isotropic material as shown in Table 8.5, and it also exhibits five independent elastic constants. Measurements of the composite C_{ij} and matrix C_{ij} then permit us to determine all the fiber C_{ij} through Eq. (8.34) (Ogi et al. 2003a). The result is shown in Table 8.6.

A direct comparison with a previous study is impractical because no study has reported a complete set of the fiber C_{ij}. But, there are reports on the tensile test Young's modulus E_3 in the fiber longitudinal direction at room temperature. The reported values range from 400 to 428 GPa (Jansson et al. 1991; Mital et al. 1994), which the EMAR fiber E_3 falls within.

Further confirmation is made by comparing measurements and calculations of C_{ij} for the cross-ply composite that contains different alignment of the fibers. Considering the cross-ply composite to consist of three phases (matrix, fibers arrayed by 0°, and fibers arrayed by 90°), the overall elastic constants are calculated using Eq. (8.35) and the fiber C_{ij} in Table 8.6. The calculated and measured coefficients are compared with each other in Table 8.7. Good agreement between them, especially for the diagonal C_{ij}, indicates the validity of the resultant fiber C_{ij}.

In the calculations, the polycrystalline Ti-alloy matrix has been assumed to be isotropic. However, it might not be the case because the composite was fabricated by the foil-fiber-foil technique; texture may have developed. In this case, the matrix

Table 8.6 Elastic constants (GPa) of the unidirectional SiC$_f$/Ti-6Al-4V composite of transverse isotropy, Ti-6Al-4V alloy of isotropy, and SiC fiber of transverse isotropy at 293 K. The fiber longitudinal direction is along the x_3-axis

	C_{11}	C_{33}	C_{12}	C_{13}	C_{44}	$^*C_{66}$	E_1	E_3	B
Composite (SiC/Ti-6Al-4V)	195	253	85.4	72.2	53.9	54.8	150	216	121
Matrix (Ti-6Al-4V)	161	161	76.2	76.2	42.7	42.4	113	113	105
SiC fiber	299	425	106	62	86.3	96.5	258	406	163

$^*C_{66} = (C_{11} - C_{12})/2$

Table 8.7 Measured and calculated elastic constants (GPa) of the cross-ply SiC_f/Ti-6Al-4Vcomposite with tetragonal symmetry at 293 K. Two principal axes of the fiber are along the x_1- and x_2-axes, respectively

	C_{11}	C_{12}	C_{13}	C_{33}	C_{44}	C_{66}
Measurement	226	81.9	78.4	189	54.7	52.1
Calculation	225	72.0	73.9	192	55.2	55.1
Difference (%)	0.4	12.1	5.7	−1.6	−0.9	−5.8

can exhibit transverse isotropy with five independent C_{ij}, whose principal axis lies in the thickness (pressure) direction. The effect of matrix anisotropy on the resultant fiber C_{ij} was investigated by simulating various transverse anisotropy of the matrix, keeping the bulk modulus unchanged (Ogi et al. 2003a). For anisotropic matrix, the **S** components can be numerically calculated as shown by Kinoshita and Mura (1971). The simplified expression for the cylindrical inclusions is

$$S_{ikmn} = \frac{1}{4\pi}\int_{c} \frac{C^M_{jlmn}\xi_l\left(\xi_k N_{ij} + \xi_i N_{kj}\right)}{D}\mathrm{d}r. \tag{8.38}$$

Here, ξ_l denotes a component of the unit vector lying normal to the fiber axis, and N_{ij} and D denote the cofactor matrix and determinant of matrix $G_{ij} = C^M_{ipjq}\xi_p\xi_q$, respectively. The integration is made along the unit circle surrounding the fiber at the center. The sensitivities of the fiber C_{ij} to the shear modulus anisotropy ($C^M_{66} - C^M_{44}$ anisotropy) and to the longitudinal modulus anisotropy ($C^M_{33} - C^M_{11}$ anisotropy) of the matrix were investigated. A 30 % shear modulus anisotropy causes errors less than 3 % in the resultant fiber C_{ij}. This anisotropy corresponds to the shear modulus anisotropy of monocrystal titanium, indicating that the fiber C_{ij} are insensitive to the shear modulus anisotropy. A 10 % longitudinal modulus anisotropy, corresponding to the longitudinal modulus anisotropy of monocrystal titanium, causes errors less than 3 %, except for the fiber C_{11}. (The fiber C_{11} is sensitive to the $C^M_{33} - C^M_{11}$ anisotropy.) The texture would cause elastic anisotropy in the matrix, but the relative magnitude would be less than a few percent, confirming the view that the deduced fiber C_{ij} are little affected by the matrix anisotropy. Thus, the assumption of matrix isotropy yields a reliable set of fiber C_{ij}.

8.4.6 EMAR at High Temperatures

Because the metal matrix composites and the SiC fiber are the materials to be used at elevated temperatures, their elastic constants are required at high temperatures, which can be measured with the EMAR system shown in Figs. 3.31 and 3.32. The static field direction is changeable by rotating the permanent magnets about the

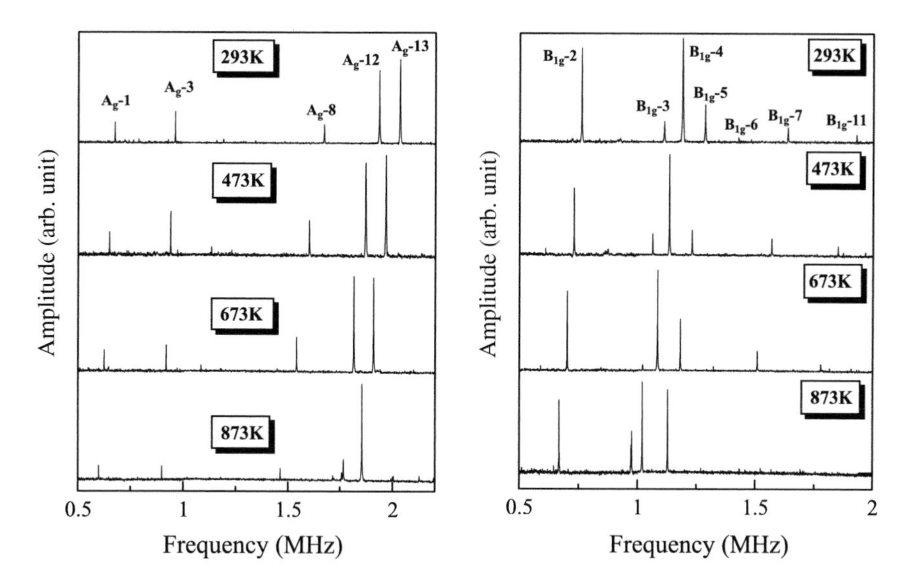

Fig. 8.13 Resonance spectra of the SiC_f/Ti-6Al-4V unidirectional composite at various temperatures for A_g vibration modes (*left*) and B_{1g} vibration modes (*right*). Reprinted form Ogi et al. (2003a), with permission from Taylor & Francis Ltd, (http://www.tandfonline.com)

cylindrical chamber axis to select the detecting vibration modes. This configuration excites and detects only two vibration groups (A_g and one of the B_g groups) independently, but they are sufficient to deduce all the five and six independent elastic constants of the unidirectional composite (Ogi et al. 2003a) and cross-ply composite (Ogi et al. 2001). (Note that the unidirectional composite shows nearly transverse isotropy.) The number of modes used in the inverse calculation was typically twenty-five.

Figure 8.13 shows EMAR spectra measured with the unidirectional composite material at various temperatures. Capability of mode selection and good signal-to-noise ratio demonstrate the EMAR's usefulness for the C_{ij} measurement at elevated temperatures. Shifts to lower frequencies indicate the elastic softening with temperature increase. Figure 8.14a shows temperature dependence of the unidirectional composite C_{ij} measured under the pressure of 10^{-4} Torr. They show linear decreases with temperature without hysteresis. The Ti alloy matrix C_{ij} also showed linear decrease up to 900 K (Fisher and Renken 1964). (Note that the α-β phase transition occurs near 1000 K in the matrix.) The rms difference between the measured and calculated resonance frequencies after the inverse calculations was typically 0.8 %. This indicates that the transverse isotropy assumption for the unidirectional composite is allowable within this error band. The rms difference for the isotropic matrix was 0.2 %. Figure 8.14b shows the temperature dependence of the SiC fiber C_{ij} determined by a micromechanics calculation (Eq. 8.34). The scattered results for C_{33}, C_{11}, and the bulk modulus B are caused by their weak contributions to the resonance frequencies.

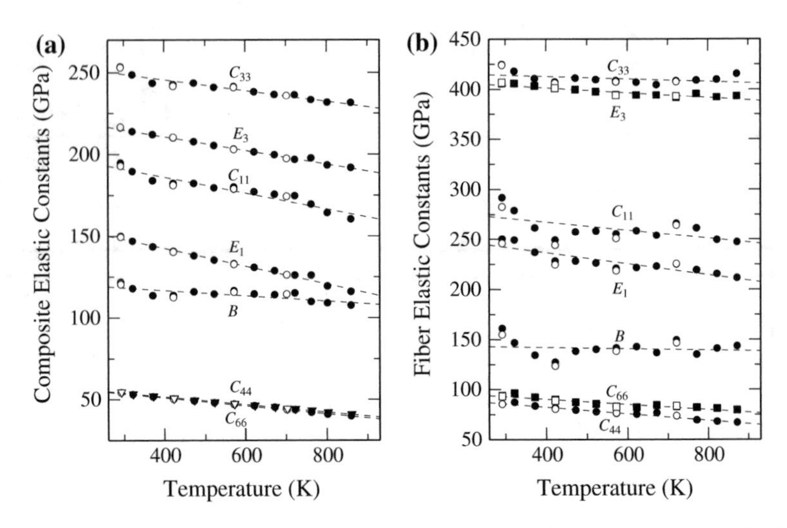

Fig. 8.14 Temperature dependences of the elastic constants of **a** the unidirectional SiC_f/Ti-6Al-4V composite and **b** the SiC fiber. *Solid* symbols denote measurements in the heating process and *open* ones in the cooling process. The fiber's longitudinal axis is in the x_3-axis. Reprinted form Ogi et al. (2003a), with permission from Taylor & Francis Ltd, (http://www.tandfonline.com)

A large difference appears in the normalized temperature derivatives of the fiber C_{ij} between the fiber's longitudinal and radial directions; the derivative for C_{33} is smaller in magnitude than that for C_{11} by a factor 4.7 and that for E_3 smaller than for E_1 by a factor 3.0. This is perhaps attributable to (i) nearly temperature-independent elastic constants of the carbon core and/or (ii) anisotropic elastic constants of the SiC and carbon. The fiber itself is a composite consisting of the SiC and carbon core. A micromechanics calculation for an isotropic SiC with an isotropic carbon core reveals that contributions of the carbon's elastic constants to the fiber C_{33} and E_3 are much larger than those to the other C_{ij}. Thus, the temperature derivatives of carbon's elastic constants highly affect those of the fiber C_{33} and E_3. When the SiC and carbon are elastically anisotropic, anisotropy in the temperature derivatives will occur.

The composite materials are used in air and sometimes suffer from oxidation at elevated temperatures. This was investigated for the cross-ply composite (Ogi et al. 2001). Figure 8.15 shows the temperature dependences of the cross-ply C_{ij} during heating. The EMAR measurements were taken in the atmosphere. Elastic constants exhibit linear decrease up to near 650 K and then show anomalies around 700 K. The composite C_{ij} changed reversibly on heating and cooling below this temperature, but they changed irreversibly once being heated beyond this point. (Note that the C_{ij} returned to the initial values by exactly tracing the heating history after being heated to 900 K, when the measurements were taken in vacuum as shown in Fig. 8.14). The irreversible process is significant in the internal friction behavior. Figure 8.16 shows the temperature dependence of internal friction of B_{2g}-2 mode

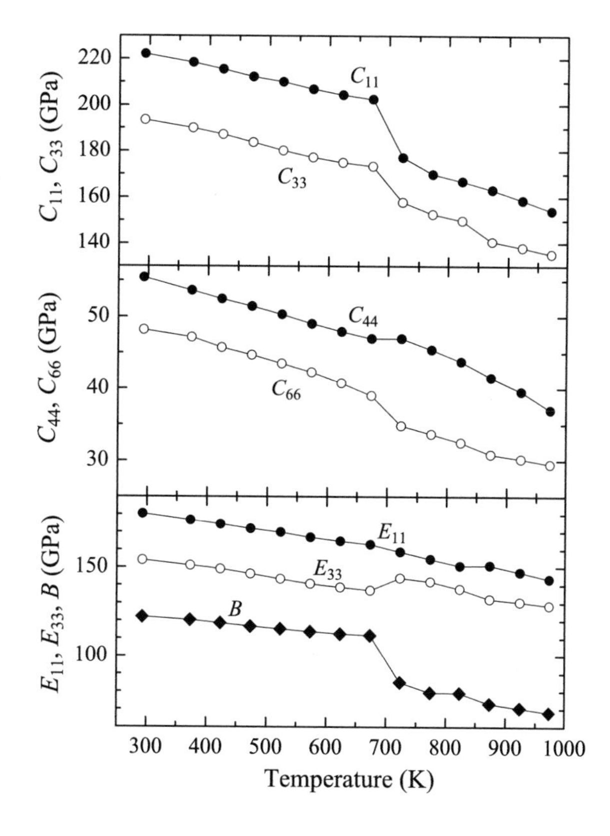

Fig. 8.15 Temperature dependence of the elastic constants of the SiC_f/Ti-6Al-4V cross-ply composite measured in the atmosphere (after Ogi et al. 2001)

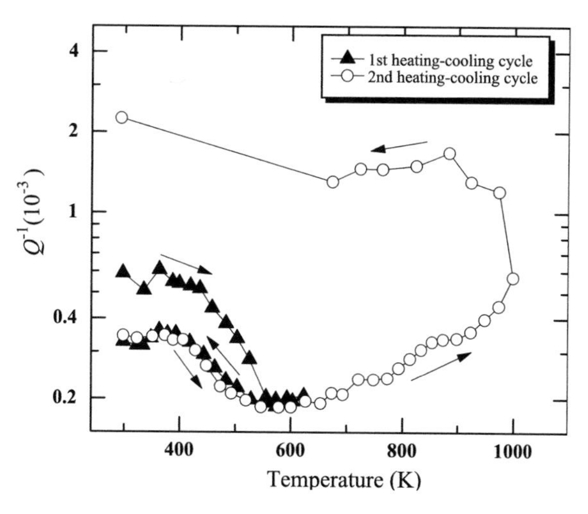

Fig. 8.16 Temperature dependence of the composite internal friction (B_{2g}-2 mode) (after Ogi et al. 2001)

($\sim$0.7 MHz). Internal friction decreased to one-third of the initial value by the first heating sequence up to 623 K. During cooling, it did not follow the heating temperature dependence. The second heating, however, caused the same temperature

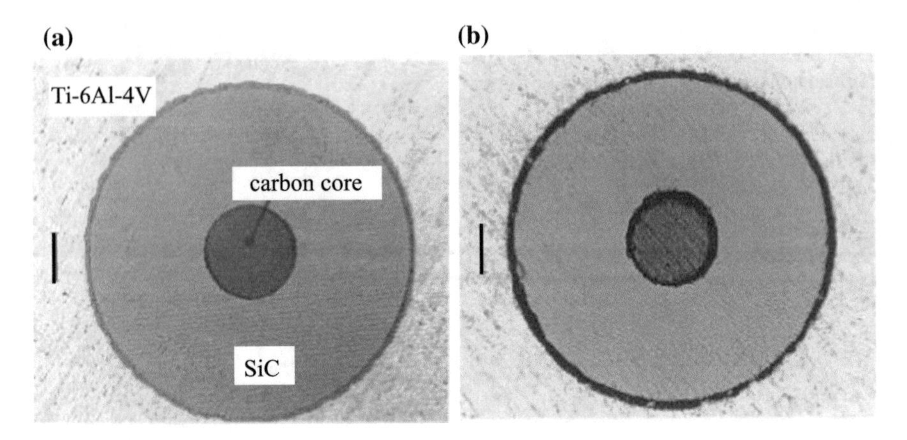

Fig. 8.17 Optical microstructure of the SiC_f/Ti-6Al-4V cross-ply composite **a** before and **b** after heating up to 1000 K. *Bars* indicate 20 μm (after Ogi et al. 2001)

dependence as that shown in the first cooling and raised the internal friction above 700 K. Internal friction increased markedly and irreversibly at 973 K. This trend was also the case with other resonance modes.

The anomalies of the elastic constants and the increase of internal friction above 700 K indicate a microstructure change caused by the exposure to high temperatures. Figure 8.17 shows the photomicrographs before and after being heated up to 1000 K. After heating, dark regions appear along the interfaces between the matrix and SiC, and between SiC and the carbon core, where carbon is involved at high concentration. Such a region is absent before heating. The matrix appears to remain the same.

Thus, the oxidation of carbon is considered to occur since the measurements were taken in air. Oxidation caused desorption of carbon atoms and partial disbonding at the interfaces, which changed the apparent macroscopic elastic stiffnesses and internal friction. In the lower-temperature region below 650 K, the same temperature dependence of internal friction was observed between the first cooling and second heating. The C_{ij} also changed reversibly in that temperature region. Thus, disbonding does not occur at low temperatures. This suggests that the practical use of this composite must be in the temperature region below 650 K.

8.5 High-Temperature Elastic Constants of Titanium through hcp-bcc Phase Transformation

The high-temperature EMAR system shown in Fig. 3.31 was used to measure the elastic constants of titanium across the hcp-bcc phase transformation (Ogi et al. 2004). The specimen was a rectangular parallelepiped of monocrystal titanium, whose

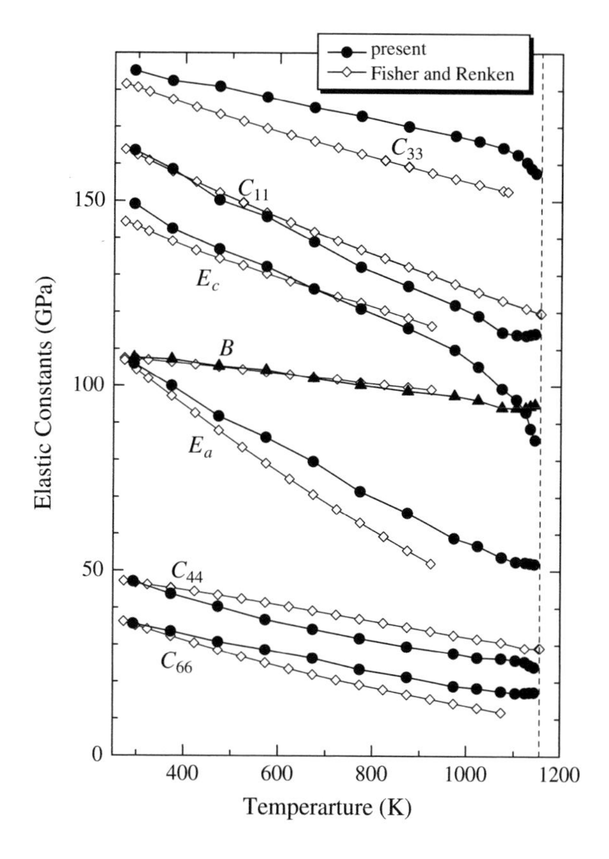

Fig. 8.18 Temperature dependences of monocrystal hcp titanium elastic constants. Reprinted from (2004). Copyright (2004), with permission from Elsevier

crystallographic axes were along the three sides, measuring 3.880 × 4.182 × 1.749 mm^3. It shows the five independent elastic constants.

Figure 8.18 shows temperature dependences of principal hcp-phase elastic constants, including the results by Fisher and Renken (1964) for comparison. B denotes the bulk modulus, and E_a and E_c denote Young's moduli along the a- and c-axes, respectively. At room temperature, the elastic anisotropy appears between the directions parallel and normal to the basal plane: C_{11} is smaller than C_{33} by 13 %, E_a smaller than E_c by 41 %, and C_{66} smaller than C_{44} by 32 %. Significant behavior occurs just below the phase transformation temperature: Elastic constants related to the c-axis (C_{33}, E_c, and C_{44}) decrease, whereas those related to the a-axis (C_{11}, E_a, and C_{66}) increase. The lattice parameter ratio c/a of hcp titanium is 1.59, smaller than the ideal value of 1.63. This indicates larger atomic distance in the basal plane (0001), yielding C_{66}, C_{11}, and E_a smaller than C_{44}, C_{33}, and E_c, respectively. The absolute values of the normalized temperature derivative are larger for C_{66} than for C_{44}, and so on, which is also caused by the larger distance among the atoms constituting the (0001) plane; a larger distance between atoms causes lower elastic stiffness and larger thermal vibration.

Approaching the phase transformation temperature at 1152 K, C_{11}, C_{66}, and E_a increase. This unusual stiffening can be attributed to premonitory behavior associated with the hcp-bcc transformation: The (0001) closed-packed plane in the hcp phase becomes a (110) plane in the bcc phase. The lattice parameter in the hcp phase is $a = 2.95$ Å at room temperature and that for the bcc phase is $a = 3.31$ Å at 1173 K, meaning that the distance between nearest atoms is 2.86 Å. Thus, the distance between the next-nearest atoms in the basal plane of the hcp phase decreases during the phase transition, which will increase C_{66}, C_{11}, and E_a. On the other hand, the c-axis-related elastic constants significantly decrease just below the transformation. This supports the idea suggested by Fisher and Renken (1964) that the direction of high-amplitude thermal vibration shifts from along the a-axis to the c-axis. The occurrence of the high-amplitude mode along the c-axis affects strongly the elastic constants C_{33}, C_{44}, and E_c so as to decrease them because of lattice anharmonicity.

8.6 Lotus-Type Porous Copper

Porous metals attract much attention as a new industrial material, because they have noticeable features such as lightweight, high capability of impact energy absorption, high damping, and low thermal conductivity. Typical of them are sphere-type porous materials. They, however, show high stress concentration around the pores when subjected to external load and fail to retain high strength. Recently, to avoid this shortcoming, lotus-type porous metals with long straight pores aligned unidirectionally have been fabricated by the unidirectional solidification method in hydrogen and argon atmosphere (Simone and Gibson 1996; Hyun et al. 2001; Hyun and Nakajima 2001). Mechanical properties of such a lotus-type porous copper show strong anisotropy about the solidification direction. Tensile strength, for example, in the solidification direction equals the specific strength determined from the relative density, but the strength normal to the direction decreases drastically with porosity (Hyun et al. 2001). One expects elastic anisotropy as well. Viewing this material as a composite consisting of matrix of a polycrystalline metal and the inclusions of aligned pores, the macroscopic symmetry should be hexagonal with the c-axis along the solidification direction and five independent elastic constants exist as shown in Eq. (8.16). The conventional ultrasonic techniques for the elastic wave velocities are inapplicable, because the pores scatter the ultrasonic waves due to the infinite mismatch of acoustic impedance. The EMAR method has then been applied to deduce all the independent elastic constants of lotus-type porous copper (Ichitsubo et al. 2002a).

The materials were prepared by the unidirectional solidification method in hydrogen and argon atmosphere. Their specifications are given in Table 8.8. Rectangular parallelepiped specimens, measuring about 9 mm on each side, were machined so as to have the surfaces normal and parallel to the solidification direction (x_3 direction) as shown in Fig. 8.19. Figure 8.20 shows microstructures for the 17.4 % porosity porous copper.

Table 8.8 Porosities, densities, and pore diameters of the four specimens of lotus-type porous copper. "ave," "dev," "max," and "min" denote the average, standard deviation, maximum, and minimum, respectively. Reprinted from Ichitsubo et al. (2002a). Copyright (2002), with permission from Elsevier

Porosity (%)	Density (kg/m^3)	Pore diameter (μm)			
		ave	dev	max	min
0	8904	–	–	–	–
17.4	7350	25.1	9.3	65.0	1.0
31.1	6135	16.6	8.3	55.3	1.8
59.1	3642	381	276	1295	14.7

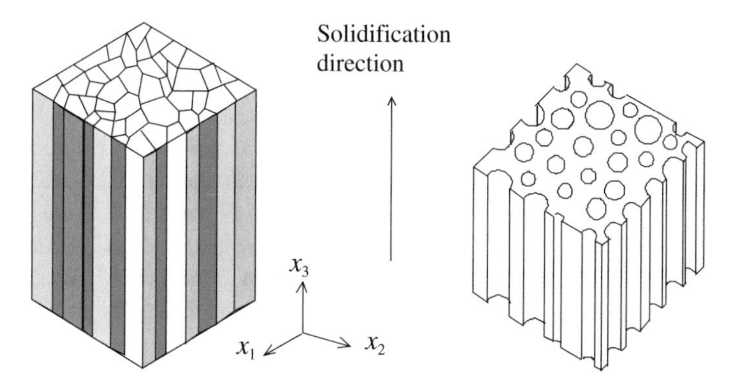

Fig. 8.19 Schematic illustration of nonporous (*left*) and porous (*right*) specimens. X-ray diffractions indicate that the crystallites' ⟨001⟩ directions are parallel to the solidification direction x_3. Reprinted from Ichitsubo et al. (2002a). Copyright (2002), with permission from Elsevier

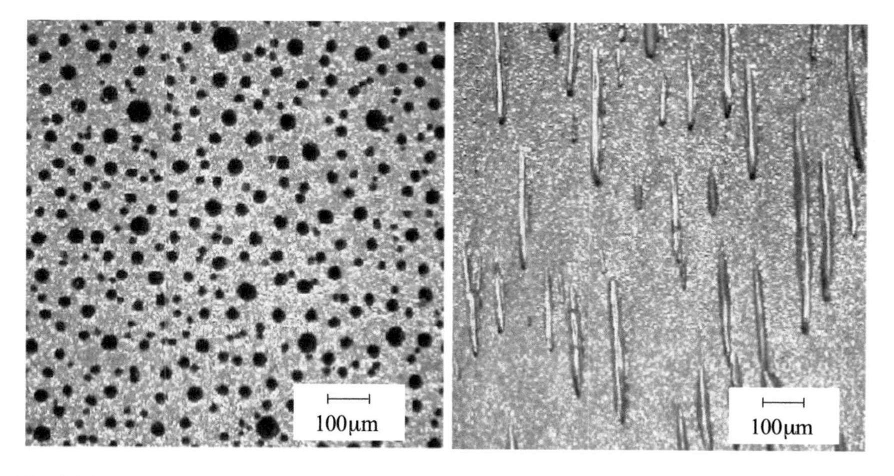

Fig. 8.20 Microstructures of 17.4 % porosity specimen seen in the solidification direction (*left*) and normal to that direction (*right*). Reprinted from Ichitsubo et al. (2002a). Copyright (2002), with permission from Elsevier

X-ray diffraction spectra indicated that the $\langle 001 \rangle$ crystallographic directions were approximately oriented in the solidification directions of all the specimens. This implies that the matrix copper is a bundle of the single-crystal rods whose $\langle 001 \rangle$ axes are oriented toward the solidification direction; rotation around the $\langle 001 \rangle$ axes is random. The overall elastic constants of the matrix phase are calculated by averaging the elastic constants of the $\langle 001 \rangle$ oriented crystal,

$$\bar{A}_{ijkl} = \frac{1}{2\pi} \int_0^{\pi} a_{pi} a_{qj} a_{rk} a_{sl} A_{pqrs} \mathrm{d}\theta, \tag{8.39}$$

where A_{ijkl} denotes either the elastic stiffness coefficients or elastic compliance coefficients, corresponding to Voigt or Reuss approximations, respectively. a_{ij} is a component of the rotation transformation by an angle θ around the x_3-axis. The summation convention is implied. The Hill approximation given by the average of the two approximations provides the matrix elastic constants shown in Table 8.9.

Figure 8.21 shows the EMAR spectrum from the 31.1 % porosity specimen. The resonance frequencies were entered into the inverse calculation to deduce the elastic constants. The results are shown in Fig. 8.22. $E_{//}$ and $E_{\perp}$ represent Young's moduli parallel and normal to the solidification direction, respectively. $E_{//}$ decreases linearly with the porosity where no stress concentration is implied. $E_{\perp}$ drops rapidly, which implies the stress concentration around the pores. Relationships $C_{11} \geq C_{33}$

Table 8.9 Hill averages of elastic constants of copper monocrystal about [001] axis (GPa). The x_3-axis is in the solidification direction. $E_{//}$ and $E_{\perp}$ denote Young's moduli parallel and normal to the solidification direction. Used cubic symmetry elastic constants of monocrystal copper are C_{11} = 168.4, C_{12} = 121.4, and C_{44} = 75.4 GPa. Reprinted from Ichitsubo et al. (2002a). Copyright (2002), with permission from Elsevier

C_{11}	C_{33}	C_{12}	C_{13}	C_{44}	$^*C_{66}$	$E_{//}$	$E_{\perp}$
187.6	168.4	102.3	121.4	75.4	42.7	66.7	97.9

$^*C_{66} = (C_{11} - C_{12})/2$

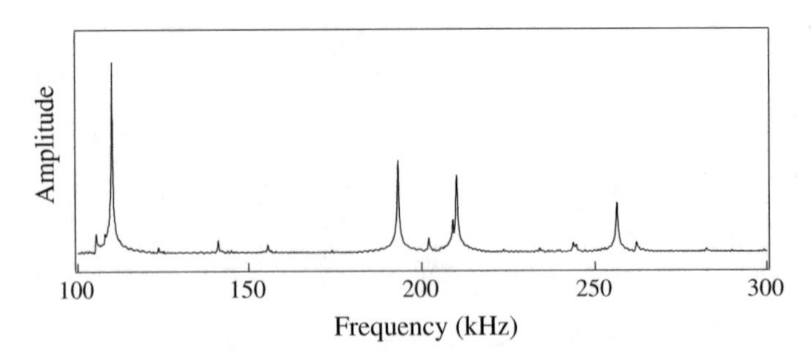

Fig. 8.21 EMAR spectrum of porous copper with 31.1 % porosity

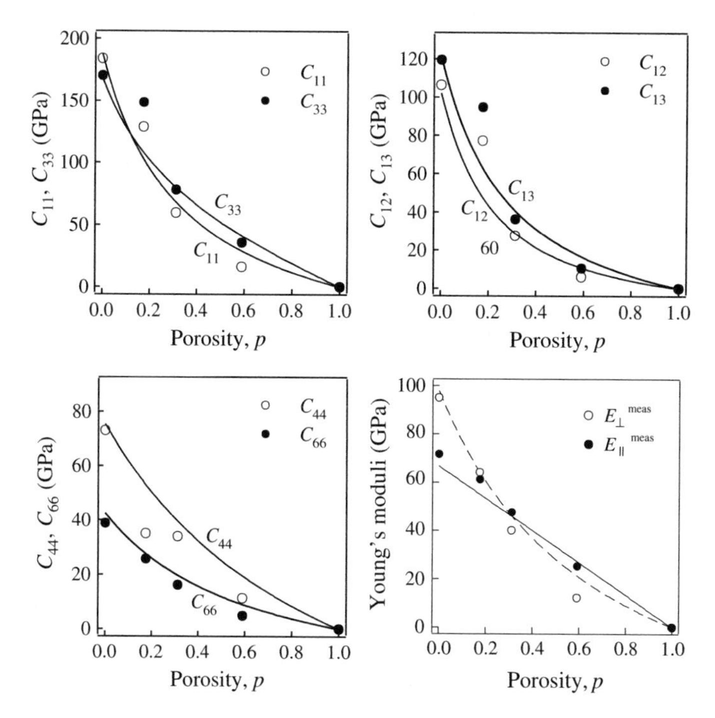

Fig. 8.22 Changes of the elastic constants of lotus-type porous copper with porosity. *Solid* and *open circles* are measurements, and *solid* and *broken lines* are calculations by the micromechanics modeling for the continuous cylindrical inclusions. Reprinted from Ichitsubo et al. (2002a). Copyright (2002), with permission from Elsevier

and $E_{\perp} > E_{//}$ hold for small porosities because of the matrix anisotropy (Table 8.9), but these relations are reversed for high porosities because of the anisotropy caused by the aligned long pores.

The elastic constants appear to be independent of the pore diameter; the primary parameter is the porosity, which a micromechanics calculation has confirmed (Ichitsubo et al. 2002a). The overall elastic constants of the lotus-type copper were computed using Eq. (8.35) by considering that it is composed of the anisotropic matrix and continuous cylindrical pores. Eshelby's tensor **S** for such a long cylindrical inclusion embedded in an anisotropic matrix is given by Eq. (8.38). The results are compared with the measurements in Fig. 8.22; favorable agreement with the EMAR measurements is observed. The calculation needs only the matrix elastic constants (Table 8.9) and the porosity.

The pores are sometimes not continuous (open) cylinders but long (closed) ellipsoids. Figure 8.23 investigates the effect of the inclusion's aspect ratio on the elastic constants. Eshelby's tensor for such an ellipsoidal inclusion is numerically obtained and tabulated by Ichitsubo et al. (2002a). The aspect ratio larger than three gives rise to the identical results as the continuous cylinder case (infinite aspect ratio); most lotus-type porous materials have pores with aspect ratio larger than

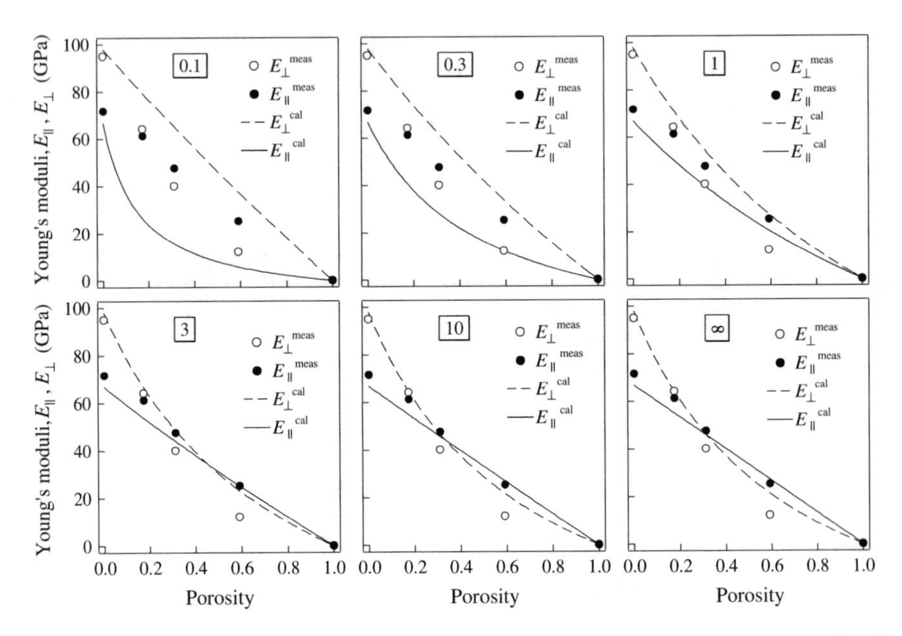

Fig. 8.23 Influence of the inclusion's aspect ratio on the resulting Young's moduli. Solid and open circles are measurements, and *solid* and *broken lines* are calculations. *The number* shown in each diagram represents the aspect ratio of the ellipsoidal inclusion pore. Reprinted from Ichitsubo et al. (2002a). Copyright (2002), with permission from Elsevier

three. Thus, the micromechanics calculation can predict all of the elastic constants of a lotus-type porous material, provided the crystallographic orientation in the solidification direction and the porosity are known. These are easily available with an X-ray analysis and optical microscopy, respectively.

8.7 Ni-Base Superalloys

Ni-base superalloy is a promising high-temperature material for turbine blades in industrial gas turbine, aeroengines, and so on. Fabrication techniques such as the directional solidification and the single-crystal solidification have remarkably improved its tolerance to plastic deformation at high temperatures. It consists of a Ni alloy face-centered-cubic (fcc) matrix of γ phase with a disordered structure and fcc cuboidal precipitates of γ' phase with an ordered $L1_2$ structure (Ni_3Al) (Fig. 8.24a). Its high resistance to deformation at high temperatures, or creep, relies on the morphological evolution of the γ' phase from the cuboidal structure into a laminar structure (directional coarsening) called rafted structure (Fig. 8.24b), which occurs at the early stage of creep life; the γ-γ' interfaces prevent dislocations from gliding. The rafted structure, however, collapses at the final stage of creep, and many γ channels appear across the γ' rafts to allow dislocation slip and climbing

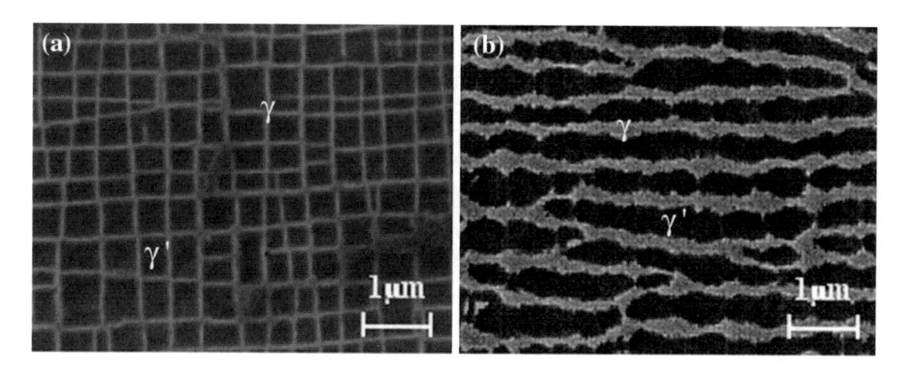

Fig. 8.24 Microstructures of **a** the virgin material consisting of the disordered γ phase matrix and cuboidal ordered γ' phase precipitations and **b** the rafted structure after uniaxial stress of 137 MPa was applied for 4 h at 1100 °C in the direction normal to the rafted planes. Reprinted from Ichitsubo et al. (2002b). Copyright (2002), with permission from Elsevier

until failure. Such a microstructure evolution has received intense study by many researchers (Sullivan et al. 1968; Tien and Copley 1971; Pineau 1976; Miyazaki et al. 1979).

Evolution of elastic properties is one of the central issues in this field of study, because the elastic misfit is the primary cause for the rafting to start. Furthermore, it could monitor the material structure with a nondestructive way. The creep behavior is closely related to the lamellar microstructure, whose evolution can be sensed through the macroscopic elastic properties. The virgin material (Fig. 8.24a) with the cuboidal γ' phase precipitates macroscopically shows cubic elastic symmetry. On the other hand, the rafted structure (Fig. 8.24b) is expected to show tetragonal elastic symmetry with a fourfold rotation axis perpendicular to the laminar planes (see Eq. 8.14).

This was investigated by Ichitsubo et al. (2002b). The material was TMS-26: the first-generation single-crystal superalloy. A single-crystal rod of the superalloy was subjected to the following heat treatment sequence: 1330 °C for 4 h, air cooling, 1120 °C for 5 h, air cooling, 870 °C for 20 h, and air cooling. Table 8.10 gives the chemical composition. The volume fraction of γ' phase was estimated to be 60–65 % from the intensities of the X-ray diffraction measurements. The material had experienced a creep test under a tensile stress of 137 MPa for 4 h at 1100 °C. Figure 8.24b shows the microstructure after this creep loading. A specimen with the rafted structure was machined into a rectangular parallelepiped with sides in the three principal directions as shown in Fig. 8.25. The sides measured 3.440, 3.433,

Table 8.10 Chemical composition of TMS-26 superalloy (at %). Reprinted from Ichitsubo et al. (2002b). Copyright (2002), with permission from Elsevier

Ni	Co	Cr	Mo	W	Al	Ta
64.8	8.82	6.34	1.26	3.97	11.98	2.84

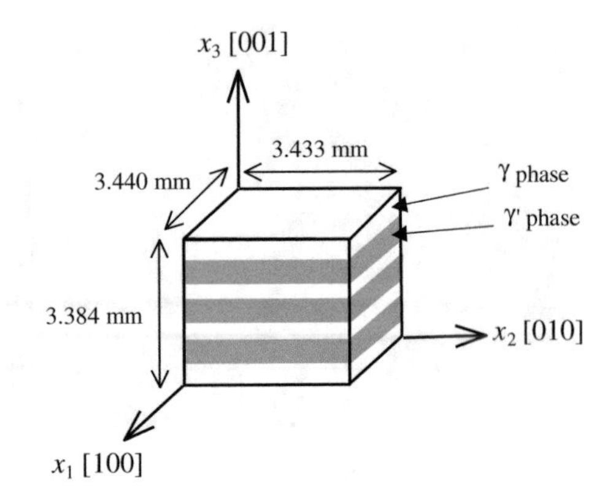

Fig. 8.25 Coordinate system and dimensions of the rafted rectangular parallelepiped specimen. The x_3-axis is in the direction perpendicular to the rafted planes. Reprinted from Ichitsubo et al. (2002b). Copyright (2002), with permission from Elsevier

Table 8.11 Tetragonal symmetry elastic constants of a rafted Ni-base superalloy (TMS-26) (GPa). E_{100} and E_{001} denote Young's moduli parallel and normal to the laminar plane, respectively. Reprinted from Ichitsubo et al. (2002b). Copyright (2002), with permission from Elsevier

C_{11}	C_{33}	C_{12}	C_{13}	C_{44}	C_{66}	E_{100}	E_{001}	B
252.5	250.4	158.5	157.2	131.8	132.4	130.6	130.2	189.0

and 3.382 mm. The mass density was 8978 kg/m^3. The crystallographic orientations were determined by a Laue back-reflection technique.

The EMAR method was applied to determine the tetragonal symmetry elastic constants. The results are shown in Table 8.11. We see small anisotropies between C_{11} and C_{33} (0.8 %); C_{12} and C_{13} (0.8 %); C_{44} and C_{66} (0.1 %); and E_{100} and E_{001} (0.3 %). This indicates that the rafted material exhibits nearly cubic symmetry despite its obvious laminar microstructure. This suggests that the elastic constants of the γ phase are close to those of the γ' phase at room temperature. Both phases show cubic symmetry.

8.8 Thin Films

8.8.1 Anisotropy of Thin Films

Many thin films are emerging as candidates for a wide variety of advanced applications such as microelectronics devices, data storage media, and microelectromechanical systems (MEMS). Understanding their mechanical properties is the key for designing applications and improving the film-processing techniques. In particular, elastic stiffness coefficients remain issues of central importance, because various film-processing techniques produce various elastic stiffnesses. Thus, developing a reliable methodology for measuring elastic stiffness coefficients of thin films is a matter of great interest.

Such a thin film normally exhibits anisotropy in the elastic properties between the in-plane (film plane) and normal (film growth) directions, even if the film material is polycrystal or amorphous. The film macroscopically shows transverse isotropy (or hexagonal symmetry) and possesses five independent elastic stiffness coefficients C_{ij} (C_{11}, C_{33}, C_{13}, C_{44}, and C_{66}) as given in Eq. (8.16), where the film thickness direction is in the x_3 direction. Several techniques have been reported for measuring the thin-film C_{ij}. The typical acoustic approach measures the flexural vibration resonance frequency of a reed composed of a film/substrate layered plate to calculate the in-plane Young's modulus (Peraud et al. 1997; Sakai et al. 1999; Papachristos et al. 2001). In this method, the mechanical contacts, needed for the acoustic transduction and supports, always diminish the sensitivity to the film's elastic constants. Huang and Spaepen (2000) developed a quasistatic uniaxial tension measurement to obtain the in-plane Young's modulus of free-standing films. They adopted a laser strain measurement to minimize the gripping influence. Another approach is to use surface acoustic wave transducers attached to the films to measure the velocities that provide some of the C_{ij} (Moreau et al. 1990; Kim et al. 1992). Brillouin scattering (Davis et al. 1991) has also been used for measuring the Rayleigh wave velocity traveling along the film surface, which is closely related to the out-of-plane shear modulus (C_{44}). To summarize, existing methods provide only a few elastic stiffness coefficients.

The EMAR method is capable of determining thin-film elastic constants because it measures resonance frequencies of specimens with quite high accuracy relying on its noncontacting nature. Free-vibration resonance frequencies of a film/substrate layered solid depend on all the film's C_{ij} and substrate's C_{ij}, as well as their dimensions and mass densities. Measurements of the resonance frequencies and use of known substrate properties allow determination of the film's C_{ij}. This methodology has been established by Ogi et al. (2002a, b). Determination of the film's C_{ij} proceeds in two steps. First, C_{33} and C_{44} are determined by measuring the thickness resonance frequencies of the layered specimen. Different dependencies of the fundamental and higher resonance modes on the film moduli and density allow their simultaneous determination. Second, the remaining three stiffnesses are determined by measuring the free-vibration resonance frequencies of the rectangular parallelepiped specimen and by performing an inverse calculation to find the most suitable C_{ij}.

8.8.2 Thickness Resonance of Layered Plate

Consider a triple-layered plate consisting of a substrate and films deposited on both surfaces of the substrate. A frequency equation for thickness resonance in such a triple-layered plate is obtainable by considering six element plane waves traveling in the thickness (x_3) direction. The resulting expression takes the form (Ogi et al. 2002a)

$$C_f k_f \tan \eta = C_s k_s \tan(\gamma - \delta), \tag{8.40}$$

where

$$\left.\begin{aligned} \cos \delta &= \cos \alpha \cos \beta + \kappa \sin \alpha \sin \beta \\ \sin \delta &= \cos \alpha \sin \beta - \kappa \sin \alpha \cos \beta \end{aligned}\right\}, \tag{8.41}$$

and

$$\begin{aligned} &\alpha = k_f d_1, \quad \beta = k_s d_1, \quad \gamma = k_s(d_1 + d_2), \\ &\eta = k_f d_3, \text{ and } \kappa = \frac{C_f k_f}{C_s k_s}. \end{aligned} \tag{8.42}$$

C_f and C_s denote the film and substrate elastic constants, respectively, either for the longitudinal mode (C_{33}) or for the shear mode (C_{44}). k_f and k_s denote the wave numbers for the film and substrate, respectively, and they are expressed by the wave velocity v and the resonance frequency f as $k = 2\pi f/v$. d_1 and d_3 denote the film thicknesses and d_2 the substrate thickness. Equation (8.40) applies also to a double-layered specimen (film/substrate) by taking $d_3 = 0$. Thus, the film elastic constants can be determined from the substrate moduli, substrate thickness, and film thickness. It is sometimes necessary to measure the resonance frequencies for the substrate alone to accurately determine the substrate C_{ij}, which is followed by the deposition of the film and then measurement of the resonance frequencies with the layered structure.

Film thickness (or film density if the thickness is known) highly affects the measurements, and it must be determined accurately. This can be done by measuring the fundamental resonance frequency. The fundamental mode shows the largest particle acceleration and inertia resistance at the film, which is sensitive to the film mass or thickness if the density is known. The fundamental mode is, however, insensitive to the film modulus because of nearly zero stress in the film region subjected to the stress-free boundary. Instead, a higher-order mode causes larger strain in the film region and is capable of detecting the film elastic constants as illustrated in Fig. 8.26.

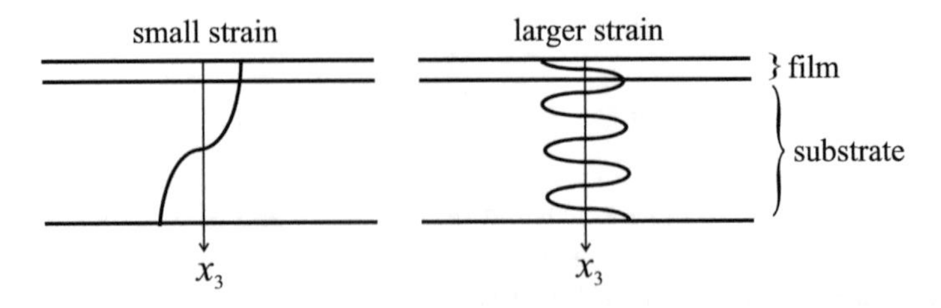

Fig. 8.26 Illustration of displacement distributions of the fundamental (*left*) and higher-order (*right*) thickness resonances in a film/substrate layered plate. The fundamental mode, which causes a small stress and the maximum displacement in the film region, is sensitive to the film's mass. Higher-order modes cause larger strain in the film region to be sensitive to the film's modulus

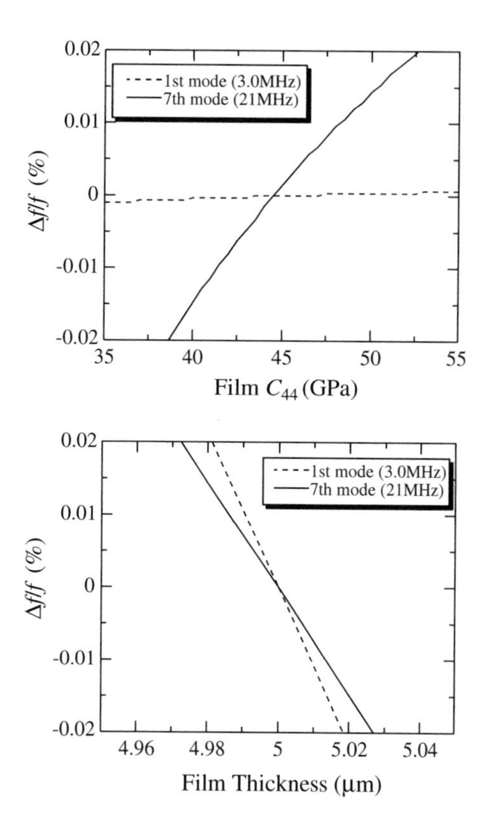

Fig. 8.27 Dependences of the shear wave resonance frequencies on the film C_{44} and thickness. The calculation is for a polycrystalline copper film of 5 μm thickness on aluminum substrate with 0.5 mm thickness

Figure 8.27 shows the sensitivity of the shear wave resonance frequencies to the film (C_{44}) and the film thickness d_1, calculated using Eq. (8.40). The calculations were made for the 5-μm Cu/0.5-mm Al/5-μm Cu triple-layered plate, supposing $C_f = 44$ GPa, $C_s = 26$ GPa, $\rho_f = 8930$ kg/m^3, and $\rho_s = 2650$ kg/m^3 (ρ's are the densities). For the fundamental resonance mode, the film modulus hardly affects the resonance frequency because of almost uniform displacement within the film (nearly a stress-free condition). As the order of the resonance becomes higher, the modulus sensitivity increases. This trend can be also found for the longitudinal wave (C_{11}). On the other hand, the thickness sensitivity is the highest with the fundamental mode because of the largest particle acceleration at the surface to increase the inertia resistance. Thus, combined use of the first-order and higher-order resonance frequencies enables us to simultaneously determine the moduli and thickness of the film.

Validity of the determination is governed by the accuracy of the resonance frequency measurement. The EMAR method was demonstrated to measure the through-thickness resonance frequencies with a high enough accuracy. At a constant temperature, the standard deviation of the measured resonance frequency is typically less than 0.01 %, which can yield the shear modulus in a 43–46 GPa range and the thickness in a 4.995–5.005 μm range for the copper film in the case of Fig. 8.27.

8.8.3 Free-Vibration Resonance of Layered Rectangular Parallelepiped

For bulk materials, we have compared the free-vibration resonance frequencies with calculations to deduce the best-fitting elastic constants as described in the previous sections. An analogous approach deduces the film C_{ij} by measuring the resonance frequencies with adequate accuracy. Because no analytical solution exists for a rectangular parallelepiped solid, the displacements in a vibrating solid have been approximated by linear combinations of the normalized Legendre functions (see Eq. 8.9) or simply of the power series (Migliori et al. 1993). For a layered rectangular parallelepiped, however, such a basis function fails to express the strain discontinuity at the interfaces because of the different elastic moduli. Heyliger (2000), then, separated the x_1-x_2 and x_3 dependences of the displacements. He used a power series ($x_1^k x_2^l$, k, l = 0, 1, 2, ...) for the x_1-x_2 dependence and Lagrangian interpolation polynomials for the x_3 dependence. For the displacement along the x_1-axis, the expression takes the form

$$u_1(x_1, x_2, x_3, t) = \sum_{i=1}^{m} \sum_{j=1}^{n} U_{ij}^{(1)}(t) \Psi_i(x_1, x_2) \overline{\Psi}_j(x_3). \tag{8.43}$$

Here, n denotes the number of the Lagrangian interpolation polynomial functions and m the number of the in-plane basis functions $\Psi_i(x_1, x_2)$. $\overline{\Psi}_j(x_3)$ denotes the one-dimensional Lagrangian interpolation polynomials (Reddy 1987). Approximations for u_2 and u_3 take similar forms. Incorporating a standard least-square procedure into Heyliger's method determines the film's C_{ij} inversely (Ogi et al. 2002b).

To study the reliability of this method, a numerical simulation was made for a titanium thin film epitaxially deposited on the (0001) surface of a monocrystal titanium rectangular parallelepiped. The resulting film C_{ij} via the inverse calculation must agree with the substrate C_{ij} because the same material was added. Figure 8.28 shows the results, where D is the substrate thickness and d the film thickness. The film C_{66} and the in-plane Young's modulus E_1 are determined within 1.5 % error for $d/D \geq 1.5$ %, while C_{33}, C_{12}, C_{13}, and the out-of-plane Young's modulus. E_3 contained large errors for $d/D < 2.5$ %. In the free vibration of a rectangular parallelepiped solid, many flexural and torsional vibrations occur causing the maximum bending and shearing stresses at the specimen surfaces where the film is deposited. The film E_1 and C_{66} can, therefore, contribute a lot to those vibration modes. However, other elastic constants contribute little. For example, C_{33} affects mostly the longitudinal mode wave that propagates in the thickness direction, and such a C_{33}-dependent mode appears at much higher frequencies. It cannot be detected in actual measurements because a large number of overlapping overtone peaks occur from other vibration modes. Thus, the free-vibration resonance method provides accurate C_{66} and E_1, but inaccurate other C_{ij}. For this reason, the C_{33} and C_{44} were separately determined using the thickness resonance method in

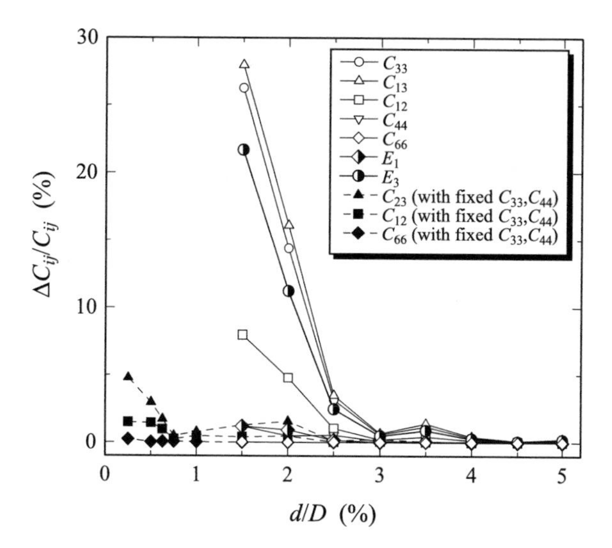

Fig. 8.28 Fractional errors of inversely determined monocrystal titanium film C_{ij} with various film thickness. The same material was assumed for the substrate. Reprinted with permission from Ogi et al. (2002a). Copyright (2002), AIP Publishing LLC

Sect. 8.8.2. Indeed, when these two are fixed, the remaining three elastic constants are determined within 5 % error for $d/D > 0.2$ % (Fig. 8.28). Exact mode identification is essential for least-square fitting between the measured and calculated resonance frequencies. This difficulty is overcome by selecting one vibration group by controlling the Lorentz force direction. (There are four vibration groups for a layered rectangular parallelepiped according to the deformation symmetry, as tabulated by Heyliger (2000).) The mode-selective principle was described in detail in Sect. 8.1.2.

8.8.4 Ni-P *Amorphous Alloy Thin Film*

The EMAR method was applied to a Ni-P amorphous alloy thin film (Ogi et al. 2002a). Ni-11.5mass%P ($Ni_{80}P_{20}$) amorphous alloy was deposited on both surfaces of polycrystalline Al-4.5mass%Mg alloy plate by an electroless-plating method. The Ni-P film was 12 μm thick, and the total thickness of the triple-layered specimen (Ni-P/Al-Mg/Ni-P) was 0.799 mm. The thickness resonance measurement gave the isotropic substrate moduli before the deposition as $C_{11} = 106.7 \pm 0.05$ GPa and $C_{44} = 26.36 \pm 0.005$ GPa. Archimedes method mass density of the substrate was 2645 kg/m^3. Film's surface roughness was less than 1 nm. Through-thickness observation of the film by transmission electron microscopy showed no defects, and a halo ring electron diffraction pattern appeared, a typical characteristic of an amorphous phase. Transverse isotropic (hexagonal) symmetry was assumed for this film.

Square specimens, measuring 30 mm on each side, were machined from a large layered plate for the thickness resonance measurements, and rectangular parallelepiped specimens measuring about $2.679 \times 2.618 \times 0.799$ mm^3 were machined for the subsequent free-vibration resonance measurements. All measurements were taken in a heat-insulated chamber, which regulated the specimen temperature to 30 ± 0.05 °C. A bulk-wave EMAT (Sect. 3.1) was used for measuring the thickness resonances. The EMAT proving area was 5 × 7 mm^2.

Figure 8.29 is a typical resonance spectrum measured by the EMAT. Reproducibility of a single resonance frequency measurement was of the order of 10^{-6}. The standard deviation among completely independent measurements was less than 0.005 %. The resulting C_{33} and C_{44} are given in Table 8.12 with the possible maximum errors estimated from the standard deviation and differences among specimens.

Figure 8.30 shows the free-vibration resonance spectra of a rectangular parallelepiped specimen measured by the EMAR system shown in Fig. 8.1,

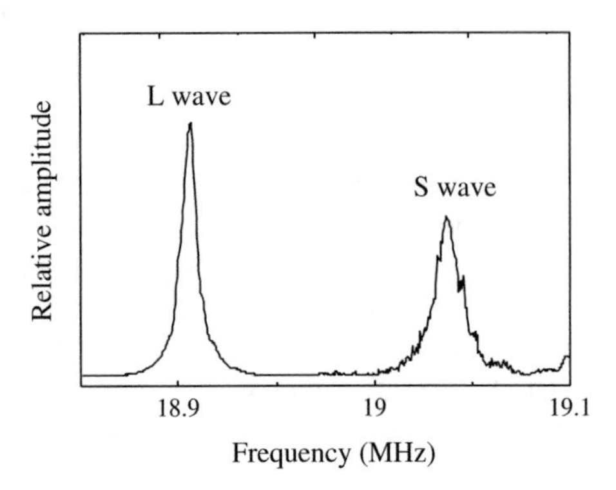

Fig. 8.29 Thickness resonance spectrum measured for the Ni-P/Al-Mg/Ni-P triple-layered plate. The fifth longitudinal-wave resonance peak and the tenth shear wave resonance peak appear. Reprinted with permission from Ogi et al. (2002a). Copyright (2002), AIP Publishing LLC

Table 8.12 Elastic stiffness coefficients (GPa) and Poisson's ratio of Ni-P amorphous alloy thin films. The film growth direction is in the x_3 direction. Micromechanics calculation explains the anisotropy of the amorphous film. Reprinted with permission from Ogi et al. (2002a). Copyright (2002), AIP Publishing LLC

	Ogi et al. (EMAR) (2002a) $Ni_{80}P_{20}$	Calculation	Barmatz and Chen (1974) $Ni_{76}P_{24}$	Logan and Ashby (1974) $Ni_{76}P_{24}$
C_{11}	140 ± 14	139	–	–
C_{33}	85 ± 3.0	86	–	–
C_{12}	64 ± 14	63	–	–
C_{13}	30 ± 15	45	–	–
C_{44}	35.2 ± 0.8	31	–	–
C_{66}^*	38.3 ± 0.5	38	–	35
E_1	107 ± 1.0	102	100	95
E_3	76 ± 8.0	66	–	–
B	65 ± 1.5	71	–	111
ν_{12}	0.41 ± 0.015	0.34	–	0.36
ν_{13}	0.21 ± 0.1	0.34	–	–
ν_{31}	0.14 ± 0.05	0.22	–	–

*$C_{66} = (C_{11} - C_{12})/2$

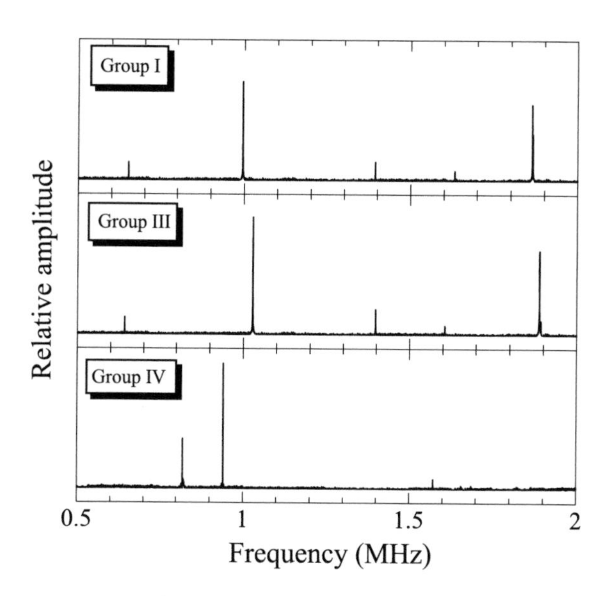

Fig. 8.30 Resonance spectra of the free vibration of the Ni-P/Al-Mg/Ni-P layered rectangular parallelepiped specimen, measuring 2.679 × 2.618 × 0.799 mm^3. Individual measurement setup excites one vibration group among the four possible groups. Reprinted with permission from Ogi et al. (2002a). Copyright (2002), AIP Publishing LLC

demonstrating that each vibration group is independently excited. (Three of the four vibration groups were detected.) The reproducibility of the measurements was of the same order as the thickness resonance measurements. Typical rms difference in measured resonance frequencies was 0.2 %, which was one order smaller than the C_{ij} contributions to the resonance frequencies, i.e., $\left(C_{ij}/f\right)\left(\partial f/\partial C_{ij}\right)$.

Table 8.12 shows the complete set of elastic constants of the Ni-P amorphous film measured by the EMAR method with those reported for Ni-P amorphous alloy thin films in the past. Barmatz and Chen (1974) used a vibrating reed method for the in-plane Young's modulus (E_1), and Logan and Ashby (1974) used tension and torsion measurements for E_1 and the in-plane shear modulus (C_{66}). Their results are comparable with the EMAR E_1 and C_{66} despite the slight difference of chemical composition. Logan-Ashby's bulk modulus B disagrees with that determined by EMAR because they assumed the film to be isotropic.

Particularly significant is the elastic anisotropy between in-plane and out-of-plane directions as shown in Fig. 8.31; C_{11} is larger than C_{33} by 49 %, E_1 larger than E_3 by 34 %, and C_{66} larger than C_{44} by 8.4 %. For the same material, Takashima et al. (2000, 2001) observed strong anisotropy in the crack growth behavior. The fracture toughness was evaluated as 4.2 and 7.3 MPa · $m^{1/2}$ for cracks propagating in the in-plane and out-of-plane directions, respectively, indicating that this material is more tolerable to crack propagation in the film growth direction. The larger C_{11} suggests stronger in-plane binding force, being consistent with their observation.

There are two possible mechanisms for the elastic anisotropy in the amorphous alloy film. First is inhomogeneous microstructure. Phosphorus-rich and

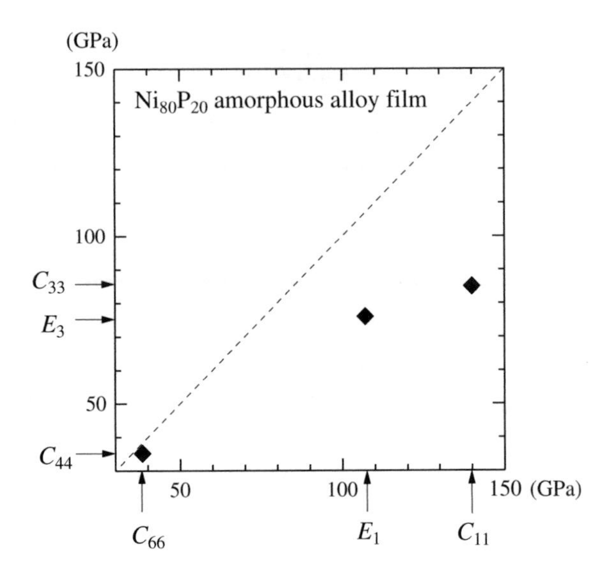

Fig. 8.31 Elastic constants related to the film-growing direction versus those to the in-plane direction. Their differences indicate transverse isotropy of Ni-P amorphous film

phosphorus-poor regions could appear periodically to make the columnar structure as actually seen in Fe-P amorphous alloy film (Armyanov et al. 1997). Also, the chemical composition could vary in the film growth direction. Both inhomogeneities may cause transverse isotropy. Second is local incomplete cohesion inside the film material (Huang and Spaepen 2000), which could occur during deposition. Incomplete cohesive regions lying in the x_1-x_2 plane will cause transverse isotropic effective stiffnesses. Considering the large anisotropy, the latter appears to be the primary cause. This effect has been investigated using a micromechanics model by replacing the incomplete cohesive regions with oblate ellipsoidal microcracks aligned parallel to the depositing face as shown in Fig. 8.32. The nonzero components of Eshelby's tensor **S** for such an inclusion are given in contracted notation by

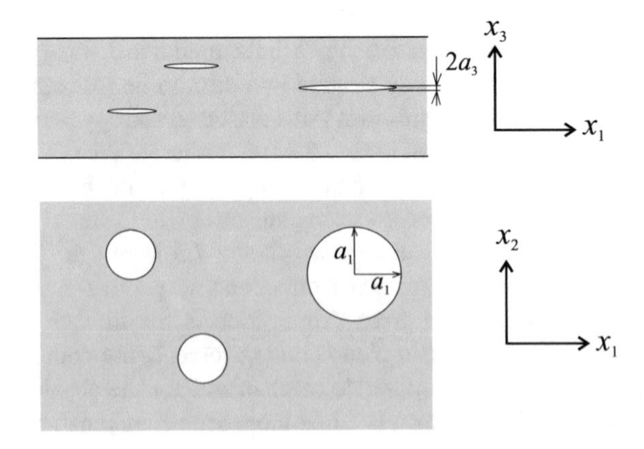

Fig. 8.32 Oblate ellipsoidal microcracks for modeling the incomplete cohesion regions within the film. Reprinted with permission from Ogi et al. (2002a). Copyright (2002), AIP Publishing LLC

$$
\begin{aligned}
S_{11} &= S_{22} = \frac{\pi(13-8\nu)}{32(1-\nu)}\cdot\frac{a_3}{a_1}, \quad S_{33} = 1 - \frac{\pi(1-2\nu)}{4(1-\nu)}\cdot\frac{a_3}{a_1},\\
S_{13} &= S_{23} = -\frac{a_3}{a_1}\cdot\frac{\pi(1-2\nu)}{8(1-\nu)}\left\{1-\frac{4}{\pi(1-2\nu)}\cdot\frac{a_3}{a_1}\right\},\\
S_{12} &= S_{21} = -\frac{\pi(1-8\nu)}{32(1-\nu)}\cdot\frac{a_3}{a_1}, \quad S_{31} = S_{32} = \frac{\nu}{1-\nu}\left\{1-\frac{\pi(1+4\nu)}{8\nu}\cdot\frac{a_3}{a_1}\right\},\\
S_{44} &= S_{55} = 1-\frac{\pi(2-\nu)}{4(1-\nu)}\cdot\frac{a_3}{a_1}, \quad S_{66} = \frac{\pi(7-8\nu)}{16(1-\nu)}\cdot\frac{a_3}{a_1}.
\end{aligned}
\tag{8.44}
$$

Equation (8.31) predicts the macroscopic elastic constants of the Ni-P film containing the inclusions. Table 8.12 compares the measurements with the calculations when a volume fraction of 5×10^{-5} and an inclusion aspect ratio of $a_1/a_3 = 10^{-4}$ are assumed. This modeling analysis predicts the effective elastic stiffness coefficients close to the measurements. Thus, thin and aligned microcracking can cause such a strong anisotropy even with an isotropic matrix.

It is difficult to detect the incomplete cohesive regions with microscopic observations, but such a heterogeneous structure would occur in many thin films. The elastic anisotropy measurement will be a key for evaluating the thin-film microstructures.

8.9 Resonance Measurements at Elevated Temperatures for Piezoelectric Materials

The antenna transmission technique shown in Sect. 3.13 is a key method for evaluating acoustic properties of piezoelectric materials at elevated temperatures, because it allows the noncontacting electromagnetic acoustic coupling. The antenna configuration as shown in Fig. 8.33 has been utilized in the high-temperature instrument in Fig. 3.31 for measuring elastic constants of α-quartz and langasite crystals, and for studying hopping conduction in Fe-doped gallium nitride.

Fig. 8.33 Setup of the antennas and specimen

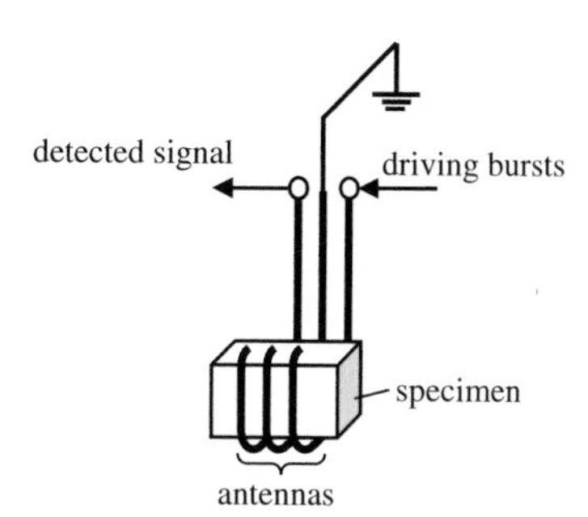

8.9.1 Langasite ($La_3Ga_5SiO_{14}$)

Piezoelectric materials such as α-quartz, lithium niobate ($LiNbO_3$), and langasite ($La_3Ga_5SiO_{14}$) are used in many applications such as resonators, actuators, and sensors. They possess many material constants: the elastic constants C_{ij}, piezoelectric coefficients e_{ij}, and dielectric coefficients ε_{ij}. Their accurate values, especially the elastic constants, are required for designing those devices. The crystallographic structure of α-quartz and langasite belongs to the trigonal system with point group 32, thus six C_{ij}, two e_{ij}, and two ε_{ij} as follows:

$$\left[C_{ij}\right] = \begin{bmatrix} C_{11} & C_{12} & C_{13} & C_{14} & 0 & 0 \\ C_{12} & C_{11} & C_{13} & -C_{14} & 0 & 0 \\ C_{13} & C_{13} & C_{33} & 0 & 0 & 0 \\ C_{14} & -C_{14} & 0 & C_{44} & 0 & 0 \\ 0 & 0 & 0 & 0 & C_{44} & C_{14} \\ 0 & 0 & 0 & 0 & C_{14} & (C_{11}-C_{12})/2 \end{bmatrix}. \tag{8.45}$$

$$\left[e_{ij}\right] = \begin{bmatrix} e_{11} & -e_{11} & 0 & e_{14} & 0 & 0 \\ 0 & 0 & 0 & 0 & -e_{14} & -e_{11} \\ 0 & 0 & 0 & 0 & 0 & 0 \end{bmatrix}, \tag{8.46}$$

$$\left[\varepsilon_{ij}\right] = \begin{bmatrix} \varepsilon_{11} & 0 & 0 \\ 0 & \varepsilon_{11} & 0 \\ 0 & 0 & \varepsilon_{33} \end{bmatrix}. \tag{8.47}$$

The dielectric coefficients are available from low-frequency capacitance measurements (IEEE 1984), but measuring all of the elastic and piezoelectric coefficients presents a formidable task as demonstrated by Smith and Welsh (1971). The conventional methods involved many independent measurements on many crystals in many orientations: pulse echo measurements or rod resonance measurements coupled with the resonance-antiresonance measurements of electric impedance. Then, one must solve a set of labyrinthine equations. Various errors easily enter, being associated with the use of different specimens, crystal misorientation, resonance frequency shifts by attaching electrodes, and so on.

Here, an advanced methodology is presented for determining all the C_{ij} and e_{ij} from a single specimen using EMAR. The acoustic spectroscopy measures the macroscopic resonance frequencies of a simple-shaped specimen. Although this method has been used to determine the elastic stiffness coefficients, one can use it to determine any property coupled to the macroscopic resonance frequencies. These include

1. shape
2. dimensions
3. mass or mass density
4. orientation

5. elastic stiffnesses, and
6. piezoelectric coefficients and/or dielectric coefficients (in the case of piezoelectric material).

The resonance spectrum is a *fingerprint* that depends on many interconnected physical properties. Measuring resonance frequencies can simultaneously determine the C_{ij} and e_{ij}. However, the contributions of e_{ij} to the resonance spectrum are much less than those of C_{ij}, and their determination requires extremely high accuracy in the resonance frequency measurements. The ideal approach to achieve this is a noncontacting measurement with the system shown in Fig. 8.33. The noncontacting measurements of the resonance frequencies and internal friction are then made possible by inserting a specimen into the antenna assembly and by applying the high-power rf burst to the generation antenna to cause the free vibrations through the converse piezoelectric effect (see Sect. 3.13). After the excitation, the detection antenna receives the vibration through the piezoelectric effect. Johnson et al. (2000) adopted this approach for studying temperature dependence of internal friction of langatate ($La_3Ga_{5.5}Ta_{0.5}O_{14}$). Ogi et al. (2003b) determined all the elastic and piezoelectric coefficients of langasite ($La_3Ga_5SiO_{14}$) with this technique, which is described in the following.

Langasite ($La_3Ga_5SiO_{14}$) attracts many researchers because of its large piezoelectric coefficients and less temperature-dependent elastic constants. These properties are especially attractive for SAW filters and allow langasite to replace quartz in a wide variety of electric devices. There are six independent C_{ij} and two e_{ij} as shown in Eqs. (8.45) and (8.46). Also, six independent internal friction Q_{ij}^{-1} exist.

Oriented rectangular parallelepiped specimens were used. Each side measured 3–10 mm. The mass density was 5724 kg/m^3. Free vibrations of an oriented rectangular parallelepiped crystal with 32 point-group symmetry fall into the four vibration groups labeled as A_g, B_g, A_u, and B_u, according to the deformation symmetry as shown in Table 8.13 (Ohno 1990). l, m, and n denote the orders of the Legendre functions used in the basis functions (Eq. (8.9)).

Table 8.13 Deformation symmetry of the four vibration groups of an oriented rectangular parallelepiped crystal with 32 point-group symmetry. u_i denotes the displacement component along the x_i-axis and ϕ the electric potential. E and O denote even and odd functions of the axis. The origin is located at the center of the rectangular parallelepiped. Vibration group notation follows Ohno (1990)

Group	Displacement	l	$m + n$	Group	Displacement	l	$m + n$
A_g	u_1	O	E	B_g	u_1	E	O
	u_2	E	O		u_2	O	E
	u_3	E	O		u_3	O	E
	ϕ	O	E		ϕ	E	O
A_u	u_1	E	E	B_{3u}	u_1	O	O
	u_2	O	O		u_2	E	E
	u_3	O	O		u_3	E	E
	ϕ	E	E		ϕ	O	O

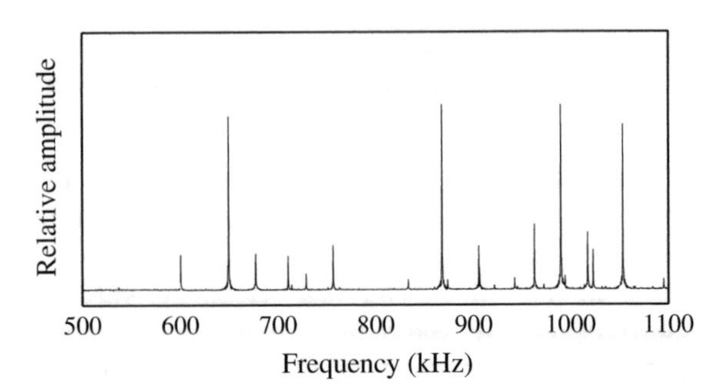

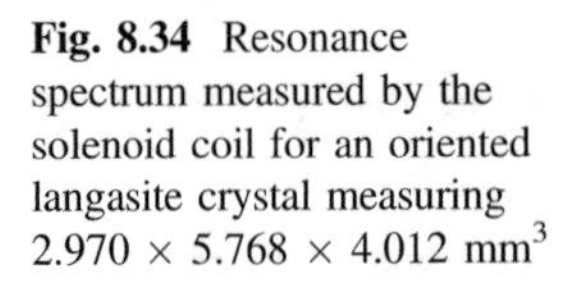
Fig. 8.34 Resonance spectrum measured by the solenoid coil for an oriented langasite crystal measuring 2.970 × 5.768 × 4.012 mm^3

Figure 8.34 shows a resonance spectrum. Good signal-to-noise ratio permitted determining the resonance frequencies with reproducibility better than 10^{-5}. The measurements were compared with the calculations for determining the C_{ij} and e_{ij} with the inverse calculation scheme (see Sect. 8.2). Choosing proper combinations of basis functions, referring to Table 8.13, can reduce the calculation time.

One cannot separately determine the e_{ij} and ε_{ij} from mechanical spectroscopy because their ratios affect the resonance frequencies. Fortunately, the ε_{ij} can be obtained by the capacitance measurements with sufficient accuracy. Several researchers have determined ε_{ij} within ±2 % band, while there are a lot of discrepancies for other coefficients (see Table 8.14). The dielectric coefficients ε_{ij} were thus fixed to the averaged values over the previous studies; $\varepsilon_{11} = 19.04\varepsilon_0$ and $\varepsilon_{33} = 50.51\varepsilon_0$, where ε_0 denotes the dielectric constant in vacuum. The converged coefficients reconstructed the resonance frequencies close to the measurements as shown in Fig. 8.35. The typical rms difference between them was 0.08 %. When the piezoelectric coefficients were neglected (purely elastic analysis), the calculated frequencies were smaller than the observations. This indicates piezoelectric stiffening (Auld 1973).

Table 8.14 compares the C_{ij} and e_{ij} determined by the EMAR with those in the literature. The C_{ij} determined by the EMAR fall within the band of the previous studies. However, the EMAR e_{ij} lies considerably outside the scattering band; e_{11} is smaller by 14–28 %, and e_{14} is larger by 31–62 % than the existing results. The EMAR measurement seems more reliable because it removes the error sources of the conventional measurements. They are (i) coupling agent and excess load for transduction, (ii) electrodes deposited on the specimen surfaces, and (iii) use of many specimens oriented in various directions.

Using the antenna, one can measure a ring-down curve of the free vibration to determine the attenuation coefficient or internal friction. Figure 8.36 presents a diagram of internal friction versus resonance frequency on three langasite crystals. No frequency dependence is seen in this range. Table 8.14 shows the internal friction tensor Q_{ij}^{-1} of langasite determined inversely (Sect. 8.2). Obviously, Q_{11}^{-1} and Q_{33}^{-1} exceed Q_{44}^{-1} and Q_{66}^{-1} by a factor of 5. The reproducibility of the Q_{ij}^{-1} measurement was less than 20 %. Thus, this difference is physically meaningful and indicates that longitudinal wave attenuation is larger than shear wave attenuation, a contrary trend to metals where dislocations' anelastic movement causes larger shear wave Q^{-1} as seen

Table 8.14 Elastic stiffness coefficients (GPa), piezoelectric coefficients (C/m^2), and the internal friction tensor of langasite ($La_3Ga_5SiO_{14}$). The dielectric coefficients are fixed in the inverse calculation as $\varepsilon_{11} = 19.04\varepsilon_0$ and $\varepsilon_{33} = 50.51\varepsilon_0$, where ε_0 denotes the dielectric constant in vacuum

	EMAR	Malocha et al. (2000)	Bungo et al. (1999)	Inoue and Sato (1998)	Sakharov et al. (1995)	Sil'vestrova et al. (1986)	Ilyaev et al. (1986)	Kaminskii et al. (1983)	EMAR $Q_{ij}^{-1}(10^{-4})$
C_{11}	189.1	188.5	189	189.5	189.3	190.2	188.9	190.9	1.73
C_{33}	261.3	261.7	268	259.9	262.4	262.1	262.2	261.9	0.95
C_{44}	53.56	53.71	53.3	53.91	53.84	53.82	53.9	52.4	0.26
C_{66}	42.3	42.21	42.4	42.4	42.16	42	42.2	43.2	0.27
C_{13}	98.87	96.88	102	97.86	95.28	91.9	96.8	104.2	1.65
C_{14}	14.38	14.15	14.1	14.64	14.93	14.7	14.3	15.2	0.16
e_{11}	−0.351	−0.402	−0.438	−0.428	−0.431	–	−0.44	−0.45	–
e_{14}	0.188	0.13	0.104	0.114	0.108	–	0.07	0.077	–

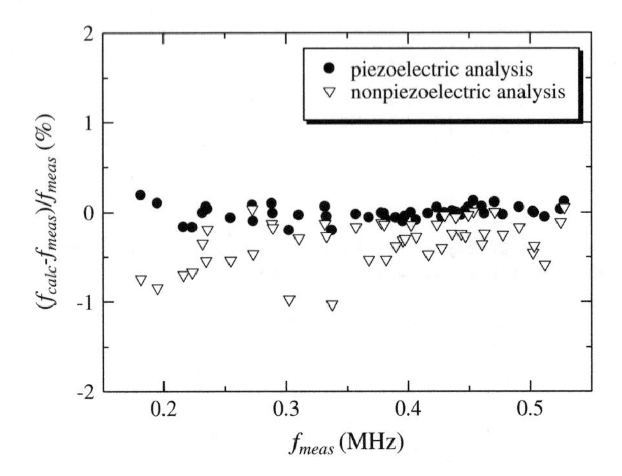

Fig. 8.35 Differences between the measured (f_{meas}) and calculated (f_{calc}) resonance frequencies of an oriented langasite crystal measuring 8.032 × 9.814 × 6.022 mm^3. Taking the piezoelectric effect into consideration, the calculations agree with the measurements within 0.08 % on average

in copper (Table 8.3), for example. Possible origin of internal friction of such a brittle material is phonon-phonon interactions. Acoustic waves break an equilibrium state of phonons because of lattice anharmonicity. The scattered thermal-mode phonons subsequently equilibrate through interaction with a low-frequency-mode (acoustic) phonons and other thermal-mode phonons, during which energy loss arises. Such energy loss can be expressed by the relaxation time τ of an acoustic phonon to relax to a thermal phonon with angular frequency of the acoustic wave ω (Mason 1965). At room temperature and in the kilohertz frequency range ($\omega\tau \ll 1$), it reduces to

$$Q_{\mathrm{phon}}^{-1} = \text{const. } \omega\tau. \tag{8.48}$$

This theory agreed with measurements of germanium, quartz, and silicon as summarized in Mason's review (1965). Especially, germanium shows the longitudinal/shear-mode internal friction ratio about 5, being very close to that of langasite. Usually, the thermal relaxation time for longitudinal waves is about twice that for shear waves because of the difference of the associated volume change. Langasite's Q_{ij}^{-1} then may be interpreted as phonon-phonon interactions. At the same time, a question remains; the loss caused by this mechanism increases with frequency, which disagrees with the result in Fig. 8.36. Such a discrepancy in the frequency dependence of Q_{ij}^{-1} has been reported for a langatate, an isomorph of langasite (Johnson et al. 2000). Thus, a modification of the phonon-phonon theory is necessary.

The elastic constants of langasite were then measured up to 1200 K for a rectangular parallelepiped (8.049 × 4.061 × 3.958 mm^3) (Nakamura et al. 2012). Figure 8.37 shows resonance spectra measured at several temperatures. Even at 1224 K, 35 resonance frequencies were measurable, which was sufficient to determine the six independent elastic constants C_{ij}, where the piezoelectric coefficients were assumed unchanged. (This assumption would cause ~1 % errors in the

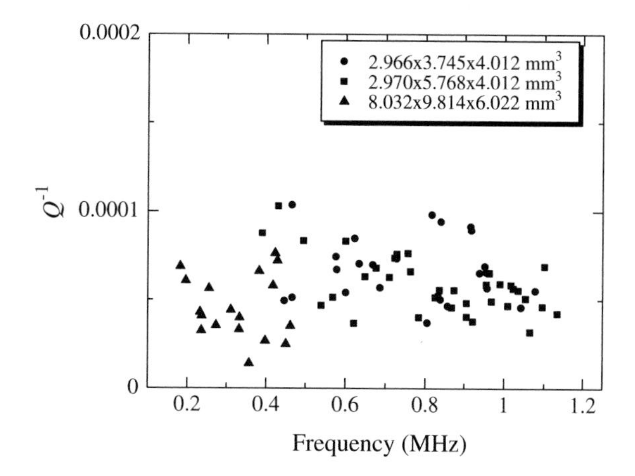

Fig. 8.36 Internal friction of langasite crystals measured by the noncontacting free-decay method

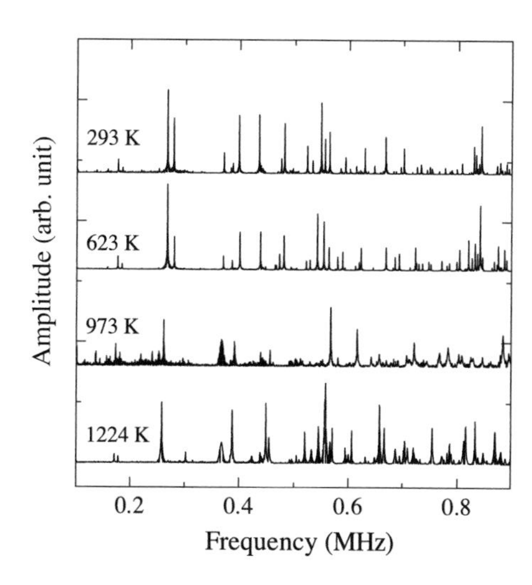

Fig. 8.37 Resonance spectra of a langasite specimen measured with the antenna transmission resonance method at several temperatures. Reprinted with permission from Nakamura et al. (2012). Copyright (2012), AIP Publishing LLC

resultant C_{ij}.) Figure 8.38 shows the temperature dependences C_{ij}, and Fig. 8.39 shows Young's moduli E_a and E_c, the bulk modulus B, and Poisson's ratios ν_{12}, ν_{13}, and ν_{31} calculated from the measured C_{ij}. The elastic constants decrease linearly as temperature increases, and their first-order temperature coefficients are listed in Table 8.15 together with C_{ij} at 293 and 1224 K.

8.9.2 α-Quartz

The elastic constants of α-quartz were also evaluated at elevated temperatures. Figure 8.40 shows the temperature dependence of the elastic constants, and Fig. 8.41 shows the temperature dependence of Young's moduli, bulk modulus,

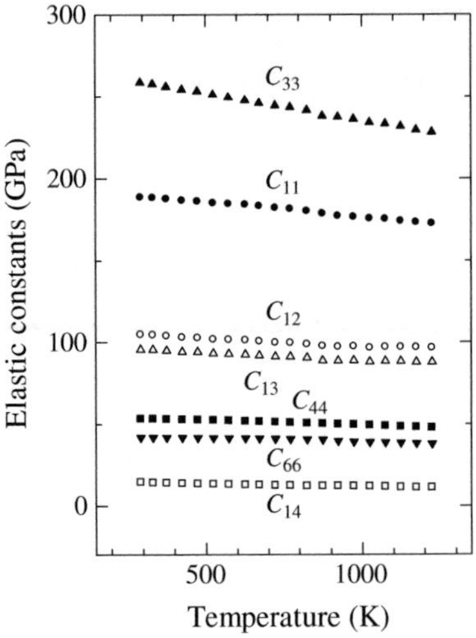

Fig. 8.38 Temperature dependences of elastic constants of langasite. Reprinted with permission from Nakamura et al. (2012). Copyright (2012), AIP Publishing LLC

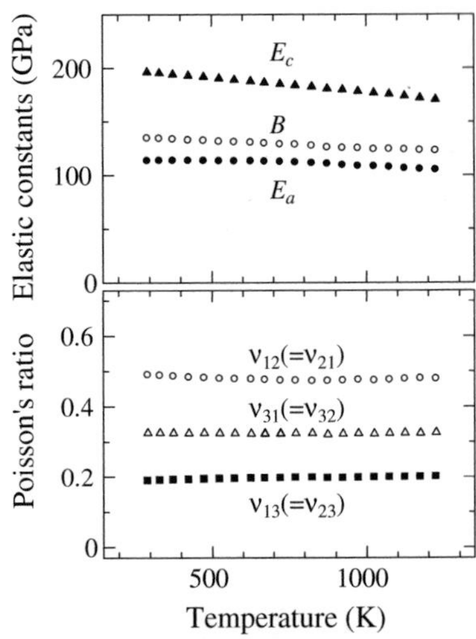

Fig. 8.39 Temperature dependence of Young's moduli, bulk modulus, and Poisson's ratios of langasite. Reprinted with permission from Nakamura et al. (2012). Copyright (2012), AIP Publishing LLC

Table 8.15 Elastic constants C_{ij} (GPa) at room temperature and elevated temperatures, and the first-order temperature coefficients a_{ij} (10^{-4} K^{-1}) of the elastic constants of langasite and α-quartz. Reprinted with permission from Nakamura et al. (2012). Copyright (2012), AIP Publishing LLC

	ij	11	12	13	14	33	44	66
Langasite	C_{ij} at 293 K	188.6	104.7	95.6	14.4	258.9	53.3	42.0
	C_{ij} at 1224 K	172.5	96.4	87.8	11.0	228.1	47.7	38.0
	a_{ij}	−0.90	−0.98	−0.99	−2.66	−1.29	−1.04	−0.85
α-quartz	C_{ij} at 294 K	86.8	7.0	11.8	−18.0	105.4	58.2	39.9
	C_{ij} at 843 K	72.0	−27.4	−3.4	−13.4	79.9	41.3	49.7
	a_{ij}	−0.49	−25.08	−6.94	0.67	−2.19	−1.67	1.66

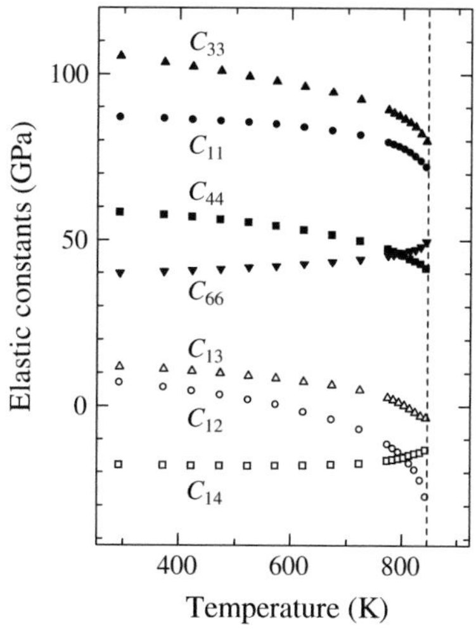

Fig. 8.40 Temperature dependence of elastic constants of α-quartz. Reprinted with permission from Nakamura et al. (2012). Copyright (2012), AIP Publishing LLC

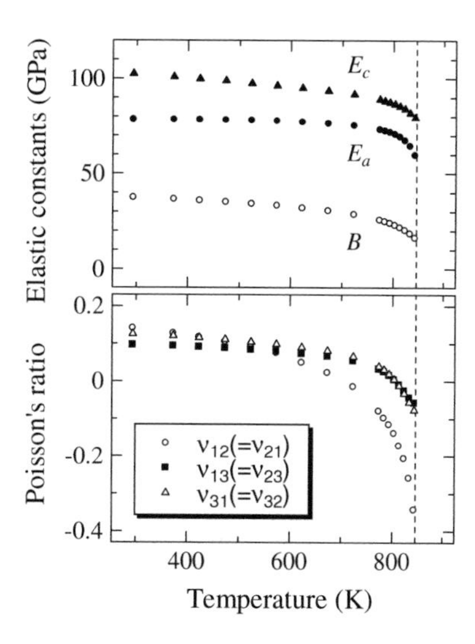

Fig. 8.41 Temperature dependence of Young's moduli, bulk modulus, and Poisson's ratios of α-quartz. Reprinted with permission from Nakamura et al. (2012). Copyright (2012), AIP Publishing LLC

and Poisson's ratios. Their temperature coefficients over the linear temperature dependence are given in Table 8.15. Quartz causes phase transition from α to β phases at 846 K, and approaching the phase transition temperature, the in-plane shear modulus C_{66} increases. This anomaly was attributed to the unchanged local in-plane shear strain at the phase transition, which is related to the in-plane shear modulus C_{66} (Nakamura et al. 2012).

8.9.3 *Phonon-Assisted Hopping Conduction of Carriers in* GaN

Gallium nitride (GaN) is a key material in various electric devices such as a high-electron-mobility transistor. The device performances, however, deteriorate with increasing temperature (Aktas et al. 1996; Maeda et al. 2001). Fe ions have been doped in GaN as deep acceptors for compensating n-type carriers to achieve a semi-insulating GaN (Monemar and Lagerstedt 1979; Maier et al. 1994), and deeply trapped carriers can be thermally activated, causing the hopping conduction (Mott and Twose 1961; Look et al. 1996), which creates the additional electron pass and affects the carrier mobility. It is thus important to understand electron transport behavior at elevated temperatures in the semi-insulating GaN.

The free carrier flow is restricted in an insulated semiconductor, and the conduction principally occurs by hopping of trapped carriers by acceptors between the sites. The hopping conduction is a thermally activated phenomenon; the phonon-assisted carrier movement is efficiently enhanced when the polarization switching rate caused by ultrasonic vibration is matched with the jump rate of a carrier from site to site. Thus, focusing a resonant mode, internal friction shows a maximum with increasing temperature because a part of the acoustic energy is spent on the carrier movement, and the frequency decrement (modulus defect) occurs at the matching temperature (Hutson and White 1962). The resonance frequency change Δf and corresponding attenuation coefficient α take the Debye-type relaxation forms (Debye 1929)

$$\left.\begin{aligned} \frac{\Delta f}{f} &= \frac{\Delta_R}{2}\frac{1}{1+(\omega\tau)^2} \\ \alpha &= \frac{\delta_R}{2}\frac{\omega^2\tau}{1+(\omega\tau)^2} \end{aligned}\right\}. \tag{8.49}$$

Here, ω is the angular frequency, and Δ_R denotes the relaxation strength. τ denotes the relaxation time for the carrier hopping and is inversely proportional to the Boltzmann factor ($\tau \propto \exp(E_h/kT)$) (Mott and Twose 1961; Look et al. 1996). An Arrhenius plot for the peak attenuation yields the activation energy E_h. An Fe-doped GaN crystal ($3.493 \times 2.990 \times 0.410$ mm^3) was used for evaluating the activation energy for hopping conduction between Fe sites.

Figure 8.42 shows temperature behavior of the attenuation coefficient for various resonance modes. The attenuation peak temperature rises as the frequency increases, and the Arrhenius plot yields the activation energy of 0.23 ± 0.05 eV (Ogi et al. 2015). Previous studies used optical spectroscopy methods for studying excited states of Fe ions in GaN (Heitz et al. 1997; Malguth et al. 2006), where the formation of an Fe^{2+} hole complex was suggested and their binding energy was estimated. However, various energy bands were superimposed in their spectra, and it was fairly difficult to identify the energy states with the phonon-assisted charge

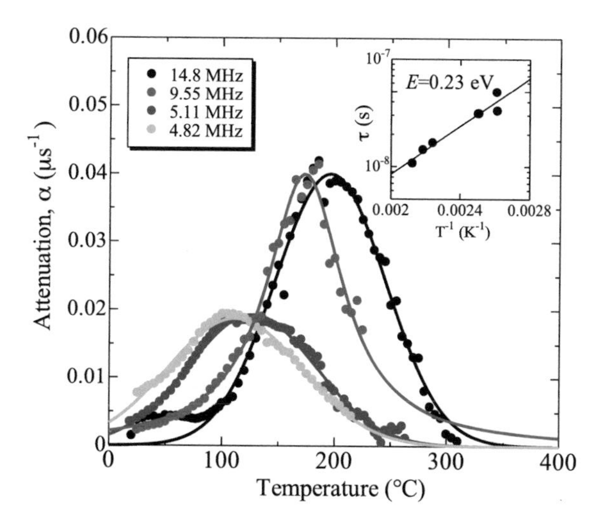

Fig. 8.42 Temperature dependence of the attenuation coefficient with various frequencies. The *inset* shows the Arrhenius plot, yielding the activation energy. Reprinted from Ogi et al. (2015). Copyright (2015), AIP Publishing LLC

transfer process from such broad spectra. For example, Heitz et al. (1997) estimated the binding energy for the complex to be 0.28 eV, whereas Malguth et al. (2006) denied this value and presented a much smaller value of 0.05 eV. Thus, the lower energy bands of the charge transfer processes coupled with phonons have not been clearly observed. In contrast, the acoustic resonance method allows the direct excitation of the phonon-assisted carrier flow and then unambiguous determination of the activation energy.

References

Aktas, O., Fan, Z. F., Mohammad, S. N., Botchkarev, A. E., & Morkoç, H. (1996). High temperature characteristics of AlGaN/GaN modulation doped field-effect transistors. *Applied Physics Letters, 69*, 3872–3874.

Armyanov, S., Vitkova, S., & Blajiev, O. (1997). Internal stress and magnetic properties of electrodeposited amorphous Fe-P alloys. *Journal of Applied Electrochemistry, 27*, 185–191.

Auld, A. B. (1973). *Acoustic Fields and Waves in Solids*. New York: Wiley.

Barmatz, M., & Chen, H. S. (1974). Young's modulus and internal-friction in metallic glass alloys from 1.5 to 300 K. *Physical Review B, 9*, 4073–4083.

Bungo, A., Jian, C., Yamaguchi, K., Sawada, Y., Uda, S., & Pisarevsky, Y. (1999). Analysis of surface acoustic wave properties of the rotated Y-cut langasite substrate. *Japanese Journal of Applied Physics, 38*, 3239–3243.

Davis, B. M., Seidman, D. N., Moreau, A., Ketterson, J. B., Mattson, J., & Grimsditch, M. (1991). Supermodulus effect in Cu/Pd and Cu/Ni superlattices. *Physical Review B, 43*, 9304–9307.

Debye, P. (1929). *Polar Molecules* (p. 93). New York: Dover Publications Inc.

Demarest, J. H, Jr. (1971). Cube resonance method to determine the elastic constants of solids. *The Journal of the Acoustical Society of America, 49*, 768–775.

Dunn, M. L., & Ledbetter, H. (1995a). Elastic moduli of composites reinforced by multiphase particles. *Journal of Applied Mechanics, 62*, 1023–1028.

Dunn, M. L., & Ledbetter, H. (1995b). Poisson's ratio of porous and microcracked solids: Theory and application to oxide superconductors. *Journal of Materials Research, 10*, 2715–2722.

Dunn, M. L., & Ledbetter, H. (1997). Elastic-plastic behavior of textured short-fiber composites. *Acta Materialia, 45*, 3327–3340.

Eer Nisse, E. P. (1967). Variational method for electroelastic vibration analysis. *IEEE Transactions on Sonics and Ultrasonics*, *SU-14*, 153–160.

Eshelby, J. D. (1957). The determination of the elastic field of an ellipsoidal inclusion and related problems. *Proceedings of the Royal Society of London A: Mathematical, Physical and Engineering Sciences, 241*, 376–396.

Fisher, E. S., & Renken, C. J. (1964). Single-crystal elastic moduli and the hcp → bcc transformation in Ti, Zr, and Hf. *Physical Review, 135*, A482–A494.

Hearmon, R. F. S. (1966). *Elastische, piezoelektrische, piezooptische und elektro-optische Konstanten von Kristallen, in Landolt-Börnstein Zahlenwerte und Funktionen aus Naturwissenschaften und Technik, Neue Serie; Gruppe 3, Kristall- und Festkorperphysik* (Vol. 1). Berlin: Springer.

Heitz, R., Maxim, P., Eckey, L., Thurian, P., Hoffmann, A., Broser, I., Pressel, K., & Meyer, K. (1997). Excited states of Fe^{3+} in GaN. *Physical Review B, 55*, 4382–4387.

Heyliger, P. (2000). Traction-free vibration of layered elastic and piezoelectric rectangular parallelepipeds. *The Journal of the Acoustical Society of America, 107*, 1235–1245.

Holland, R. (1968). Resonant properties of piezoelectric ceramic rectangular parallelepipeds. *The Journal of the Acoustical Society of America, 43*, 988–997.

Huang, H., & Spaepen, F. (2000). Tensile testing of free-standing Cu, Ag and Al thin films and Ag/Cu Multilayers. *Acta Materialia, 48*, 3261–3269.

Hutson, A. R., & White, D. L. (1962). Elastic wave propagation in piezoelectric semiconductors. *Journal of Applied Physics, 33*, 40–47.

Hyun, S. K., Murakami, K., & Nakajima, H. (2001). Anisotropic mechanical properties of porous copper fabricated by unidirectional solidification. *Materials Science and Engineering A, 299*, 241–248.

Hyun, S. K., & Nakajima, H. (2001). Fabrication of lotus-structured porous iron by unidirectional solidification under nitrogen gas, In *Cellular Metals and Metal Forming Technology* (pp. 181–186). Bremen: MIT-Verlag.

Ichitsubo, T., Tane, M., Ogi, H., Hirao, M., Ikeda, T., & Nakajima, H. (2002a). Anisotropic elastic constants of lotus-type porous copper: Measurements and micromechanics modeling. *Acta Materialia, 50*, 4105–4115.

Ichitsubo, T., Ogi, H., Hirao, M., Tanaka, K., Osawa, M., Yokokawa, T., et al. (2002b). Elastic constant measurement of ni-base superalloy with the RUS and mode selective EMAR methods. *Ultrasonics, 40*, 211–215.

IEEE. (1984). *IEEE Standard on Piezoelectricity*, Part II, *IEEE Transactions on Sonics and Ultrasonics*, SU31.

Ilyaev, A., Umarov, B., Shabanova, L., & Dubovik, M. (1986). Temperature dependence of electromechanical properties of LGS crystals. *Physica Status Solidi (a), 98*, K109–K114.

Inoue, K., & Sato, K. (1998). Propagation characteristics of surface acoustic waves on langasite. *Japanese Journal of Applied Physics, 37*, 2909–2913.

Isaak, D., Carnes, J., Anderson, O., & Oda, H. (1998). Elasticity of fused silica spheres under pressure using resonant ultrasound spectroscopy. *The Journal of the Acoustical Society of America, 104*, 2200–2206.

Jansson, S., Dève, H. E., & Evans, A. G. (1991). The anisotropic mechanical properties of a Ti matrix composite reinforced with SiC fibers. *Metallurgical Transactions A, 22*, 2975–2984.

Johnson, W., Kim, S., & Lauria, D. (2000). Anelastic loss in langatate, In *Proceedings of the 2000 IEEE/EIA International Frequency Control Symposium and Exhibition* (pp. 186–190).

Kaminskii, A., Silvestrova, I., Sarkisov, S., & Denisenko, G. (1983). Investigation of trigonal $(La_{1-x}Nd_x)_3Ga_5SiO_{14}$ crystals. *Physica Status Solidi (a), 80*, 607–620.

Kim, J. O., Achenbach, J. D., Shinn, M., & Barnett, S. A. (1992). Effective elastic constants and acoustic properties of single-crystal TiN/NbN superlattices. *Journal of Materials Research, 7*, 2248–2256.

Kinoshita, N., & Mura, T. (1971). Elastic fields of inclusions in anisotropic media. *Physica Status Solidi (a), 5*, 759–768.

Kröner, E. (1978). Self-consistent scheme and graded disorder in polycrystal elasticity. *Journal of Physics F: Metal Physics, 8*, 2261–2267.

Kuokkala, V.-T., & Schwarz, R. B. (1992). The use of magnetostriction film transducers in the measurement of elastic moduli and ultrasonic attenuation of solids. *Review of Scientific Instruments, 63*, 3136–3142.

Ledbetter, H., Dunn, M., & Couper, M. (1995c). Calculated elastic constants of alumina-mullite ceramic particles. *Journal of Materials Science, 30*, 639–642.

Ledbetter, H., Fortunko, C. M., & Heyliger, P. (1995a). Orthotropic elastic constants of a boron-aluminum fiber-reinforced composite: An acoustic-resonance-spectroscopy study. *Journal of Applied Physics, 78*, 1542–1546.

Ledbetter, H., Fortunko, C. M., & Heyliger, P. (1995b). Elastic constants and internal friction of polycrystalline copper. *Journal of Materials Research, 10*, 1352–1353.

Ledbetter, H. (2003). Acoustic Studies of Composite-Materials Interfaces, in Nondestructive Characterization of Materials, Vol. 11, 689–696.

Lei, M., Ledbetter, H., & Xie, Y. (1994). Elastic constants of a material with orthorhombic symmetry: An alternative measurement approach. *Journal of Applied Physics, 76*, 2738–2741.

Leisure, R. G., & Willis, F. A. (1997). Resonant ultrasound spectroscopy. *Journal of Physics: Condensed Matter, 9*, 6001–6029.

Logan, J., & Ashby, M. F. (1974). Mechanical properties of 2 metallic glasses. *Acta Metallurgica, 22*, 1047–1054.

Look, D. C., Reynolds, D. C., Kim, W., Aktas, Ö., Botchkarev, A., Salvador, A., & Morkoç, H. (1996). Deep-center hopping conduction in GaN. *Journal of Applied Physics, 80*, 2960–2963.

Maeda, N., Tsubaki, K., Saitoh, T., & Kobayashi, N. (2001). High-temperature electron transport properties in AlGaN/GaN heterostructures. *Applied Physics Letters, 79*, 1634–1636.

Maier, K., Kunzer, M., Kaufmann, U., Schneider, J., Monemar, B., Akasaki, I., & Amano, H. (1994). Iron acceptors in gallium nitride (GaN). *Materials Science Forum, 143–147*, 93–98.

Malguth, E., Hoffmann, A., Gehlhoff, W., Gelhausen, O., Phillips, M. R., & Xu, X. (2006). Structural and electronic properties of Fe^{3+} and Fe^{2+} centers in GaN from optical and EPR esxperiments. *Physical Review B, 74*, 165202.

Malocha, D., Cunha, M., Adler, E., Smythe, R., Frederick, S., Chou, M., Helmbold, R., & Zhou, Y. (2000). Recent measurements of material constants versus temperature for langatate, langanite and langasite, In *Frequency Control Symposium and Exhibition, 2000. Proceedings of the 2000 IEEE/EIA International* (pp. 200–205).

Mason, W. P. (1965). Effect of impurities and phonon process on the ultrasonic attenuation of germanium, crystal quartz, and silicon, In *Physical Acoustics* (Vol. 3B, pp. 235–286). New York: Academic Press.

Maynard, J. (1992). The use of piezoelectric film and ultrasound resonance to determine the complete elastic tensor in one measurement. *The Journal of the Acoustical Society of America, 91*, 1754–1762.

Maynard, J. (1996). Resonant Ultrasound Spectroscopy. *Physics Today, 49*, 26–31.

Migliori, A., & Sarrao, J. (1997). *Resonant Ultrasound Spectroscopy*. New York: Wiley-Interscience.

Migliori, A., Sarrao, J., Visscher, M. W., Bell, T., Lei, M., Fisk, Z., & Leisure, R. (1993). Resonant ultrasound spectroscopy techniques for measurement of the elastic moduli of solids. *Physica B: Condensed Matter, 183*, 1–24.

Mital, S. K. (1994). Micro-fracture in high-temperature metal-matrix laminates. *Composite Science and Technology, 50*, 59–70.

Miyazaki, T., Nakamura, K., & Mori, H. (1979). Experimental and theoretical investigations on morphological changes of γ' precipitates in Ni-Al single-crystals during uniaxial stress-annealing. *Journal of Materials Science, 14*, 1827–1837.

Mochizuki, E. (1987). Application of group theory to free oscillations of an anisotropic rectangular parallelepiped. *Journal of Physics of the Earth, 35*, 159–170.

Monemar, B., & Lagerstedt, O. (1979). Properties of VPE-grown GaN doped with Al and some iron-group metals. *Journal of Applied Physics, 50*, 6480–6491.

Moreau, A., Ketterson, J. B., & Huang, J. (1990). Three methods for measuring the ultrasonic velocity in thin films. *Materials Science and Engineering A, 126*, 149–154.

Mori, T., & Tanaka, K. (1973). Average stress in matrix and average elastic energy of materials with misfitting inclusions. *Acta Metallurgica, 21*, 571–574.

Mott, N. F. & Twose, W. D. (1961). The theory of impurity conduction. *Advances in Physics, 10* (7), 107–163.

Mura, T. (1987). *Micromechanics of Defects in Solids* (2nd ed.). The Hague: Martinus Nijhoff.

Nakamura, N., Sakamoto, M., Ogi, H., & Hirao, M. (2012). Elastic constants of langasite and α quartz at high temperatures measured by antenna transmission acoustic resonance. *Review of Scientific Instruments, 83*, 073901.

Ogi, H., Ledbetter, H., Kim, S., & Hirao, M. (1999a). Contactless mode-selective resonance ultrasound spectroscopy: Electromagnetic acoustic resonance. *The Journal of the Acoustical Society of America, 106*, 660–665.

Ogi, H., Takashima, K., Ledbetter, H., Dunn, M. L., Shimoike, G., Hirao, M., & Bowen, P. (1999b). Elastic constants and internal friction of an SiC-fiber-reinforced Ti-alloy-matrix crossply composite: Measurement and theory. *Acta Materialia, 47*, 2787–2796.

Ogi, H., Dunn, M., Takashima, K., & Ledbetter, H. (2000). Elastic properties of a SiC_f/Ti unidirectional composite: Acoustic resonance measurements and micromechanics predictions. *Journal of Applied Physics, 87*, 2769–2774.

Ogi, H., Ledbetter, H., Takashima, K., Shimoike, G., & Hirao, M. (2001). Elastic properties of a crossply SiC_f/Ti composite at elevated temperatures. *Metallurgical and Materials Transactions A: Physical Metallurgy and Materials Science, 32*, 425–429.

Ogi, H., Shimoike, G., Hirao, M., Takashima, K., & Higo, Y. (2002a). Anisotropic elastic-stiffness coefficients of an amorphous Ni-P film. *Journal of Applied Physics, 91*, 4857–4862.

Ogi, H., Shimoike, G., Takashima, K., & Hirao, M. (2002b). Measurement of elastic-stiffness tensor of an anisotropic thin film by electromagnetic acoustic resonance. *Ultrasonics, 40*, 333–336.

Ogi, H., Kai, S., Ichitsubo, T., Hirao, M., & Takashima, K. (2003a). Elastic-stiffness coefficients of a silicon-carbide fiber at elevated temperatures: Acoustic spectroscopy and micromechanics modeling. *Philosophical Magazine, 83*, 503–512.

Ogi, H., Nakamura, N., Sato, K., Hirao, M., & Uda, S. (2003b). Elastic, anelastic, and piezoelectric coefficients of langasite ($La_3Ga_5SiO_{14}$): Resonance ultrasound spectroscopy with laser-doppler interferometry. *IEEE Transactions on Ultrasonics Ferroelectrics, and Frequency Control, UFFC-50*, 553–560.

Ogi, H., Kai, S., Ledbetter, H., Tarumi, R., Hirao, M., & Takashima, K. (2004). Titanium's high-temperature elastic constants through the hcp-bcc phase transformation. *Acta Materialia, 52*, 2075–2080.

Ogi, H., Tsutsui, Y., Nakamura, N., Nagakubo, A., Hirao, M., Imade, M., et al. (2015). Hopping conduction and piezoelectricity in Fe-doped GaN studied by non-contacting resonant ultrasound spectroscopy. *Applied Physics Letters, 106*, 091901.

Ohno, I. (1976). Free vibration of a rectangular parallelepiped crystal and its application to determination of elastic constants of orthorhombic crystals. *Journal of Physics of the Earth, 24*, 355–379.

Ohno, I. (1990). Rectangular parallelepiped resonance method for piezoelectric crystals and elastic constants of alpha-quartz. *Physics and Chemistry of Minerals, 17*, 371–378.

Papachristos, V. D., Panagopoulos, C. N., Christoffersen, L. W., & Markaki, A. (2001). Young's modulus, hardness and scratch adhesion of Ni-P-W multilayered alloy coatings produced by pulse plating. *Thin Solid Films, 396*, 173–182.

Peraud, S., Pautrot, S., Villechaise, P., Mendez, J., & Mazot, P. (1997). Determination of Young's modulus by a resonant technique applied to two dynamically ion mixed thin films. *Thin Solid Films, 292*, 55–60.

Pineau, A. (1976). Influence of uniaxial stress on morphology coherent precipitates during coarsening-elastic energy considerations. *Acta Metallurgica, 24*, 559–564.

Reddy, J. N. (1987). A generalization of two-dimensional theories of laminated composite plates. *Communications in Applied Numerical Methods, 3*, 173–181.

Sakai, S., Tanimoto, H., & Mizubayashi, H. (1999). Mechanical behavior of high-density nanocrystalline gold prepared by gas deposition method. *Acta Materialia, 47*, 211–217.

Sakharov, S., Senushencov, P., Medvedev, A., & Pisarevsk, Y. (1995), New data on temperature stability and acoustical losses of langasite crystals, in *49th Proceedings of the 1995 IEEE International Frequency Control Symposium* (pp. 647–652).

Sil'vestrova, I., Pisarevskii, Y., Senyushchenkov, P., & Krupnyi, A. (1986). Temperature dependence of the properties of $La_3Ga_5SiO_{14}$ single crystal. *Soviet Physics Solid State, 28*, 1613–1614.

Simone, A. E., & Gibson, L. J. (1996). The tensile strength of porous copper made by the GASAR process. *Acta Metallurgica, 44*, 1437–1447.

Smith, R., & Welsh, F. (1971). Temperature dependence of the elastic, piezoelectric, and dielectric constants of lithium tantalate and lithium niobate. *Journal of Applied Physics, 42*, 2219–2230.

Sullivan, C. P., Webster, G. A., & Piearcey, B. J. (1968). Effect of stress cycling on creep behaviour of a wrought nickel-base alloy at 955 °C. *Journal of the Institute Metals, 96*, 274–281.

Sumino, Y., Ohno, I., Goto, T., & Kumazawa, M. (1976). Measurement of elastic constants and internal friction in single-crystal MgO by rectangular parallelepiped resonance. *Journal of Physics of the Earth, 24*, 263–273.

Takashima, K., Ogura, A., Ichikawa, Y., Shimojo, M., & Higo, Y. (2000). Anisotropic fracture behavior of electroless deposited Ni-P amorphous alloy thin films, In *MRS 2000 Fall Meeting Abstract* (pp. 574–579).

Takashima, K., Shimojo, M., Higo, Y., & Swain, M. V. (2001). Fracture behavior of micro-sized specimens with fatigue pre-crack prepared from a Ni-P amorphous alloy thin film, in *ASTM STP1413*. (pp. 72–81).

Tanaka, K., & Koiwa, M. (1996). Single-crystal elastic constants of intermetallic compounds. *Intermetallics, 4*, S29–S39.

Tanaka, K., Okamoto, K., Inui, H., Minonishi, Y., Yamaguchi, M., & Koiwa, M. (1996a). Elastic constants and their temperature dependence for the intermetallic compound Ti_3Al. *Philosophical Magazine A, 73*, 1475–1488.

Tanaka, K., Ichitsubo, T., Inui, H., Yamaguchi, M., & Koiwa, M. (1996b). Single-crystal elastic constants of γ-TiAl. *Philosophical Magazine Letters, 73*, 71–78.

Tien, J. K., & Copley, S. M. (1971). Effect of uniaxial stress on periodic morphology of coherent gamma prime precipitations in Nickel-base superalloy crystals. *Metallurgical Transactions, 2*, 215–219.

Wadsworth, J., & Froes, F. H. (1989). Developments in metallic materials for aerospace applications. *JOM, 41*, 12–19.

Chapter 9
Resonance Ultrasound Microscopy

Abstract The antenna transmission technique (Sect. 3.13) is closely related to the EMAT phenomenon, where dynamic electric field is coupled with the deformation of piezoelectric material. It allows noncontacting excitation of intended vibrational mode in a piezoelectric material and has been applied for developing piezoelectric probes for measuring local Young's modulus quantitatively. This technique is called resonance ultrasound microscopy, or RUM. This chapter describes the physical principle of the quantitative local Young's modulus measurement and some experimental results with RUM.

Keywords Biasing force · Contact stiffness · Electron backscattering pattern (EBSP) · Grain boundary · Green function · Hertzian contact · Langasite oscillator · Diamond tip · Poisson's ratio · Silicon-carbide fiber · Tungsten-carbide ball bearing

9.1 Local Elastic Stiffness

The elastic stiffness of materials reflects interatomic strength for bond stretching and bond bending. It also involves information on inclusions, noncohesive grain bonds, and oriented heterogeneous phases. In the case of defective inclusions, the stiffness decreases because the stress concentration around the defects causes larger deformation. Thus, macroscopic elastic stiffness can be used for nondestructive evaluation tool of defects. The elastic stiffness at a local area is expected to be more sensitive to defects because the volume fraction of defects increases. Also, local stiffness provides us with bond strength at interfaces in composite materials. The local stiffness can be a measure for material reliability, especially for composites.

To measure the local elastic stiffness, vibrational measurements have been successfully incorporated into atomic force microscopy (AFM), where the resonance frequency change of an AFM cantilever contacting the material is detected (Yamanaka et al. 2000; Rabe et al. 2002). However, quantitative evaluation of the local stiffness is difficult because the cantilever is not isolated acoustically. Namely, many ambiguous components participate in the resonator system, not only the

M. Hirao and H. Ogi, *Electromagnetic Acoustic Transducers*,
Springer Series in Measurement Science and Technology,
DOI 10.1007/978-4-431-56036-4_9

cantilever but also the piezoelectric element to excite the vibration, the gripping condition, and the biasing force, which is variable depending on the surface roughness of the specimen. As an alternative method, resonance ultrasound microscopy (RUM) has been invented (Ogi et al. 2006, 2008). It uses a bar-shaped langasite oscillator. The oscillator is acoustically isolated from any other materials except for the contact point with specimen, and the vibration is excited and detected by a noncontacting line antenna. The elastic properties of the specimen are then extracted unambiguously, realizing the quantitative measurement of materials' local stiffness, although the spatial resolution is inferior to the ultrasonic AFM. The most important goal of this point-contact scanning measurement is to map the stiffness distribution quantitatively. This demand poses three challenges. First, the measurement should be independent of surface roughness. Second, it should be independent of temperature. Third, the vibrating probe should be isolated from any other components except for the examined specimen.

9.2 Isolation of Oscillator

Figure 9.1 shows the schematic of the langasite oscillator, which is intended to achieve a high spatial resolution ($\sim$200 nm). A monocrystal rectangular parallelepiped langasite (6.0 $\times$ 0.68 $\times$ 0.68 mm^3) is supported by a nylon fixture at the nodal point of the fundamental mode of longitudinal resonance (see Sect. 8.9.1 for langasite). The longitudinal direction of the crystal is selected to be in the x-direction of the trigonal system for three reasons. First, the x-direction Young's modulus, $E_{[100]}$, of langasite takes a minimum value ($E_{[100]}$ = 114 GPa and $E_{[001]}$ = 198.6 GPa), providing the highest sensitivity to the contact stiffness as shown below. Second, the temperature derivative of Young's modulus is smaller in the x direction (($dE_{[100]}/dT)/E_{[100]}$ = 2.8 $\times$ 10^{-5} K^{-1} and ($dE_{[001]}/dT)/E_{[001]}$ = 1.1 $\times$ 10^{-4} K^{-1} (Tarumi et al. 2011)),

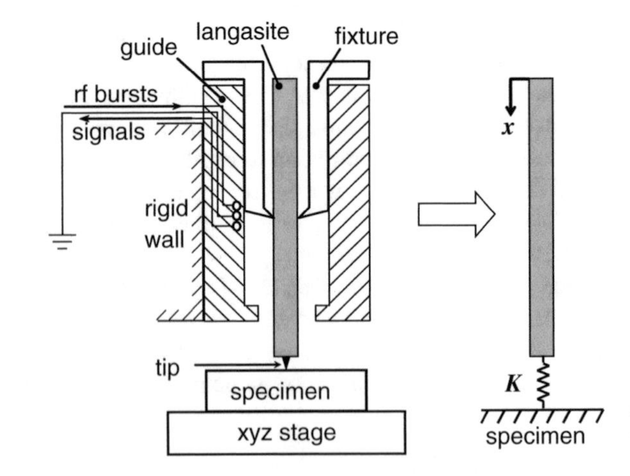

Fig. 9.1 Schematic of the langasite probe with a diamond tip (*left*) and the vibrational model with the contact spring representing the contact stiffness K between the tip and the specimen (*right*). Reprinted from Ogi et al. (2008), Copyright (2008), with the permission from AIP Publishing LLC

yielding stable resonance frequency and resulting in a calibration-free measurement for the temperature change. Third, the longitudinal vibration in the x direction is easily excited by applying the quasistatic electric field in the x direction, owing to the large absolute value of the piezoelectric coefficient, e_{11} (see Table 8.14). A conical monocrystal diamond tip, whose radius is 2 μm, is attached at the center of the bottom surface of the oscillator to make contact with the specimen, where the vibrational amplitude takes a maximum (antinode). A cylindrical guide surrounding the fixture allows only vertical movement of the probe. Thus, the biasing force for the contact is kept unchanged, being independent of surface roughness; it principally equals the weight from the oscillator mass together with the fixture (0.59 mN).

The line antenna consists of generation wire, detection wire, and a grounding wire. It is embedded in the guide. We apply tone-burst signal ($\sim$50 V_{pp} amplitude and 100 μs duration) to the generation wire to radiate the quasistatic electric field in the x direction at the nodal line of the side face of the crystal, where the maximum stress occurs for an intended resonance. The electric field in the longitudinal direction predominantly causes longitudinal vibrations (A_g vibration group in Table 8.13). After the excitation, the detection wire picks up resonance vibration through the piezoelectric effect. A frequency scan yields the resonance spectrum contactlessly, and the Gaussian function fitting provides the resonance frequency. Figure 9.2 shows resonance spectra measured before and after contact with a low-carbon steel. It is observed that the resonance frequency increases with the mechanical constraint caused by the contact and also the peak height (and Q-value) decreases due to the energy leakage into the specimen. Because this frequency scanning requires relatively longer time, we monitor the phase of the received signal at a fixed frequency to determine the resonance frequency change from the linear relationship between the phase and frequency near the resonance frequency.

At each measuring point, resonance frequency at a noncontacting state is first measured by the superheterodyne spectroscopy method (see Sect. 5.2); resonance frequency at the contacting condition is measured by lifting the specimen up to make a contact, and the frequency change between them is recorded. Then, the probe is separated from the specimen, and the stage moves the specimen to the next measuring point. The probe does not scratch the specimen surface.

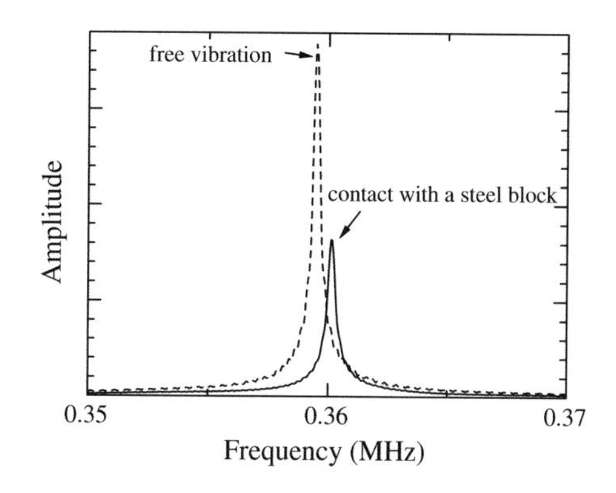

Fig. 9.2 Resonance spectra measured by the line antenna at a noncontacting state and in contact with a low-carbon steel specimen. Reprinted from Ogi et al. (2008), Copyright (2008), with the permission from AIP Publishing LLC

9.3 Vibrational Analysis

The contact problem of the isolated oscillator can be modeled as a resonator of a bar oscillator connected with the rigid wall through a spring as shown in Fig. 9.1. The spring constant is characterized by the contact stiffness, K, which depends on elastic constants of the tip and specimen, the biasing force, and the tip radius. Assuming a simple bar resonance and solving the one-dimensional wave equation with the boundary conditions, one derives a frequency equation of the oscillator in contact with a specimen (Ogi et al. 2008),

$$kL\ \tan(kL) = \frac{K}{K_{\mathrm{osc}}} \equiv p. \tag{9.1}$$

Here, $k = 2\pi f\sqrt{\rho/E_{\mathrm{osc}}}$ is the wavenumber and $K_{\mathrm{osc}} = E_{[100]}A/L$ is the equivalent spring constant of the oscillator for a static load. ρ and E_{osc} are mass density and Young's modulus of the oscillator. A and L are the cross-sectional area and the length of the langasite crystal, respectively. p indicates the contribution of the contact stiffness to the resonator system; larger p causes larger frequency shift. As mentioned, we selected the x axis of the langasite crystal to be vertical because the x-direction Young's modulus $E_{[100]}$ of langasite is the smallest to provide the largest p value. Figure 9.3 shows the dependence of the frequency shift Δf on the contact stiffness K for the langasite oscillator used in this study. The broken line is the three-dimensional solution for the A_g-1 mode by the Rayleigh-Ritz method (Sect. 8.2), and the solid line is the one-dimensional solution from Eq. (9.1). Their difference is insignificant, and Eq. (9.1) allows us to inversely determine the contact stiffness K from the measured frequency shift Δf.

Hertzian contact model with isotropic bodies (Johnson 1985) connects K to the effective Young's modulus, E^*_{iso}, via

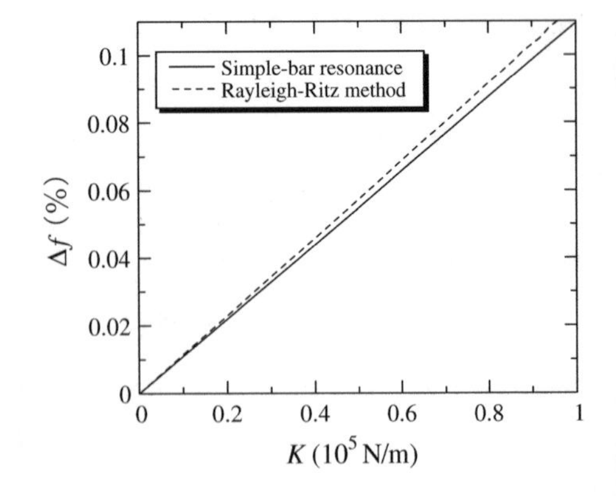

Fig. 9.3 Relationship between the frequency change and the contact stiffness for langasite crystal of $0.68 \times 0.68 \times 6.0\ \mathrm{mm}^3$. Reprinted from Ogi et al. (2008), Copyright (2008), with the permission from AIP Publishing LLC

$$E^{*}_{\mathrm{iso}} = \sqrt{\frac{K^{3}}{6FR}} = \left(\frac{1 - \nu^{2}_{\mathrm{spe}}}{E_{\mathrm{spe}}} + \frac{1 - \nu^{2}_{\mathrm{tip}}}{E_{\mathrm{tip}}} \right)^{-1}, \tag{9.2}$$

where E and ν are Young's modulus and Poisson's ratio; the subscripts "spe" and "tip" indicate quantities of the specimen and tip, respectively. R is the tip radius, and F is the biasing force. Thus, we can determine the effective Young's modulus of the specimen at the local contact area by substituting K in Eq. (9.2).

One complication arises from the small contact area of RUM, which is comparable to or less than the grain size of polycrystalline materials, and the elastic anisotropy comes into measurement. Willis (1966) and Swanson (2004) analyzed the elastic field caused by contact between an indenter and an anisotropic half-space solid using Fourier transformation. Following their analytical method, we derive the contact stiffness for anisotropic solids and study the effect of elastic anisotropy on the measured Young's modulus.

We take a Cartesian coordinate system, where the x_3-axis lies in the depth direction toward the half-space of the contacting specimen. The x_1-x_2 plane defines the specimen surface. The origin is located on the specimen surface and corresponds to the center of the contact area. The contact area is assumed to be an ellipse, whose major and minor axes are denoted by a_1 and a_2, respectively. The pressure distribution at the contact interface is assumed to be expressed by $p_0\sqrt{1 - x_1^2 a_1^2 - x_2^2 a_2^2}$ with the maximum pressure p_0 at the center. Distributions of displacements are calculated by solving the equilibrium equation with respect to boundary conditions of the free surface and the normal biasing force on the plane of $x_3 = 0$, performing Fourier transformation for displacements about the x_1 and x_2 directions. The resultant expression for the relative vertical displacement of the contact interface, δ, is

$$\delta = \sqrt[3]{\frac{9 I_0^3 F^2}{128 R I_1}}, \tag{9.3}$$

$$\begin{aligned} I_0 &= \sum_{\lambda}^{\mathrm{spe,tip}} \int_0^{2\pi} G_3^{\lambda}(\varepsilon \cos\theta, \sin\theta)\, d\theta, \\ I_1 &= \sum_{\lambda}^{\mathrm{spe,tip}} \int_0^{2\pi} G_3^{\lambda}(\varepsilon \cos\theta, \sin\theta) \cos^2\theta \, d\theta, \end{aligned} \tag{9.4}$$

with $\varepsilon = a_1/a_2$, which is very close to unity for most materials, and $\varepsilon = 1$ is postulated. G_3^{λ} is the Green function for the surface displacement in the x_3 direction caused by the unit point force at the origin. It is given by

$$G_3^{\lambda}(\xi_1, \xi_2) = \frac{C^{\lambda}}{\sqrt{\xi_1^2 + \xi_2^2}}, \tag{9.5}$$

in the case of contact between transversely isotropic bodies; C^{λ} is a constant determined from the elastic constants. Equation (9.3) shows a nonlinear relation between the elastic indentation δ and the applied force F.

For a dynamic measurement with small amplitude, we can define the contact stiffness K by taking perturbations of δ and F as

$$K = \sqrt[3]{\frac{48RI_1}{I_0^3}} \tag{9.6}$$

to the first-order approximation. Combining Eqs. (9.2) and (9.6), we can define the effective Young's modulus for anisotropic materials as

$$E^*_{\text{aniso}} = \sqrt{\frac{8I_1}{I_0^3}}. \tag{9.7}$$

Our final goal of developing RUM is to make a calibration-free measurement, that is, measurement of a material's Young's modulus from only the resonance frequency shift without using reference specimens. The measured frequency shift Δf can be converted to the contact stiffness K unambiguously using Eq. (9.1). Then, the effective Young's modulus E^*_{iso} is determined by Eq. (9.2), which is used to

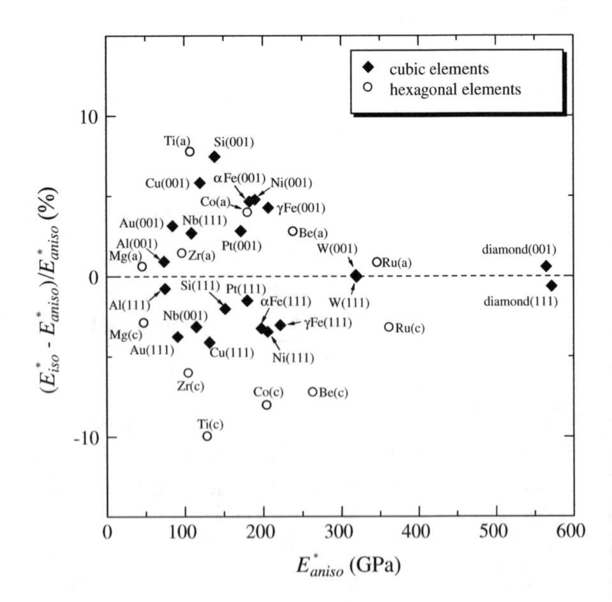

Fig. 9.4 Difference between E^*_{aniso} and E^*_{iso} for cubic and hexagonal elements. The notations (a) and (c) for hexagonal elements denote planes perpendicular and parallel to the basal plane, respectively. Reprinted from Ogi et al. (2008), Copyright (2008), with the permission from AIP Publishing LLC

deduce Young's modulus E_{spe} assuming Poisson's ratio (ambiguity of Poisson's ratio is insignificant as shown in Eq. (9.2)). The above analysis indicates that E^*_{iso} differs from E^*_{aniso} in general. Their difference is calculated and plotted in Fig. 9.4 for the principal crystallographic planes of cubic and hexagonal elements using E_{tip} = 1132 GPa and ν_{tip} = 0.067 for the isotropic diamond tip. It is observed that the difference is negligible in the case of tungsten and diamond, although up to 10 % error could be included for the materials of larger elastic anisotropy such as titanium and silicon.

9.4 Young's Modulus Imaging

The above RUM is capable of imaging the Young's modulus distribution of polycrystalline copper, which is compared with the calculated one based on the electron backscattering pattern (EBSP) measurements. The studied material was an oxygen-free high-conductivity copper. Purity exceeded 99.95 %. The EBSP measurement was taken before the elastic stiffness mapping. After annealing at 800 °C for 2 h, the surface layer (~30 μm thick) was removed chemically to obtain a nonstrained surface, and Kikuchi patterns were obtained for every 7 μm. Figures 9.5a, b compare an image obtained by the scanning electron microscopy with a crystallographic orientation image determined by the EBSP measurement. We measured the resonance frequency for every 5 μm on the same area examined by EBSP. At each measuring point, the change of the resonance frequency between before and after making contact was recorded. Figure 9.5c shows the resonance frequency image, showing clearly the difference of the elastic stiffness depending on the crystallite orientation. The resonance frequency appears nonuniform even within a single grain, indicating elastic constant distribution.

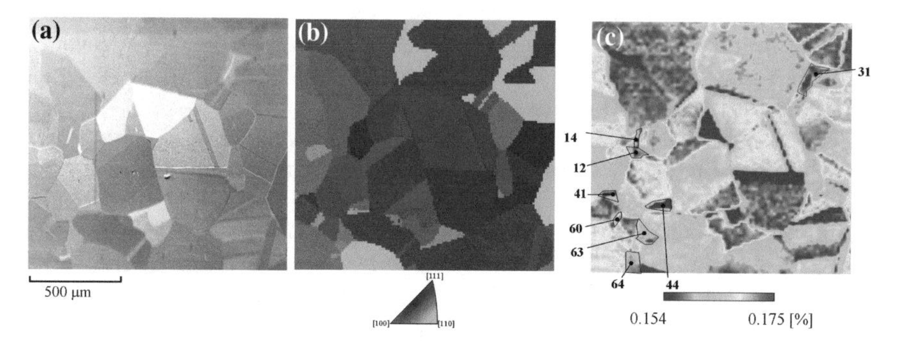

Fig. 9.5 **a** Scanning electron microscopy image, **b** crystallographic orientation image for the normal direction to the surface measured by the EBSP method, and **c** resonance frequency image at F = 0.019 N. Numbered grains indicate smaller stiffnesses than those estimated by grain orientations. Reprinted with permission from Ogi et al. (2006), Copyright (2006) by the American Physical Society. (http://dx.doi.org/10.1103/PhysRevB.73.174107)

Since Eq. (9.7) applies only to orthorhombic materials, it is necessary to average the elastic constant tensor over the rotational angle in the surface plane at each point using Hill's averaging method (Nakamura et al. 2004) and crystallographic orientations determined by EBSP measurements. The averaged elastic stiffness tensor $<C_{ijkl}>$ is given by

$$< C_{ijkl} > = \frac{1}{2}\left\{\frac{1}{2\pi}\int_0^{2\pi} C_{ijkl}(\theta)d\theta + \left(\frac{1}{2\pi}\int_0^{2\pi} S_{ijkl}(\theta)d\theta\right)^{-1}\right\}, \tag{9.8}$$

where S_{ijkl} is the elastic compliance tensor. This procedure yields transverse isotropic (hexagonal) elastic symmetry to accept the above analysis. Figure 9.6 compares the effective normal Young's modulus determined by the observed frequency shift and that calculated by the EBSP measurement for individual grains; there were 74 grains in the scanned area and they are numbered. The effective Young's modulus determined by the RUM fails to show a perfect agreement with that by EBSP because of the simplification of the contact stiffness calculation in order to accept the Willis-Swanson analysis. They nevertheless show a good correlation as shown by the solid line, and the overall error limit of 15 % is achieved for the high elastic anisotropy of copper; it is noted that no fitting parameters are included in the calculation of the effective modulus from the resonance frequency shift. The comparison demonstrates the validity of the contact stiffness model and that the RUM possesses quantitative sensitivity to a material's local elastic modulus.

Some grains show smaller stiffnesses than the prediction by EBSP measurements, exceeding the possible error caused by the elastic anisotropy in Fig. 9.4. These are smaller grains as labeled in Figs. 9.5c and 9.6. This discrepancy could be explained by two factors. First, elastic softening at the grain boundaries is caused by lattice distortion. The volume fraction of such softened regions increases in a small grain. Figure 9.7 shows a linear trace of the resonance frequency along the horizontal broken line, which indicates the stiffness decrease across grain boundaries. The top figure in Fig. 9.7 shows a low quality factor of EBSP at the grain boundaries, indicating low crystallinity.

Second is the elastic strain (or lattice anharmonicity). Phillips et al. (2004) found by using X-ray microdiffraction measurements that elastic strains in some small grains (not all) are much larger than those in large grains in Al-Cu polycrystalline thin films: The maximum resolved shear strain reaches 0.6 %. In small grains, less plastic deformation occurs because dislocations cannot move freely because of the grain boundary back stress. The modulus decrement is estimated roughly to be 3–8 % from the correlation line in Fig. 9.6. We estimate the elastic strain that may cause this decrement from the temperature dependence of the elastic constants. Copper's Young's modulus is 135.7 GPa at 100 K and 127.0 GPa at 300 K (Overton and Gaffney 1955); it decreases by 6.4 % by the 200-K temperature increase, corresponding to the 1 % volume change, assuming the constant thermal

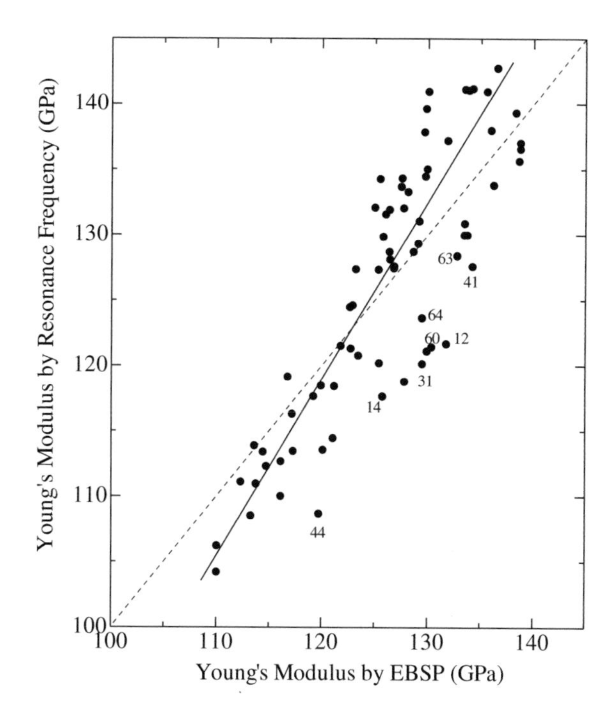

Fig. 9.6 A comparison between the effective Young's modulus determined by RUM and calculation with the EBSP measurement. The slope of the broken line is 1. The solid line indicates the strong correlation between them. Numbered grains correspond to those in Fig. 9.5c. Reprinted with permission from Ogi et al. (2006), Copyright (2006) by the American Physical Society. (http://dx.doi.org/10.1103/PhysRevB.73.174107)

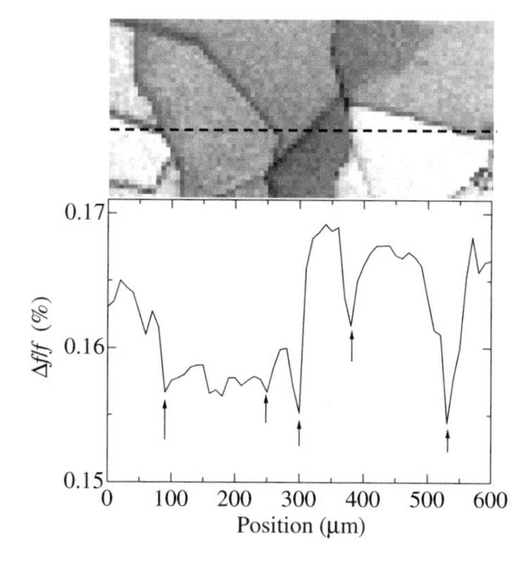

Fig. 9.7 An image of the quality factor in the EBSP measurement (*top*) and the line trace of the resonance frequency shift along the broken line shown in the image (*bottom*). Arrows indicate grain boundaries. Reprinted with permission from Ogi et al. (2006), Copyright (2006) by the American Physical Society. (http://dx.doi.org/10.1103/PhysRevB.73.174107)

expansion coefficient of 18×10^{-6} K^{-1}. Thus, to explain a decrease of Young's modulus in small grains, we expect about 0.3 % uniaxial strain. Young's modulus is highly affected by the shear modulus because it is about eight times as sensitive to the shear modulus as to the bulk modulus. Zener (1949) showed that the variation of the shear modulus with the strain energy caused by thermal vibrations (thermal expansion) is identical with that caused by the strain energy due to residual stresses. Considering the large residual resolved shear strain observed in an Au-Cu alloy, a decrease of Young's modulus by several percent is possible. Smaller grains may include residual strains large enough to change the elastic constant, approximately by 10^{-2}. The stiffness distribution within a single grain can, therefore, be interpreted as arising from the residual strain distribution.

The RUM can depict the stiffness distribution on a cross section of a silicon-carbide fiber (SCS-6) embedded in Ti-alloy matrix (see Sect. 8.4). As shown in Fig. 9.8, where the stiffness images are overlapped on the optical image of the cross section, it principally consists of four components: a carbon core, inner carbon coating, CVD (chemical vapor deposition) deposited β-SiC, and outer carbon coating. Figure 9.9 presents a line trace of the effective Young's modulus in the radial direction, showing larger stiffnesses in inner and outer coating regions. The blocks of turbostratic carbon, being similar to graphite (Ning and Pirouz 1991), are randomly oriented toward the core, while their *c* axes are predominantly oriented toward the radius in the inner and outer coatings. RUM images demonstrate their difference of the overall elastic moduli.

It is also observed that the SiC region shows lower Young's modulus than expected. We calculate the averaged-over-direction Young's modulus of 447 GPa and the effective modulus of 327 GPa from the monocrystal elastic constants of β-SiC (Li and Wang 1999). RUM result provides the effective Young's modulus smaller than 300 GPa, being around 200 GPa. This is attributable to the nano- and microdefects, which can be introduced during the CVD deposition process. In the RUM measurement, the volume of the probing region is estimated to be smaller

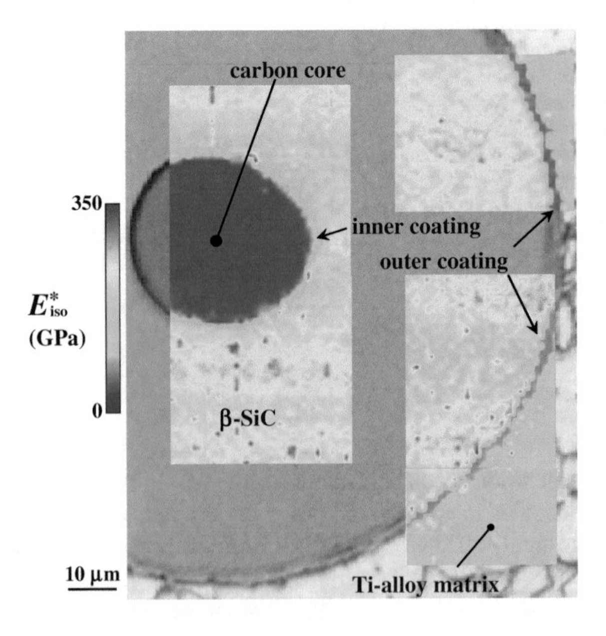

Fig. 9.8 Images of effective Young's modulus mapping on a single embedded SiC fiber overlapped on the optical microscopy image. Reprinted with permission from Ogi et al. (2008), Copyright (2008), AIP Publishing LLC

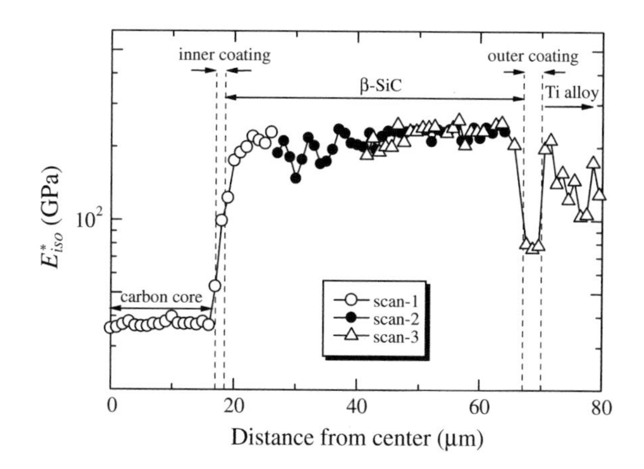

Fig. 9.9 Variation of the effective Young's modulus in the radial direction of the fiber. Three independent measurements are combined. Reprinted with permission from Ogi et al. (2008), Copyright (2008), AIP Publishing LLC

than 1 μm³. The local stress field increases the degree of the stress concentration near defects if any and makes the apparent stiffness decrease more significantly than in the case of the application of a uniform stress field.

9.5 Portable Young's Modulus Tester

Young's modulus is the most important engineering elastic stiffness of a solid, and it is indispensable for designing any structures. Various methods were proposed to measure Young's modulus. The tensile test and bending test are classical methods for measuring Young's modulus, where relationships between applied force and deformations are measured to deduce Young's modulus. They require specimens to be machined into specific shapes and they are inapplicable to small-size solids. The acoustic methods, such as bar resonance methods, pulse echo methods, and resonant ultrasound spectroscopy (RUS, see Chap. 8), allow more accurate determination of the elastic constants with small specimens. However, they again need well-shaped specimens and high-skilled techniques. Indentation tests allow estimation of Young's modulus on solids, but they involve many ambiguous parameters, such as shape of indenter, elastic-plastic behavior of the material, and more. Furthermore, all the previous methods require measurements of at least two quantities for deducing Young's modulus: force and deformation in the tensile, bending and indentation tests, resonance frequency and dimension in the acoustic resonance methods, ultrasonic round-trip time and thickness in the pulse echo method. These requirements have prevented us from making quick and easy evaluation of Young's modulus for arbitrary geometry of solids.

The principle of RUM is further mechanized to realize a portable instrument for quick evaluation of Young's modulus of a solid just by touching a probe to the solid surface. Only one quantity, the resonance frequency of the oscillator, yields Young's modulus without using any other parameters.

There are a number of key mechanical elements, with which such an instrument is made available:

(i) A square-column-shaped monocrystal langasite oscillator with $L = 20$ mm and $A = 2.0 \times 2.0$ mm^2 is used. The oscillator is supported by an O-ring at the four nodal points (apexes of square cross section at the center) of the fundamental rod resonance mode (see Fig. 9.10a) so as to minimize the influence of contacts for support on the frequency response.
(ii) Antenna is attached close to the oscillator for exciting and detecting the resonance frequency of the electrodeless oscillator contactlessly.
(iii) The tip shape must be a perfect sphere for adopting Eq. (9.2). A tungsten-carbide ball bearing of 2 mm radius is used for this purpose, whose sphericity is 0.64 μm in diameter. (The diameter fluctuation is smaller than 0.016 %.) The ball bearing was carefully polished to a thin tip (0.3 mm height). The tip weight is much smaller than that of the oscillator (less than

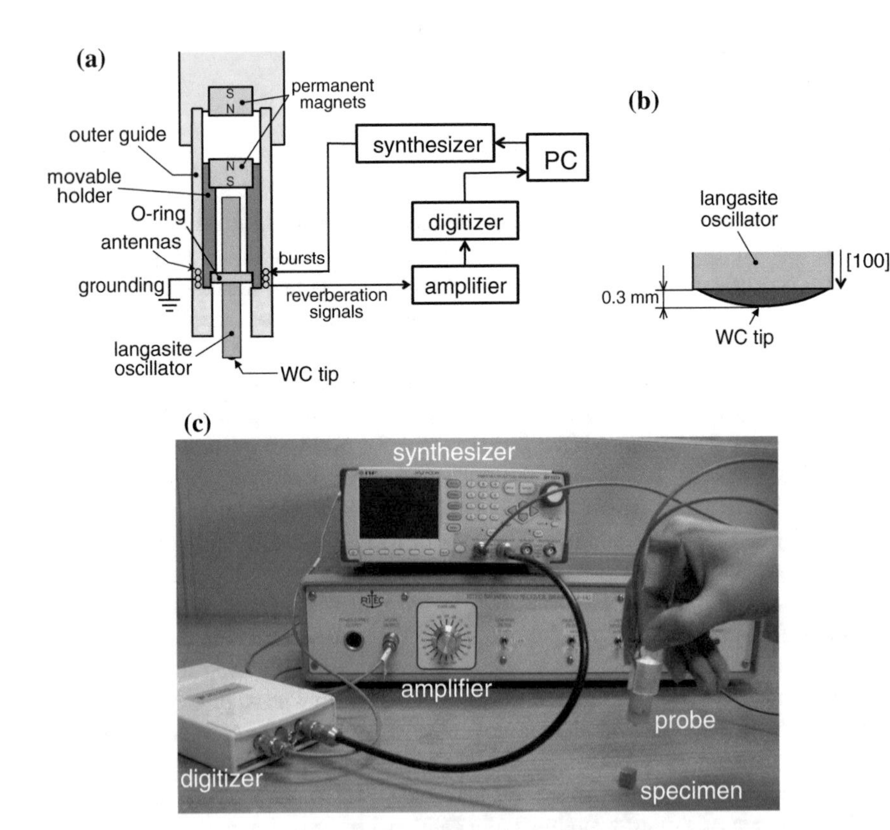

Fig. 9.10 **a** Measurement system of Young's modulus tester, **b** a close-up of the tungsten carbide tip, and **c** appearance of the whole setup. Reprinted from Ogi et al. (2014), Copyright (2014), with permission from Elsevier

2 %), affecting little the resonance frequency. It is bonded on the bottom surface of the langasite oscillator (Fig. 9.10b).

(iv) A constant biasing force is maintained using the repulsive force of permanent magnets. As shown in Fig. 9.10a, the holder, in which the O-ring and the permanent magnet are attached, smoothly slides along the outer guide together with the oscillator when the probe is pressed onto specimen, and it stops when the outer guide contacts the specimen surface. The distance between the two magnets is kept constant, giving a constant biasing force of $F = 0.455$ N. This force is much larger than the weight of the holder involving the O-ring, magnet, and oscillator (0.017 N), and it is nearly constant even in the case of inclined objects as shown in Fig. 9.11.

Tone bursts of amplitude of 20 V_{pp} and duration of 10 ms are fed to the generating antenna to excite the resonance. After the excitation, the reverberating signals are detected by the receiving antenna. After amplified by 32 dB, the waveform is digitally acquired with a portable digitizer, and Fourier transformation is performed to obtain the amplitude at the exciting frequency.

Figure 9.12 shows resonance spectra measured in contact with various materials, clearly displaying that the resonance frequency increases as the specimen stiffens. At the noncontacting state, the resonance frequency is near 106 kHz, while the theoretical value is estimated to be 118 kHz using the elastic constants and mass density shown in Table 8.14. Just as in Fig. 9.2, the peak amplitude decreases with Young's modulus of specimen. This occurs due to the leakage of vibrational energy of the oscillator into the specimen because the acoustic impedance of the specimen approaches that of the tungsten-carbide tip.

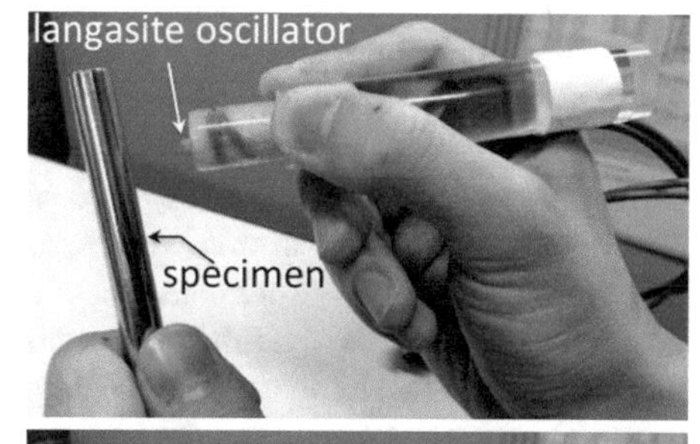

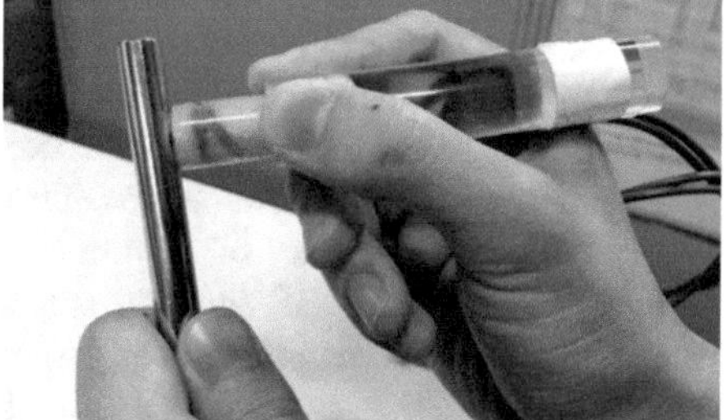

Fig. 9.11 Ballpoint pen probe contacting a specimen. Reprinted from Ogi et al. (2014), Copyright (2014), with permission from Elsevier

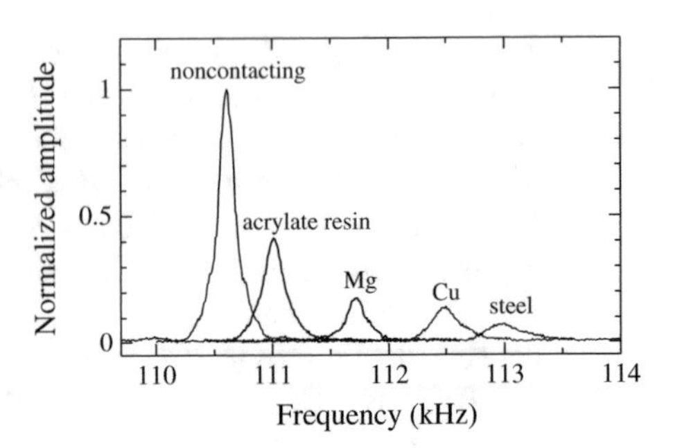

Fig. 9.12 Resonance spectra measured in contacts with various materials. Reprinted from Ogi et al. (2014), Copyright (2014), with permission from Elsevier

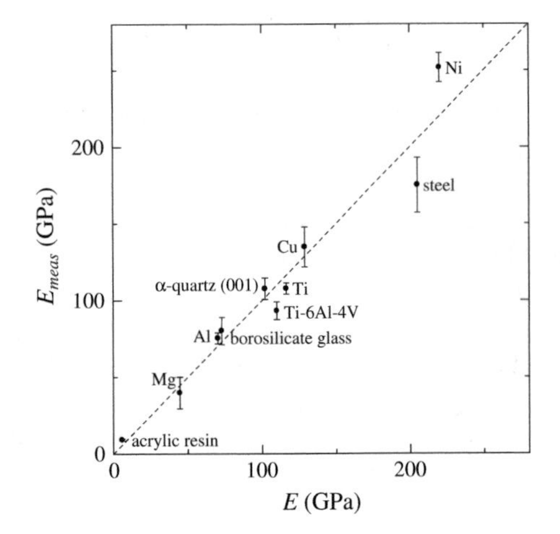

Fig. 9.13 Young's modulus measured by the present method (E_{meas}) and reported values (E). Reprinted from Ogi et al. (2014), Copyright (2014), with permission from Elsevier

The elastic properties of langasite (see Sect. 8.9.1) and those of tungsten carbide (E_{tip} = 630 GPa and ν_{tip} = 0.21) are substituted in Eq. (9.2). For the specimen's Poisson's ratio, we assume a fixed value of ν_{spe} = 0.25. We can, then, determine the specimen's Young's modulus E_{spe} just by measuring the resonance frequency at the contact condition without using any other parameters. Figure 9.13 compares Young's moduli thus measured with reported values for various materials. Despite the calibration-free measurement, they agree with each other within 10 % for materials of Young's modulus lower than 150 GPa. The difference, however, increases up to 15 % for materials of larger Young's modulus, because the probe sensitivity deteriorates with the specimen's Young's modulus. Conventional methods, such as the pulse echo method, will provide more accurate stiffness, but they require well-shaped specimens, large and expensive instruments, and a long period of time for surface preparation. The present Young's modulus tester allows a quick and pointwise measurement even for specimens with arbitrary configurations. It only needs a flat surface of 10 mm diameter or more.

Error bars in Fig. 9.13 indicate standard deviation for independent measurements on five different points. The larger error bars for polycrystalline materials (Cu, carbon steel, and Ni) are caused by different effective Young's modulus

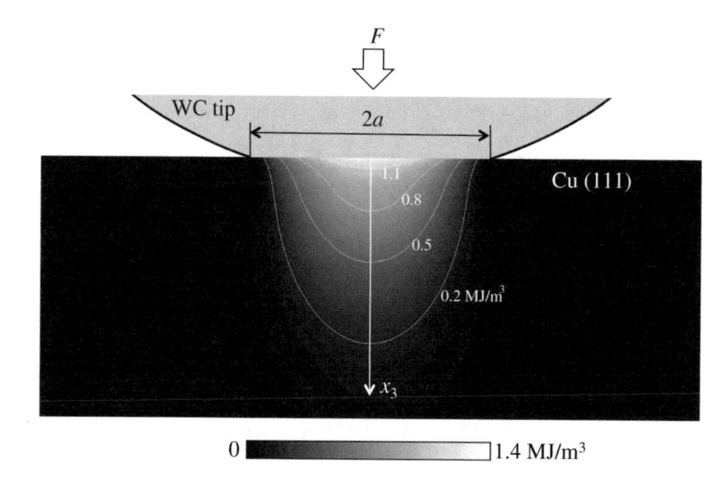

Fig. 9.14 Distribution of the strain energy caused by a contact with a (111) face of Cu. a (= 17.7 lm) is the contact radius. Parameters used are E_{tip} = 630 GPa, ν_{tip} = 0.21, E_{osc} = 118 GPa, ν_{spe} = 0.25, and F = 0.455 N. Reprinted from Ogi et al. (2014), Copyright (2014), with permission from Elsevier

depending on the crystallographic orientation. For example, Young's moduli of Cu in [100], [110], and [111] directions are 65, 129, and 189 GPa, respectively. (The aggregated value is 126 GPa.) Although the present method does not represent the normal Young's modulus directly, the anisotropic effect could alter the effective stiffness by 10 % at most. Measured Young's modulus is thus affected by orientation of grains involved in the detected region in such a locally anisotropic material.

Using the Willis approach for Hertzian contact with an anisotropic specimen, we calculated the distribution of strain energy in the transverse-isotropically oriented (111) Cu specimen. The result is shown in Fig. 9.14. The contact radius is a = 17.7 μm in this case, and the strain-energy penetration depth is close to the contact radius. The probe is then sensitive to a volume of $\sim \pi a^3$ at the surface region, and the measurement is affected by any stiffness variation in this volume. It is possible to enlarge the measuring volume by increasing the biasing force or the tip radius.

References

Johnson, K. (1985). *Contact Mechanics*. Cambridge: Cambridge University Press.

Li, W., & Wang, T. (1999). Elasticity, stability, and ideal strength of β-SiC in plane-wave-based ab initio calculations. *Physical Review B, 59*, 3993.

Nakamura, N., Ogi, H., & Hirao, M. (2004). Elastic constants of chemical-vapor-deposition diamond thin films: Resonance ultrasound spectroscopy with laser-doppler interferometry. *Acta Materialia, 52*, 765–771.

Ning, X. J., & Pirouz, P. (1991). The microstructure of SCS-6 SiC fiber. *Journal of Materials Research, 6*, 2234–2248.

Ogi, H., Hirao, M., Tada, T., & Tian, J. (2006). Elastic-stiffness distribution on polycrystalline copper studied by resonance ultrasound microscopy: Young's modulus microscopy. *Physical Review B, 73*, 174107.

Ogi, H., Inoue, T., Nagai, H., & Hirao, M. (2008). Quantitative imaging of Young's modulus of solids: A contact-mechanics study. *Review of Scientific Instruments, 79*, 053701.

Ogi, H., Sakamoto, Y., & Hirao, M. (2014). Calibration-free portable young's-modulus tester with isolated langasite oscillator. *Ultrasonics, 54*, 1963–1966.

Overton Jr, W. C., & Gaffney, J. (1955). Temperature variation of the elastic constants of cubic elements. I. Copper. *Physical Review, 98*, 969–977.

Phillips, M. A., Spolenak, R., Tamura, N., Brown, W. L., MacDowell, A. A., Celestre, R. S., Padmore, H. A., Batterman, B. W., Arzt, E., & Patel, J. R. (2004). X-ray microdiffraction: Local stress distributions in polycrystalline and epitaxial thin films. *Microelectronic Engineering, 75*, 117–126.

Rabe, U., Kopycinska, M., Hirsekorn, S., Saldana, J. M., & Schneider, G. A. (2002). High-resolution characterization of piezoelectric ceramics by ultrasonic scanning force microscopy techniques. *Journal of Physics D: Applied Physics, 35*, 2621.

Swanson, S. R. (2004). Hertzian contact of orthotropic materials. *International Journal of Solids and Structures, 41*, 1945–1959.

Tarumi, R., Nitta, H., Ogi, H., & Hirao, M. (2011). Low-temperature elastic constants and piezoelectric coefficients of langasite ($La_3Ga_5SiO_{14}$). *Philosophical Magazine, 91*, 2140–2153.

Willis, J. R. (1966). Hertzian contact of anisotropic bodies. *Journal of the Mechanics and Physics of Solids, 14*, 163–176.

Yamanaka, K., Tsuji, T., Noguchi, A., Koike, T., & Mihara, T. (2000). Nanoscale elasticity measurement with in situ tip shape estimation in atomic force microscopy. *Review of Scientific Instruments, 71*, 2403.

Zener, C. (1949). Relation between residual strain energy and elastic moduli. *Acta Crystallographica, 2*, 163–166.

Chapter 10
Nonlinear Acoustics for Microstructural Evolution

Abstract Acoustic nonlinearity holds the potential of becoming the primary means of characterizing microstructural evolution caused by, for instance, fatigue of metals, because it is capable of probing the processes of dislocation movement (Hikata et al. (1966); Hikata and Elbaum (1966); Cantrell and Yost (1994)) and crack nucleation and growth (Buck et al. (1978); Richardson (1979); Morris et al. (1979); Nagy (1998)). The sensitivity to microstructural attributes during the incubation period of aging is often higher than that of the linear properties (velocity and attenuation) . However, there are several other nonlinearity sources being related to mechanical contact between material's surface and the transducers. The EMAT has then been expected for the nonlinearity measurement to isolate material's nonlinearity, but it has failed to excite high enough vibration amplitude in the material for inducing higher harmonics. Electromagnetic acoustic resonance (EMAR) again has overcome this problem by superimposing many ultrasonic waves coherently, and as shown later, three methods for nonlinear acoustics have been emerged.

Keywords Creep · Dislocations · EMAR · Higher harmonics · Fatigue · Second power law

10.1 Nonlinear Elasticity

In solid materials, acoustic nonlinearity arises from three sources of diverse scales. They are lattice anharmonicity, energy absorption, and imperfect interfaces. In the atomic scale, the interatomic potential contains nonparabolic terms of interatomic distance, being specific to the crystal. Those terms are the origin of non-Hookean behaviors of solids, including acoustoelastic effects (Chap. 12) and three-wave interaction (Jones and Kobett 1963), and also account for the thermal expansion.

M. Hirao and H. Ogi, *Electromagnetic Acoustic Transducers*,
Springer Series in Measurement Science and Technology,
DOI 10.1007/978-4-431-56036-4_10

If the solid has one or more mechanisms of acoustic damping, the dynamic stress-strain curve shows a hysteresis loop, whose area determines the rate of the energy dissipation per cycle. This implies that the energy absorption accompanies nonlinear phenomena, but not necessarily vice versa. In a scale comparable or larger than grain size, solids sometimes contain imperfect interfaces such as small cracks caused by fatiguing (see Chap. 16) and the separation at interphase boundaries in natural rocks, for instance. The opposite faces are in contact with each other but not well tightly. They open, close, or grind against each other responding to the applied stress and acoustic fields. Compressive phase of longitudinal wave travels across such a weakly bonded interface, but a part of tensile phase does not. This rectifying effect can distort the waveform causing the nonlinearity.

10.2 Second Harmonic Generation

The axial-shear-wave EMAT (Sect. 3.7) on cylindrical specimens was found to be capable of the nonlinearity measurement (Ogi et al. 2001). Note that the axial-shear-wave resonance occurs at uneven frequency intervals (see Figs. 13.4 and 16.10a), and the second harmonics cannot be detected at the double frequency as is usually the case of second harmonic measurement. The procedure taken is as follows:

(i) Measure the resonance frequency f_r and peak height A_1 of the first resonance mode as shown in Fig. 10.1.
(ii) Drive the EMAT coil (meander-line coil or solenoid coil) at half the resonance frequency ($f_r/2$), keeping the input power unchanged. In this case, the driving

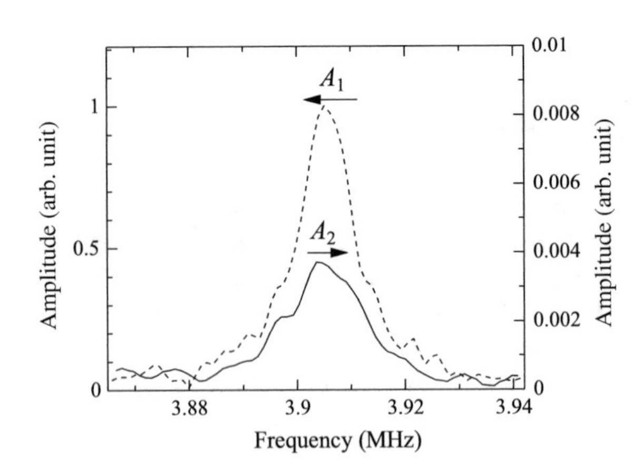

Fig. 10.1 Resonance spectra for the fundamental and the second harmonic components of the axial shear wave (n = 49) (see Sect. 3.7). The specimen was a low-carbon steel rod with 14 mm diameter. The magnetostriction-type EMAT was used with a meander-line coil of 0.9-mm period. Reprinted with permission from Ogi et al. (2001), Copyright (2001), AIP Publishing LLC

frequency does not satisfy the resonance condition, and the fundamental component does not make a detectable signal. However, the second harmonic component having the double frequency (f_r) satisfies the resonance condition and remains with a measurable amplitude. Swept frequency operation on received signal while exciting at $f_r/2$ detects the resonance spectrum that contains a peak of height A_2 at the original resonance frequency.

(iii) Define the amplitude ratio A_2/A_1 as the nonlinearity and use it for materials characterization.

This measuring principle eliminates unwanted nonlinearity, which may otherwise arise from the contacting interfaces between the specimen and transducer. Thus, the nonlinearity only from the material is detected.

The second power law has confirmed this nonlinear measurement. In metals without flaws, the higher harmonics arises from nonlinear elasticity due to lattice anharmonicity and anelasticity due to dislocation movement. These two effects are inseparable in actual nonlinear measurements. Both generate the higher harmonics, among which the second harmonic usually dominates. The second harmonics amplitude should then be proportional to the square of the fundamental wave amplitude (Hikata et al. 1965). This is called *second power law*.

The present definition of nonlinearity A_2/A_1 is based on the second power law and the proportionality between $A_1(f_r)$ and the *true* fundamental amplitude, say $A_1'(f_r/2)$. The second harmonic amplitude A_2 is of course proportional to the square of A_1', that is, $A_2 \propto A_1'^2$. The A_1' component lasts in a very short time after excitation and vanishes through mutual interference in the off-resonance condition. Within this short time, it generates the second harmonics following the second power law. But one cannot measure A_1'. Instead, one can measure the relation between A_1 and A_2, both at f_r, by changing the driving voltage. Figure 10.2 shows the results for an aluminum alloy and 0.25 mass %C steel before and after applying tensile plastic deformation. Both of them demonstrate the linear relationship between A_2 and A_1^2, proving that $A_1 \propto A_1'$. The larger slopes after applying plastic deformation are associated with the dislocation multiplication. The A_1^2-A_2 slope is proportional to the dislocation density Λ times the fourth power of the loop length L (Hikata et al. 1965). Normalization of A_2 in terms of A_1 removes the influences of liftoff, frequency dependence of the transfer efficiency, and other anomalies. Measurement of A_2/A_1 is capable of sensing the metal's fatigue progress as will be discussed in Chap. 16.

10.3 Collinear Three-Wave Interaction

Most often selected method for acoustic nonlinearity relies on the spectral analysis of a pulse signal received after traveling a certain distance through the material and detects the second harmonic generation. Intensity of fundamental harmonics, A_1, and that of the second harmonics of double frequency, A_2, are measured to calculate A_2/A_1^2, which indicates the material's nonlinearity to the first order and is often

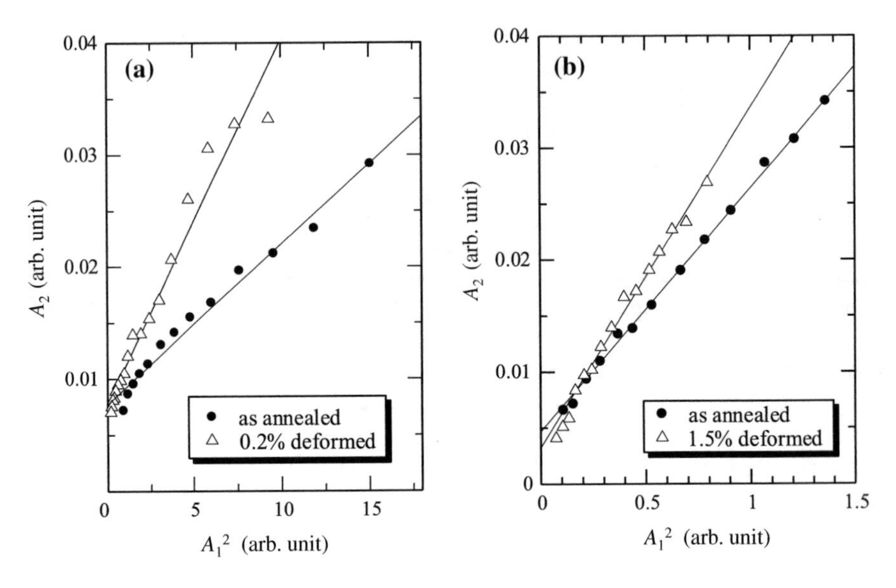

Fig. 10.2 Verification of the second power law before and after plastic elongation for **a** an aluminum alloy rod with 14 mm diameter and **b** a low-carbon steel rod with 14 mm diameter. A Lorentz force-type EMAT (Fig. 3.18) and magnetostriction-type EMAT (Fig. 3.17) were used for **a** and **b**, respectively. Reprinted with permission from Ogi et al. (2001), Copyright (2001), AIP Publishing LLC

called *nonlinear parameter.* Similar measurement is basically available with the thickness resonance, because it occurs at equal intervals of frequency, unlike the axial-shear-wave resonance aforesaid, provided that the medium is homogeneous through the plate thickness. Higher harmonics is then measurable by exciting the resonant oscillation of high enough intensity at one of the resonance frequencies and detecting the amplitude at the frequency small integer times the excitation frequency.

It should be noted that this type of nonlinear acoustic measurements includes an unavoidable problem, whichever contact or noncontact transducer is used, and whichever pulse signal or thickness resonance is detected. Distortion of waveform is more or less caused at the contact between the transducer and the target material, by the instruments in use, and during the signal processing in addition to the material's nonlinearity. These unwanted effects cannot be easily removed as their phases are different from each other. For this reason, absolute elastic nonlinearity is difficult to determine with such a conventional second harmonic measurement.

Ohtani et al. (2015) show a new method of EMAR for acoustic nonlinearity, which uses two shear-wave EMATs placed face to face on both sides of a sample plate to excite thickness resonance at two resonant frequencies, f_m and f_n, at the same position (Fig. 10.3): $f_m = mC_s/(2d)$ with shear wave velocity C_s and thickness d. The resonant orders, m and n, are chosen so that $m \pm n$ do not coincide with any multiple or divisor of m and n. The second-order elastic nonlinearity, if any, generates a third resonance of the order $m \pm n$, in which one of the EMATs can receive

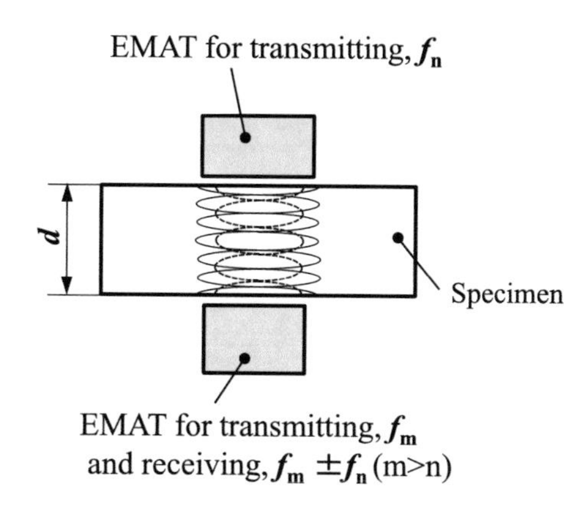

Fig. 10.3 Collinear three-wave interaction of thickness shear resonances. After Ohtani et al. (2015)

after the excitation, while it remains in the plate as reverberation. $A_3/(A_1A_2)$ is the measure of nonlinearity in this case, where A_1 and A_2 are the amplitudes of source resonances, and A_3 is that of the third resonance produced by nonlinear mixing of them. This operation is a collinear version of three-wave interaction worked out by Jones and Kobett (1963) and serves to eliminate unwanted higher harmonics from the nonlinear response.

They applied this measurement to trace the through-life microstructural evolution of crept high-temperature steel alloy (ASME Gr. 122 steel, 11Cr-0.4Mo-2W-CuVNb) and compared with the linear measurements and also with EBSD, SEM, and TEM observations. This alloy steel has been used for boiler components in ultra-supercritical (USC) power plants, which are exposed to steam temperature of 600 °C. The creep life of the welded joints decreases as the result of TypeIV damage at the heat-affected zone (HAZ) due to the finer grain size than the base material. This type of damage develops interior of the metal, and ultrasound is expected to detect the damage state for insuring safety. The resonant spectrum is

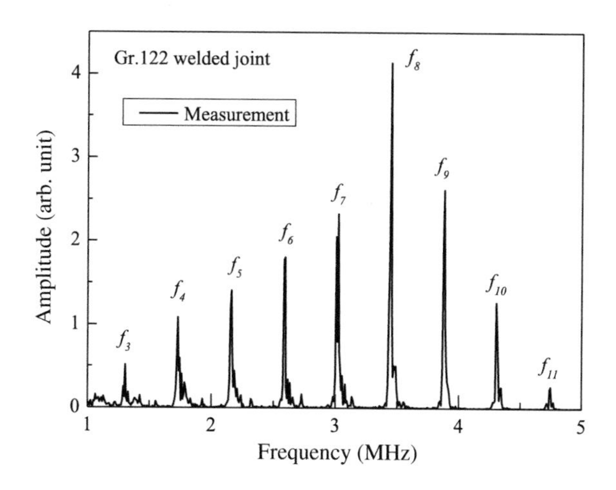

Fig. 10.4 Resonant spectrum at HAZ of ASME Gr. 122 steel specimen of 3.5 mm thick. After Ohtani et al. (2015)

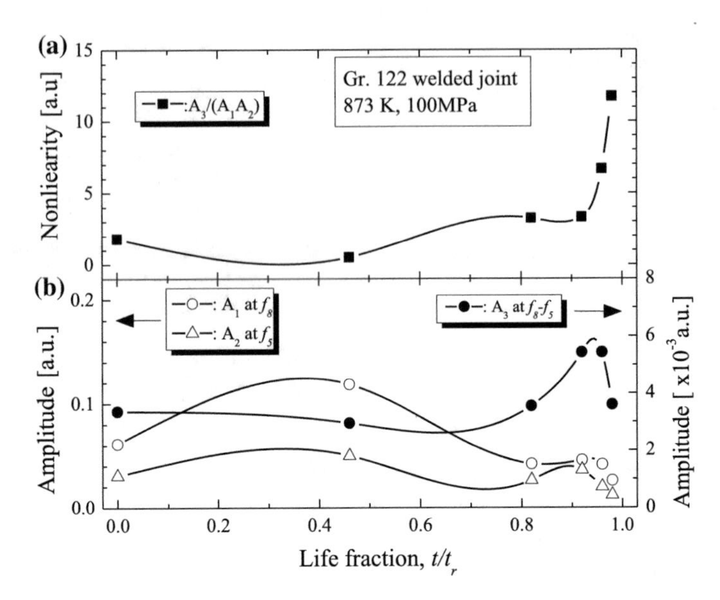

Fig. 10.5 Evolution of **a** $A_3/(A_1A_2)$ and **b** individual resonant amplitudes, A_1, A_2, and A_3 through the creep life of HAZ of ASME Gr. 122 steel specimen. After Ohtani et al. (2015)

shown in Fig. 10.4, where up to eleventh resonance is visible for 3.5-mm-thick specimen. Measurement was done by masking the EMAT coil with copper foil to give a 2×10 mm^2 active aperture for focusing on the HAZ close to weldment. Creep loading of 100 MPa was applied at 600 °C, and the rapture time was 16,340 h. They chose the resonances of $m = 8$ and $n = 5$ as sources and measured the intensity of the third resonance of difference frequency ($m - n = 3$) to find $A_3/(A_1A_2)$. The polarization of shear waves is aligned parallel to the applied stress. A_3 appeared two order of magnitude smaller than A_1 and A_2, but $A_3/(A_1A_2)$ jumped to six times larger value than the original at the final stage, indicating the approach of creep rapture (Fig. 10.5). Other measurements of velocity, attenuation, and NRUS $\Delta f/f$, another nonlinear phenomenon outlined below, are much less sensitive to the creep damage than $A_3/(A_1A_2)$. Dislocation recovery, formation of voids, and initiation of small cracks are the causes of these nonlinear observations.

10.4 Nonlinear Resonant Ultrasound Spectroscopy

Van Den Abeele et al. (2000) have set forth the feasibility of nonlinear resonant ultrasound spectroscopy (NRUS), which is another acoustic resonance method in audible frequency range for detecting material's nonlinearity. Thin rectangular beams of artificial slate used for roofing, sized $200 \times 20 \times 4$ mm^3, were put into resonance in the lowest bending mode by a speaker located 2 cm away from the middle (antinodal) point. Typical resonance frequency was around 300 Hz. The beam samples were suspended at two nodal points with thin nylon wires, which is a similar way of support to Fig. 1.2. An accelerometer attached to one end of the beam, which is another antinode, measured the out-of-plane oscillation, while the excitation frequency was discretely changed over the resonant peak. Their experiments indicated that the resonance curves scaled linearly with the applied amplitude in case of intact samples. The resonance frequency was nearly independent of the excitation voltage. In case of samples damaged by hammer impact, the resonance peak was skewed significantly as the voltage increased. The resonance frequency depended on amplitude and decreased linearly with the amplitude, meaning a softening. Relative frequency shift more than 0.01 was demonstrated. Phenomenological model based on the hysteretic nonlinearity explains the shift of resonance frequency as well as the nonlinear attenuation and harmonic generation observed with damaged quasibrittle material.

Ohtani et al. (2013) extended NRUS measurement to evaluate creep damage in austenitic stainless steel (JIS-SUS304) by incorporating EMAR technique with a shear-wave EMAT. Figure 10.6 compares the 5th resonance peaks around 1.5 MHz between undamaged and damaged samples of 5 mm thick. Creep load of 120 MPa was applied at 600 °C in air; rapture time was 298 h. The driving voltage was changed from 156 to 1560 $V_{p\text{-}p}$ to obtain these resonance curves. The resonance frequency shift induced by the amplitude increase, Δf, was observed even with undamaged sample (Fig. 10.6a), indicating the presence of nonlinear elasticity in the original state. Normalized frequency shift, $\Delta f/f$, gradually increased as the creep proceeded and took a maximum value of 0.0013 at 50 % of creep life, to which the response in Fig. 10.6b corresponds. It then decreased to minimum of 0.0003 around

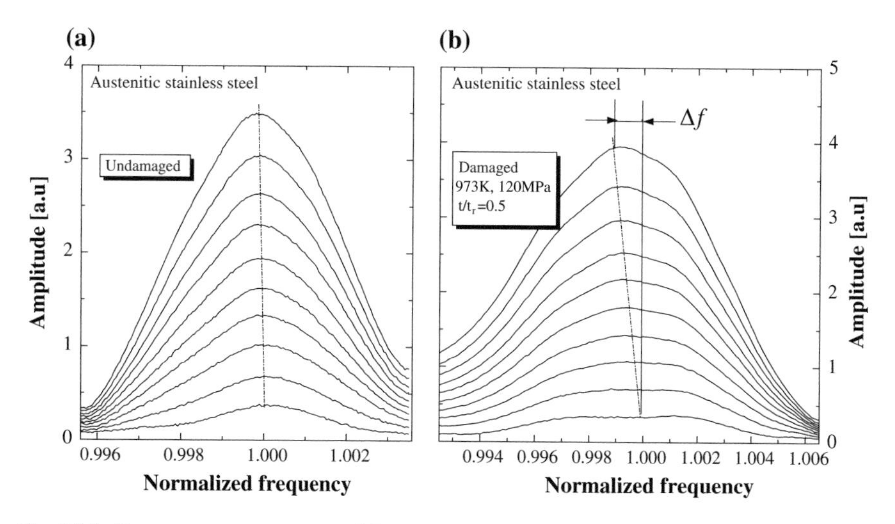

Fig. 10.6 Resonance spectra at different excitation levels **a** from undamaged and **b** creep-damaged austenitic stainless steel. The fifth resonance around 1.5 MHz was detected. Reprinted with permission from Ohtani et al. (2013), AIP Publishing LLC

70 % creep life, turned to increase, and kept increasing till rapture. This behavior was synchronized with the change of attenuation coefficient, suggesting the same mechanism behind. Indeed, TEM observation confirms that the dislocation mobility and restructuring are responsible for the evolution of $\Delta f/f$ and attenuation.

References

Buck, O., Morris, W. L., & Richardson, J. M. (1978). Acoustic harmonic generation at unbonded interfaces and fatigue cracks. *Applied Physics Letters, 33*, 371–373.

Cantrell, J. H., & Yost, W. T. (1994). Acoustic harmonic generation from fatigue-induced dislocations dipoles. *Philosophical Magazine A, 69*, 315–326.

Hikata, A., Chick, B., & Elbaum, C. (1965). Dislocation contribution to the second harmonic generation of ultrasonic waves. *Journal of Applied Physics, 36*, 229–236.

Hikata, A., & Elbaum, C. (1966). Generation of ultrasonic second and third harmonics due to dislocations. I. *Physical Review, 144*, 469–477.

Hikata, A., Sewell, F. A., Jr., & Elbaum, C. (1966). Generation of ultrasonic second and third harmonics due to dislocations. II. *Physical Review, 151*, 442–449.

Jones, G. L., & Kobett, D. R. (1963). Interaction of elastic waves in an isotropic solid. *The Journal of the Acoustical Society of America, 35*, 5–10.

Morris, W. L., Buck, O., & Inman, R. V. (1979). Acoustic harmonic generation due to fatigue damage in high-strength aluminum. *Journal of Applied Physics, 50*, 6737–6741.

Nagy, P. B. (1998). Fatigue damage assessment by nonlinear ultrasonic materials characterization. *Ultrasonics, 36*, 375–381.

Ogi, H., Hirao, M., & Aoki, S. (2001). Noncontact monitoring of surface-wave nonlinearity for predicting the remaining life of fatigued steels. *Journal of Applied Physics, 90*, 438–442.

Ohtani, T., Kusanagi, Y., & Ishii, Y. (2013). Noncontact nonlinear resonant ultrasound spectroscopy to evaluate creep damage in an austenitic stainless steel. In *Review of Progress in Quantitative Nondestructive Evaluation* (Vol. 32, pp. 1227–1233).

Ohtani, T., Honma, T., Ishii, Y., Tabuchi, M., Hongo, H., & Hirao, M. (2015). Evolutions of nonlinear acoustics and microstructural change during creep in Gr. 122 Steel welded joint. *Journal of the Society of Material Science, Japan, 64*, 80–87 (in Japanese).

Richardson, J. M. (1979). Harmonic generation at an unbonded interface. I. Planar interface between semi-infinite elastic media. *International Journal of Engineering Science, 17*, 73–85.

Van Den Abeele, K.-A., Carmeliet, J., Ten Cate, J. A., & Johnson, P. A. (2000). Nonlinear elastic wave spectroscopy (NEWS) techniques to discern material damage, part II: Single-mode nonlinear resonance acoustic spectroscopy. *Research in Nondestructive Evaluation, 12*, 31–42.

Part IV
Industrial Applications

Part IV deals with a variety of individual topics encountered in industrial applications, for which the EMATs are believed to be the best solutions. The last chapter reviews a number of field applications of EMAT techniques.

Chapter 11
Online Texture Monitoring of Steel Sheets

Abstract This chapter describes an EMAT technique for continuously monitoring the drawability ($\bar{r}$-value) of low-carbon steel sheets in a production mode. This is possible because the texture is the common source of the elastic anisotropy (ultrasonic velocities) the plastic anisotropy ($\bar{r}$-values). Traditionally, the r-value has been determined by measuring plastic strains of tensile specimens (Lankford et al. 1950) or resonance frequencies with ribbon coupons (Stickels and Mould 1970). They are destructive measurements. A nondestructive method hopefully a robust tool for online usage has long been needed.

Keywords Cold-rolled steel sheet · Crystallites' orientation distribution function (CODF) · Elastic/plastic anisotropy · Magnetostriction · Meander-line-coil EMAT · Monocrystal elastic stiffness · S_0 mode · Voigt/Reuss/Hill average

11.1 Texture of Polycrystalline Metals

Ultrasonic velocities reveal anisotropy in polycrystalline metals. They change with the propagation and polarization directions relative to the rolling and normal directions in case of a rolled plate, showing an orthorhombic symmetry. The magnitude of the anisotropy is usually small, and the fractional velocity differences are observed on the order of 10^{-2} at most. But, it is large enough to obscure the acoustoelastic effect, which refers to the ultrasonic velocity changes and their anisotropy caused by stress and is capable of measuring residual stress nondestructively (Chap. 12). One of major subjects of acoustoelastic studies has been the separation of the stress-induced anisotropy from the intrinsic anisotropy.

Elastic anisotropy stems from the metal's microstructures. Among others, nonrandom distribution of crystallite orientation, called *texture* or preferred orientation, is the largest cause of intrinsic anisotropy of ultrasonic propagation velocities. During thermomechanical processes, crystallites rotate during crystal slip and the crystal orientations tend to be partially aligned in a certain way. Other aspects including grain size, shape, precipitation particles, and even lattice defects are less

M. Hirao and H. Ogi, *Electromagnetic Acoustic Transducers*,
Springer Series in Measurement Science and Technology,
DOI 10.1007/978-4-431-56036-4_11

influential to the anisotropy when lower measurement frequencies are used. For stress measurements, texture effect should be removed somehow. But, it is natural to take advantage of the sensitivity for sensing the texture.

Any polycrystalline metal contains texture, whether natural or induced. There are several metallic materials where the chemical composition and manufacturing conditions are carefully controlled to develop appropriate texture for specific purposes. One good example is the grain-oriented Fe-3%Si alloy sheets, which have $\langle 001 \rangle$ easy directions of magnetization closely aligned in the rolling direction. Because of the strong magnetic polarization (Morris and Flowers 1981), these electrical sheets are used as core materials for power transformers. Another example is cold-rolled steel sheets used for automobile body parts, beverage cans, etc. Because the sheets are pressed and stretched to form parts, they should have such a plastic anisotropythat they deform easily in the sheet plane with little thickness reduction. Otherwise, they would break during deep drawing and stretching (Davies et al. 1972). An important measure of formability is the $\bar{r}$-value, which is proportional to the limiting drawing ratio. The $\bar{r}$-value is the in-plane average of plastic strain ratios defined by

$$\bar{r} = (r_0 + 2r_{45} + r_{90})/4. \tag{11.1}$$

Here, r_γ is the ratio of plastic strains in the width and thickness directions of a tensile specimen cut and elongated at an angle γ from the rolling direction. Cold rolling followed by annealing tends to make the $\{111\}$ crystallographic planes lie parallel to the rolling plane, resulting in the desired formability, i.e., high $\bar{r}$-value.

11.2 Mathematical Expressions of Texture and Velocity Anisotropy

The polycrystalline aggregate is regarded as a composite material made up of crystallites, whose crystallographic orientations may distribute at random or cluster around particular orientations with respect to the forming geometry. The monocrystal properties and the crystallite orientation distribution determine the gross physical properties. In the harmonic method (Bunge 1982; Roe 1965), the texture is quantified by crystallites' orientation distribution function (CODF). It is a probability density function and defines how the crystallites are oriented statistically in the bulk of a polycrystal. The independent variables are three Euler angles defining the relationship of two Cartesian coordinate systems, one being fixed to a crystallite and the other to the rolled sheet (Fig. 11.1). Being analogous to the Fourier expansion, the CODF is expanded in terms of generalized spherical harmonics (Roe 1965, 1966; Morris and Heckler 1968),

$$w(\xi, \psi, \varphi) = \sum_{l=0}^{\infty} \sum_{m=-l}^{l} \sum_{n=-l}^{l} W_{lmn} Z_{lmn}(\xi) \exp(-im\psi) \exp(-in\varphi), \tag{11.2}$$

$$W_{lmn} = \frac{1}{4\pi^2} \int_0^{2\pi} \int_0^{2\pi} \int_{-1}^{1} w(\xi, \psi, \varphi) Z_{lmn}(\xi) \times \exp(im\psi) \exp(in\varphi) \mathrm{d}\xi \mathrm{d}\psi \mathrm{d}\varphi. \tag{11.3}$$

Here, $Z_{lmn}(\xi)$ is the generalized Legendre function of $\xi = \cos\theta$. The expansion coefficients W_{lmn} represent the contribution of individual harmonics and are called orientation distribution coefficients (ODCs). If account is taken for the orthorhombic symmetry (rolling texture) and the cubic symmetry of constituent crystallites, a large number of ODCs vanish and many others are mutually dependent. The independent ODCs are $W_{000} = 1/\left(4\sqrt{2}\pi^2\right)$, W_{4m0} (m = 0, 2, 4), W_{6m0} (m = 0, 2, 4, 6), etc. The basic ODC, W_{000}, determines the isotropic property. All others contribute to anisotropy.

The CODF is used to weight the orientation-dependent monocrystal quantities and integrate them over the entire space of Euler angles to obtain the macroscopic counterparts of a textured metal. The elastic stiffness tensor is a fourth-rank tensor, whose weighted average C_{ijkl} can be given by

$$C_{ijkl} = \iiint c_{pqrs} a_{pi} a_{qj} a_{rk} a_{sl} w(\psi, \theta, \varphi) \mathrm{d}\psi \mathrm{d}\theta \mathrm{d}\varphi. \tag{11.4}$$

Here, c_{pqrs} is the monocrystal elastic stiffness tensor. $a_{pi}(\psi, \theta, \varphi)$ is the transformation tensor from the specimen coordinate system O-$x_1x_2x_3$ to the crystallite coordinate system O-$X_1X_2X_3$ and has components proportional to cosine and sine functions of the Euler angles. Because of the orthogonality of trigonometric functions, terms of higher order than l = 4 vanish in the integration of Eq. (11.4) and a weighted average of elastic stiffness tensor contains the W_{lmn} coefficients up to the fourth order (l = 4) as independent texture parameter: W_{400}, W_{420}, and W_{440} for cubic systems (Morris 1969; Davies et al. 1972).

The above is the Voigt averaging procedure for the aggregate's (second-order) elastic constants (Sayers 1982). To derive the Reuss average, the monocrystal elastic compliance tensor is weight-averaged in the analogous way and converted to the stiffness. The conversion involves linearization using the fact that the magnitude of the ODCs is on the order of 10^{-2} or less even for highly textured polycrystals. The mathematical expressions of aggregate elastic constants with Voigt, Reuss, and also Hill averages take the same form (Hirao et al. 1987). With either averaging scheme, the elastic constants C_{ij}, in Voigt's two-suffix notation, can be expressed as the sum of Lamé constants (λ and μ) and the anisotropic parts, which are linear combinations of W_{4m0} (m = 0, 2, 4):

$$
\begin{aligned}
C_{11} &= \lambda + 2\mu + \frac{12\sqrt{2}\pi^2 c}{35}\left(W_{400} - \frac{2\sqrt{10}}{3}W_{420} + \frac{\sqrt{70}}{3}W_{440}\right),\\
C_{22} &= \lambda + 2\mu + \frac{12\sqrt{2}\pi^2 c}{35}\left(W_{400} + \frac{2\sqrt{10}}{3}W_{420} + \frac{\sqrt{70}}{3}W_{440}\right),\\
C_{33} &= \lambda + 2\mu + \frac{32\sqrt{2}\pi^2 c}{35}W_{400},\\
C_{44} &= \mu - \frac{16\sqrt{2}\pi^2 c}{35}\left(W_{400} + \sqrt{\frac{5}{2}}W_{420}\right),\\
C_{55} &= \mu - \frac{16\sqrt{2}\pi^2 c}{35}\left(W_{400} - \sqrt{\frac{5}{2}}W_{420}\right),\\
C_{66} &= \mu + \frac{4\sqrt{2}\pi^2 c}{35}\left(W_{400} - \sqrt{70}W_{440}\right),\\
C_{23} &= \lambda - \frac{16\sqrt{2}\pi^2 c}{35}\left(W_{400} + \sqrt{\frac{5}{2}}W_{420}\right),\\
C_{31} &= \lambda - \frac{16\sqrt{2}\pi^2 c}{35}\left(W_{400} - \sqrt{\frac{5}{2}}W_{420}\right),\\
C_{12} &= \lambda + \frac{4\sqrt{2}\pi^2 c}{35}\left(W_{400} - \sqrt{70}W_{420}\right).
\end{aligned}
\tag{11.5}
$$

Here, x_1 is the rolling direction (RD), x_2 the transverse direction (TD), and x_3 the normal direction to the sheet plane (ND) (Fig. 11.1). c stands for the crystal's anisotropy factor, which is a negative quantity for many common cubic crystals. Because W_{4m0} is a small quantity and c has a comparable magnitude with λ and μ, the elastic anisotropy is usually weak. The expressions of λ, μ, and c in terms of the monocrystal moduli depend on the averaging method (V, R, or H); their explicit forms appear in Hirao et al. (1987):

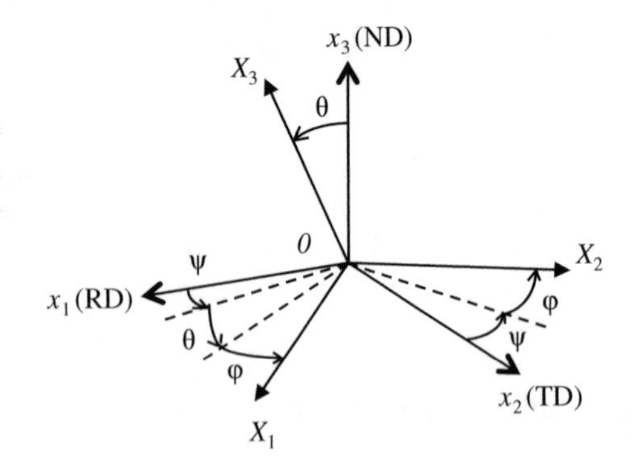

Fig. 11.1 Roe's definition of Euler angles connecting the specimen coordinate system O-$x_1x_2x_3$ to the crystallite coordinate system O-$X_1X_2X_3$

$$\left.\begin{array}{l}
\text{Voigt average:}\\
\quad \lambda+2\mu = c_{11} - 2(c_{11} - c_{12} - 2c_{44})/5\\
\quad \mu = c_{44} + (c_{11} - c_{12} - 2c_{44})/5\\
\quad c = c_{11} - c_{12} - 2c_{44}\\
\text{Reuss average:}\\
\quad \lambda+2\mu = \dfrac{2[s_{11} + s_{12} - (s_{11} - s_{12} - s_{44}/2)/5]}{(s_{11} + 2s_{12})[s_{44} + 4(s_{11} - s_{12} - s_{44}/2)/5]}\\
\quad \mu = [s_{44} + 4(s_{11} - s_{12} - s_{44}/2)/5]^{-1}\\
\quad c = -4\mu^2(s_{11} - s_{12} - s_{44}/2)\\
\text{Hill average:}\\
\quad (\lambda+2\mu)_{\text{Hill}} = \left[(\lambda+2\mu)_{\text{Voigt}} + (\lambda+2\mu)_{\text{Reuss}}\right]/2\\
\quad \mu_{\text{Hill}} = \left(\mu_{\text{Voigt}} + \mu_{\text{Reuss}}\right)/2\\
\quad c_{\text{Hill}} = \left(c_{\text{Voigt}} + c_{\text{Reuss}}\right)/2
\end{array}\right\}, \tag{11.6}$$

For α-Fe, the Hill average gives λ = 277.0 GPa, μ = 81.7 GPa, and c = −136.3 GPa. The same results were obtained by Bunge (1982) using a different notation for ODCs. Li and Thompson (1990a) extended the analysis to textured hexagonal metals.

All the components of C_{ij} take different values in Eq. (11.5), showing orthorhombic symmetry, which is consistent with many measurements, but only six (λ, μ, c, W_{400}, W_{420}, and W_{440}) are independent. It should be noted that three ODCs govern the elastic anisotropy of polycrystalline cubic metals. The dependence of elastic wave velocities on texture had been thought to be more complicated. However, within this range of approximation, the texture contributes to the elastic anisotropy only through three independent fourth-order ODCs. Simplicity is an advantage. We can easily discuss the texture effects on elastic constants and ultrasonic velocities. On the other hand, ultrasonic measurements can provide only rough characteristics of texture pertaining to the three ODCs.

11.3 Relation between ODCs and *r*-Values

The sheet formability encompasses many aspects of mechanical properties such as yield stress, elongation, strain-hardening factor (so-called *n*-value), and plastic anisotropy (*r*-value). The drawing performance is closely related to the *r*-values, which are determined by texture. Basically, the *r*-values depend on all W_{lmn} coefficients. But, the influence diminishes rapidly as l increases. Davies et al. (1972) and Bunge (1982) concluded that the *r*-value is substantially determined through W_{000} and W_{4m0}. The W_{000} coefficient gives r = 1.0 for all directions, over which W_{4m0} superimposes anisotropy.

The $\bar{r}$-value is linked to W_{400}. Both $\bar{r}$ and W_{400} represent the normal anisotropy, or the difference between the in-plane and out-of-plane properties. They have

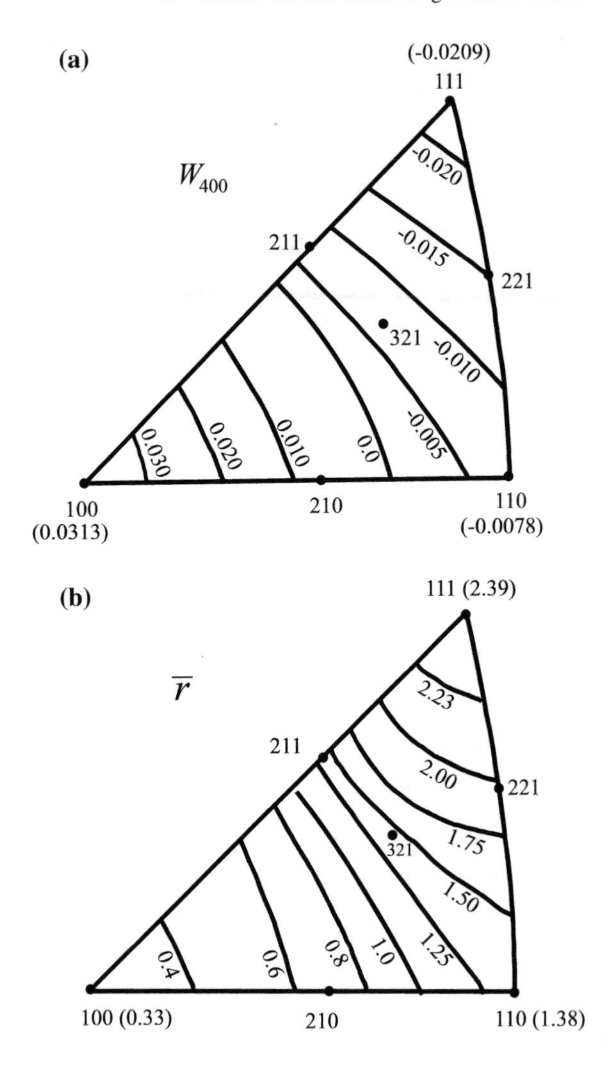

Fig. 11.2 Dependence of W_{400} and $\bar{r}$ on crystallographic plane parallel to the rolling plane in monocrystal α-Fe

nothing to do with the planar anisotropy. Figure 11.2 shows how $\bar{r}$ and W_{400} depend on the crystallographic plane parallel to the rolling plane. For calculating W_{400}, the CODF w is expressed as a combination of 24 delta functions for individual texture components and then substituted into Eq. (11.3). We have adopted the $\bar{r}$-value calculated by Nagashima et al. (1965). They calculated the r-values of monocrystal plates assuming 48 slip systems, all having <111> slip directions, with equal critical resolved shear stress. We observe that $\bar{r}$ and W_{400} relate closely. The patterns of variation are just reversed. When the {111} planes lie within the rolling plane, W_{400} takes its minimum and $\bar{r}$ is largest. Contrarily, for {100} planes, W_{400} takes the maximum and $\bar{r}$ is smallest. Moreover, the isotropic contours of $W_{400} = 0$ and $\bar{r} = 1$ are located at nearly the same place in the unit stereographic triangle.

Because W_{4m0} governs both elastic and plastic anisotropies, the r-values can be estimated by ultrasonic techniques. Any quantity can then be a tool for $\bar{r}$ measurement, provided it isolates W_{400}. Ultrasonic velocities and their combinations available for extracting W_{400} of sheets are as follows (Sayers 1982; Hirao et al. 1988; Hirao and Fukuoka 1989; Li and Thompson 1990b):

(a) Longitudinal velocity traveling in ND:

$$\rho V_{\mathrm{L}}^2 = \lambda + 2\mu + \frac{32\sqrt{2}\pi^2 c}{35} W_{400}. \tag{11.7}$$

(b) Average of shear wave velocities polarized in RD and TD, both traveling in ND:

$$\rho \frac{V_{\mathrm{S1}}^2 + V_{\mathrm{S2}}^2}{2} = \mu - \frac{16\sqrt{2}\pi^2 c}{35} W_{400}. \tag{11.8}$$

(c) Angular dependence of fundamental shear horizontal ($\mathrm{SH_0}$) wave velocity:

$$\rho V_{\mathrm{SH0}}(\gamma) = \mu + \frac{4\sqrt{2}\pi^2 c}{35}\left(W_{400} - \sqrt{70} W_{440} \cos 4\gamma\right). \tag{11.9}$$

(d) Angular dependence of fundamental symmetrical Lamb ($\mathrm{S_0}$) wave velocity:

$$\begin{aligned} V_{\mathrm{S0}}(\gamma) &= V_0 (1 - \Delta)^{1/2} \\ &\quad + \frac{c}{\rho V_0}\left(s_0 W_{400} + s_2 W_{420} \cos 2\gamma + s_4 W_{440} \cos 4\gamma\right). \end{aligned} \tag{11.10}$$

In Eqs. (11.9) and (11.10), the angle γ measures the propagation direction of guided plate modes from the RD. $V_0 = \sqrt{4\mu(\lambda + \mu)/(\lambda + 2\mu)\rho}$ (ρ: density) is the velocity of $\mathrm{S_0}$-mode wave at the low-frequency limit and $\Delta = [\lambda/(\lambda + 2\mu)]^2 (kd)^2/3$ is the correcting term for the velocity decrease caused by dispersion, that is, the finite thickness/wavelength ratio (k wavenumber, d sheet thickness). s_0 and s_2 are functions of Poisson's ratio and $s_4 = 4\pi^2/\sqrt{35} = 6.67$; for steels, $s_0 = 6.08$ and $s_2 = -9.18$. Equations (11.7)–(11.10) can be derived by substituting Eq. (11.5) into the equation of motion and, where necessary, solving it with the boundary condition specific to each propagation mode. They are correct in the first-order approximation regarding the ODCs and Δ.

There are two different approaches for the ultrasonic online $\bar{r}$-value evaluation. First is to use the close relationship between $\bar{r}$ and the velocities (or combination thereof) that contain only W_{400}. A regression curve must be calibrated beforehand from destructive tests. This is more practical, but less rigorous, than the other. The

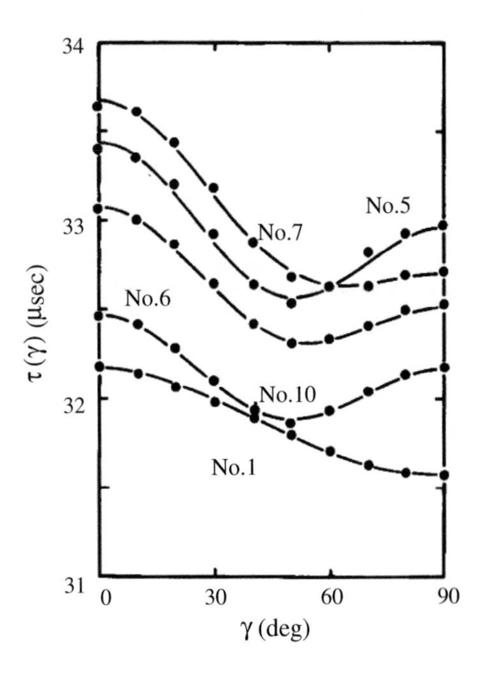

Fig. 11.3 Angular dependence of S_0-mode transit time within the rolling plane of five steel sheets of various $\bar{r}$-values. Curves are the fits with Eq. (11.10) (after Hirao et al. 1989)

main advantage is the fast measurement, a requisite for online applications. In what follows, we describe the method of using transit times of an S_0-mode plate wave instead of velocities to minimize the measuring time. The second approach aims at accurately predicting individual r-values, not only the $\bar{r}$-value, after estimating higher-order ODCs up to $l = 12$ (Bunge 1982; Sakata et al. 1990). It involves decomposing the texture into the principal elements and, via a rule-of-mixture calculation, finding the ODCs that satisfy the CODF-velocity relation based on the elastic energy method. This consumes time.

The close relation between $\bar{r}$ and W_{400} (Fig. 11.2) was experimentally confirmed by measuring the transit times in the S_0 mode. Figure 11.3 presents the angular dependence of the transit time $\tau(\gamma)$ measured with two identical Lorentz force-type EMATs (shown below in Fig. 11.5a) separated 145 mm on commercially produced sheets (Hirao et al. 1989). The operating frequency was 0.7 MHz. The solid lines are fits of the functional form of Eq. (11.10). Good agreement is apparent, which verifies the approximate analysis leading to Eq. (11.10). The planar average of them, τ_{ave}, after being corrected for the dispersion effect, should be proportional to W_{400}:

$$\tau_{ave} = \frac{L}{V_0}\left(1 - \frac{cs_0}{\rho V_0^2} W_{400}\right) + \tau_0. \tag{11.11}$$

Here, L is the propagation distance and τ_0 the constant delay. Figure 11.4a demonstrates the dependence of the $\bar{r}$-value on τ_{ave} and in turn on W_{400}. Figure 11.4b provides X-ray diffraction data for five selected poles, giving a

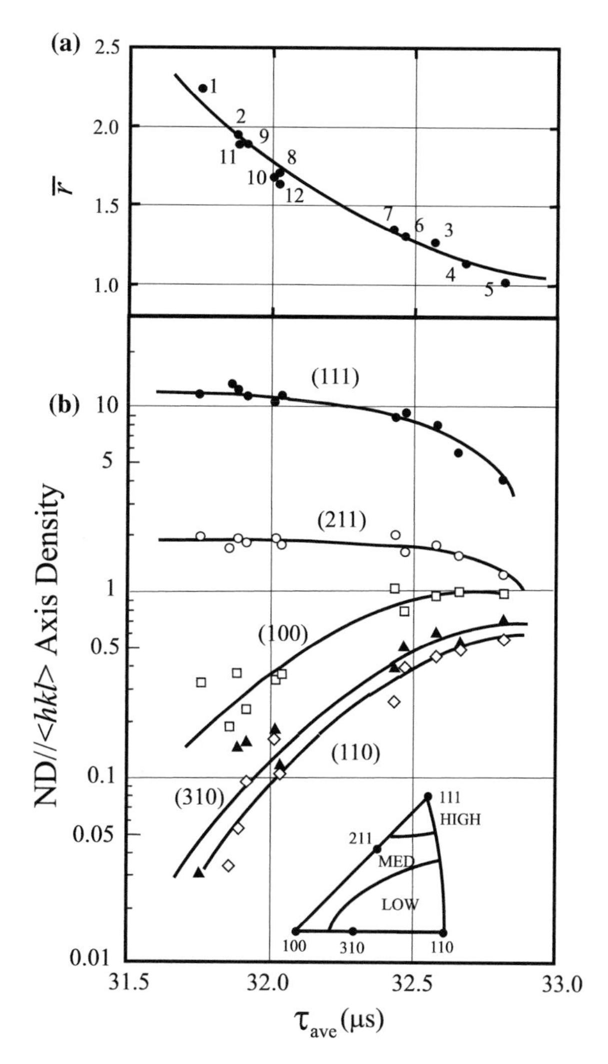

Fig. 11.4 **a** Relation between average transit time τ_{ave} and $\bar{r}$. **b** Relation between τ_{ave} and ND//$\langle hkl \rangle$ axis densities measured with X-ray diffraction technique. The thickness ranged from 0.677 to 0.901 mm (after Hirao et al. 1989)

metallurgical proof of pole density dependence of τ_{ave} or W_{400}. Diffraction intensity has been normalized with the corresponding background obtainable from a standard specimen of random orientation. The intensities of desirable texture components, {111} and {211}, are much larger than those of undesirable components. As τ_{ave} increases and $\bar{r}$ decreases to 1 at the same time, the distribution of sheet normals tends to be random, approaching the isotropic limit. Thus, τ_{ave} detects W_{400}, or the essential feature of the texture, which represents the whole crystallite orientation and is required for the formability characterization.

It might be expected that W_{440} is likewise related to the planar anisotropy $\Delta r = (r_0 - 2r_{45} + r_{90})/2$. However, a theoretical comparison is unavailable between W_{440} and Δr. Measurements imply that they are weakly interrelated (Hirao

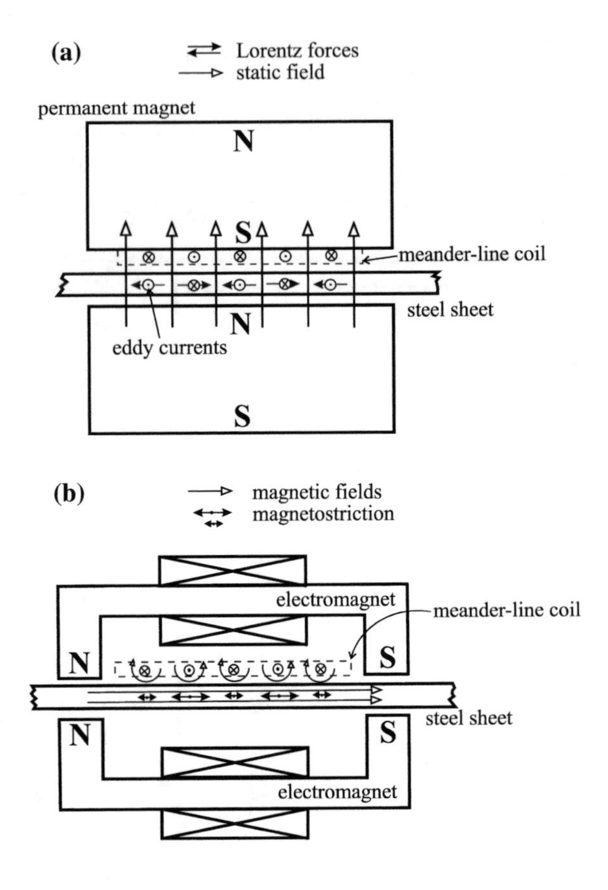

Fig. 11.5 Configuration of **a** Lorentz-force-type EMAT and **b** magnetostrictive-type EMAT to measure the velocity in S_0-mode plate wave along a steel sheet

et al. 1989). This is partly because W_{400} dominates in magnitude for Al-killed and interstitial-free steel sheets, for which the $\bar{r}$-value is one of the major issues of quality assurance. Other possible reasons include the mechanical constraints among deformed grains and the participation of higher-order ODCs.

11.4 Online Monitoring with Magnetostrictive-Type EMATs

Equations (11.7)–(11.10), or equivalent, are the basis of online ultrasonic $\bar{r}$ evaluation. EMATs are chosen commonly to measure the velocities. An EMAT exerts an electromagnetic force to the moving sheets to transmit and receive elastic waves. It allows a clearance between the EMAT face and the sheet surface. Kawashima (1990) and Clark et al. (1992) developed thickness resonance EMAT system, using a dual-mode EMAT with single magnet and a round coil (see Fig. 3.3) to determine V_L, V_{S1}, and V_{S2} simultaneously from a resonance spectrum. W_{400} is given by the bulk-wave velocity ratio $V_L/(V_{S1} + V_{S2})$. An advantage exists in using a single EMAT rather than using three pairs of them to measure the transit times of a plate mode. A compact EMAT allows detecting the inhomogeneous $\bar{r}$ values across the sheet width. However, it needs a long measuring time for sweeping the frequency and seems incompatible with the operation on fast-moving sheets. Use of the SH_0 mode requires only two measurements to obtain W_{400} (see Eq. (11.9)), but it is inadequate in a real production line of ferrous sheets. When the PPM EMAT (Fig. 3.10) is in use, the domain magnetization changes periodically under the PPM assembly when the sheet travels. Rapid movement of domain walls causes a Barkhausen noise, which induces unwanted currents in the sensing coil. To use the magnetostrictive-type EMAT (Fig. 3.12) for the SH_0 mode, we need a strong magnetization along the sheet, which is difficult to be realized for the moving sheet with a liftoff of several millimeters.

A practical EMAT system then relies on using the meander-line-coil EMATs to measure the S_0-mode traveling times at $\gamma = 0°$, 45°, and 90° and to calculate τ_{ave}. Once L and τ_0 in Eq. (11.11) are calibrated, τ_{ave} directly gives the $\bar{r}$-value without calculating the velocities. Using the same coil geometry, Fujisawa et al. (1991) compared a number of EMATs including those in Fig. 11.5; some operate with the Lorentz force and the others with the magnetostrictive force. Both generate a periodic pattern of tensile and compression strains parallel to the sheet plane. The line spacing of the coil is adjusted to equal the S_0-mode's half wavelength at a frequency of 280 kHz. They concluded that the magnetostrictive-type EMAT (Fig. 11.5b) with electromagnets is superior to others, because of three factors:

(i) The signal amplitude becomes maximum at a medium current to the electromagnets.
(ii) The maximum efficiency is nearly independent of the sheet thickness in the 0.56–1.80 mm range.
(iii) This EMAT exhibits sufficient S/N ratio with 5-mm liftoff for the sheet thickness from 0.5 to 2.5 mm. The magnetizing current has to be adjusted to obtain the maximum amplitude.

An online measurement system has been developed using three pairs of magnetostrictive-type EMATs with electromagnets (Fujisawa et al. 1991; Hirao et al. 1993). The EMATs are assembled on a fixture block. This block and another one carrying electromagnets, for strengthening the field and balancing the magnetic forces, hold the center of the sheet from top and bottom with a 3-mm air gap. The meander-line coils lie 5 mm above the sheet surface. Transmitter and receiver EMATs are placed 140 mm apart for $\gamma = 0°$ and 90°; they are separated 290 mm for $\gamma = 45°$. (No correction is made for errors caused by diffraction.) Two pairs of rollers are placed upstream and downstream of the measuring site to stabilize the sheet movement.

Each of the three channels has its own driving and receiving circuits; they are turned on one after another. The transit time is measured by referring to the zero-crossing point in the middle of the received RF signal without signal averaging. The average transit time for three directions is correlated with the $\bar{r}$-value

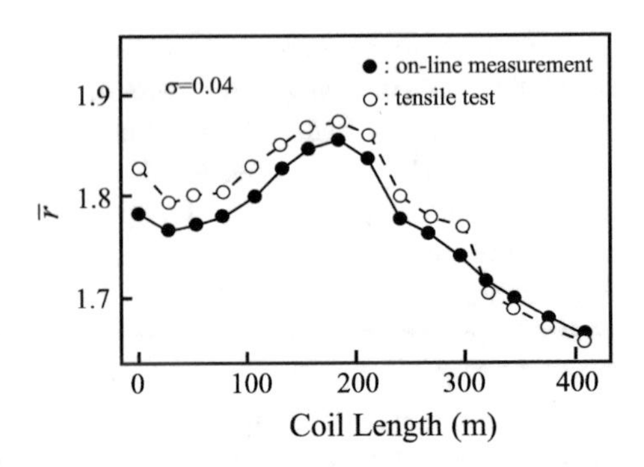

Fig. 11.6 Comparison between online measurement and tensile test of a test sheet (0.8 mm thick). Annealing condition varies along the length

using a regression curve similar to the one in Fig. 11.4a. The measuring time for a channel is 5 ms. The calculation of $\bar{r}$ needs another 5 ms, the total being 20 ms for an output $\bar{r}$. At a line speed of 300 m/min, the sheet travels 10 cm to finish a set of measurements. Usually, 16 measurements are averaged for the final evaluation. Thus, measured $\bar{r}$ is memorized in the controlling computer and displayed real time on a chart or CRT.

The environmental influences in the manufacturing condition are also studied. They are the tensile load applied to the sheet, the sheet temperature, and the change in liftoff. The $\bar{r}$-value changes by −0.001 with 10 MPa tensile stress, −0.1 with 30° temperature rise, and 0.08 with a liftoff increase of 1 mm. These influences were removed for further improved accuracy. Also, the ambient magnetic noise is negligible. It was then concluded that the industrial environment does not affect the $\bar{r}$ evaluation.

The measurement was independent of the sheet speed (and its change) up to 325 m/min, which exceeds the maximum speed in the continuous annealing line of 300 m/min. Standard deviation of 0.055 was obtained from the comparison between the online ultrasonic and tensile values of $\bar{r}$ with various grades of sheets. Figure 11.6 presents another promising result for installation in a production line. It compares the online evaluation and tensile test of $\bar{r}$, when the annealing condition is continuously changed along the length of a test sheet. The two results agree with each other within $\bar{r}$ = 0.04. This noncontact, online $\bar{r}$-value evaluation system was installed at the end of a continuous annealing line, just before the coiling stage in Kashima Steel Works, Sumitomo Metal Industries Ltd., and has been proven satisfactory for operation.

References

Bunge, H. J. (1982). *Texture Analysis in Materials Science*. London: Butterworths.

Clark, A. V., Fortunko, C. M., Lozev, M. G., Schaps, S. R., & Renken, M. C. (1992). Determination of sheet steel formability using wide band electromagnetic-acoustic transducers. *Research in Nondestructive Evaluation, 4*, 165–182.

Davies, G. J., Goodwill, D. J., & Kallend, J. S. (1972). Elastic and plastic anisotropy in deep-drawing steels. *Metallurgical Transactions, 3*, 1627–1631.

Fujisawa, K., Murayama, R., Fukuoka, H., & Hirao, M. (1991). Development of EMAT monitoring system of formability in cold-rolled sheets. In *Nondestructive Characterization of Materials* (Vol. 5, pp. 623–634).
Hirao, M., Aoki, K., & Fukuoka, H. (1987). Texture of polycrystalline metals characterized by ultrasonic velocity measurements. *The Journal of the Acoustical Society of America, 81*, 1434–1440.
Hirao, M., & Fukuoka, H. (1989). Dispersion relations of plate modes in anisotropic polycrystalline sheets. *The Journal of the Acoustical Society of America, 85*, 2311–2315.
Hirao, M., Fukuoka, H., Fujisawa, K., & Murayama, R. (1989). Characterization of formability in cold-rolled steel sheets using electromagnetic acoustic transducers. *Metallurgical Transactions A, 20*, 2385–2392.
Hirao, M., Fukuoka, H., Fujisawa, K., & Murayama, R. (1993). On-line measurement of steel sheet $\bar{r}$-value using magnetostrictive-type EMAT. *Journal of Nondestructive Evaluation, 12*, 27–32.
Hirao, M., Hara, N., Fukuoka, H., & Fujisawa, K. (1988). Ultrasonic monitoring of texture in cold-rolled steel sheets. *The Journal of the Acoustical Society of America, 84*, 667–672.
Kawashima, K. (1990). Nondestructive characterization of texture and plastic strain ratio of metal sheets with electromagnetic acoustic transducers. *The Journal of the Acoustical Society of America, 87*, 681–690.
Lankford, W. T., Snyder, S. C., & Bauscher, J. A. (1950). New criteria for predicting the press performance of deep drawing sheets. *Transactions ASM, 42*, 1197–1232.
Li, Y., & Thompson, R. B. (1990a). Relations between elastic constants C_{ij} and ODCs for hexagonal materials. *Journal of Applied Physics, 67*, 2663–2665.
Li, Y., & Thompson, R. B. (1990b). Influence of anisotropy on the dispersion characteristics of guided ultrasonic plate modes. *The Journal of the Acoustical Society of America, 87*, 1911–1931.
Morris, P. R. (1969). Averaging fourth-rank tensors with weight functions. *Journal of Applied Physics, 40*, 447–448.
Morris, P. R., & Flowers, J. W. (1981). Texture and magnetic properties. *Texture of Crystalline Solids, 4*, 129–141.
Morris, P. R., & Heckler, A. J. (1968). Crystallite orientation analysis for rolled cubic materials. In *Advances in X-ray Analysis* (Vol. 11, pp. 454–472). New York: Plenum Press.
Nagashima, S., Taketi, H., & Kato, H. (1965). Plastic anisotropy of polycrystalline iron with rolling texture. *Journal of the Japan Institute of Metals, 29*, 393–398 (in Japanese).
Roe, R.-J. (1965). Description of crystalline orientation in polycrystalline materials. III. General solution to pole figure inversion. *Journal of Applied Physics, 36*, 2024–2031.
Roe, R.-J. (1966). Inversion of pole figures for materials having cubic crystal symmetry. *Journal of Applied Physics, 37*, 2069–2072.
Sakata, K., Daniel, D., Jonas, J. J., & Bussière, J. F. (1990). Acoustoelastic determination of the higher-order orientation distribution function coefficients up to l = 12 and their use for the on-line prediction of r-value. *Metallurgical Transactions A, 21*, 697–706.
Sayers, C. M. (1982). Ultrasonic velocities in anisotropic polycrystalline aggregates. *Journal of Physics. D: Applied Physics, 15*, 2157–2167.
Stickels, C. A., & Mould, P. R. (1970). The use of young's modulus for predicting the plastic-strain ratio of low-carbon steel sheets. *Metallurgical Transactions, 1*, 1303–1312.

Chapter 12
Acoustoelastic Stress Measurements

Abstract Within the framework of linear elasticity, the principal of superposition holds and the elastic wave velocity is independent of the stress acting in the solids. But, this is an approximation after all and solid materials exhibit nonlinear responses; the velocities vary in proportion to stress although in very small magnitude in general. This phenomenon, called acoustoelasticity, originates from the anharmonic interatomic potential. This chapter provides the theoretical background of acoustoelasticity and a number of measurement result with EMATs on applied and residual stresses of industrial importance.

Keywords Elastic wave velocity · Photoelasticity · Shear wave polarization · Second-/third-order elastic constants · Texture effect

12.1 Higher-Order Elasticity

We refer to the stress-induced change of elastic wave velocities as *acoustoelasticity*. It originates from nonlinear elasticity or lattice anharmonicity, where the interatomic potential is no longer a parabolic function of interatomic distance. Within linear elasticity, the principle of superposition holds. The elastic wave propagation is independent of the stress field and the velocities would never change with stress application.

Including the higher-order elasticity, the elastic strain energy U is expanded into polynomials of strain ε as

$$U = \frac{1}{2}a\varepsilon^2 + \frac{1}{6}b\varepsilon^2 + \ldots, \tag{12.1}$$

which renders a non-Hookean stress–strain relation:

$$\sigma = a\varepsilon + \frac{1}{2}b\varepsilon^2. \tag{12.2}$$

M. Hirao and H. Ogi, *Electromagnetic Acoustic Transducers*,
Springer Series in Measurement Science and Technology,
DOI 10.1007/978-4-431-56036-4_12

The first term represents linear elasticity with the second-order elastic constant a. The second term represents the deviation from the linearity and the third-order elastic constant b determines the degree of deviation. (Actually, there are two second-order elastic constants and three third-order elastic constants in isotropic solids; three and six independent elastic constants in cubic crystals.) Equation (12.2) yields the strain-dependent elastic constant

$$\frac{\partial^2 U}{\partial \varepsilon^2} = a + b\varepsilon + \ldots, \tag{12.3}$$

from which the elastic wave velocity V is approximated to be

$$V = V_0(1 + C_{\mathrm{E}}\sigma). \tag{12.4}$$

V_0 is the stress-independent velocity. The coefficient C_{E}, expressible as a linear combination of a and b, is the stress sensitivity of velocity. In this range of approximation, nonlinear elasticity causes a linear dependence of V on stress σ; it is the case in actual measurements. Since the strain ε is on the order of 10^{-3} at the elastic limit of common metals, the acoustoelastic effect is small in magnitude.

The above simplified explanation relates the acoustoelastic response to the material's nonlinearity. A complete theory should contain other sources of nonlinearity. They include the square terms in the strain tensor expressed with displacement gradient (geometrical nonlinearity), the density change with the mean normal stress, and, where necessary, the boundary conditions to be satisfied at deformed conditions. Thorough theoretical description of acoustoelasticity can be found in Hughes and Kelly (1953), Thurston (1964, 1974), Eringen and Şuhubi (1975), and Pao et al. (1984). For the definition of the third-order elastic constants, Murnaghan (1951), Toupin and Bernstein (1961), and Brugger (1964) made major contributions. Readers requiring information on nonlinear elastic waves in general should consult the monographs by Landau and Lifshitz (1956) and Green (1973). Green gave a comprehensive review on the topics, including measurement results prior to the early 1970s.

Interest in acoustoelasticity is twofold. In solid-state physics, the third-order elastic constants are the fundamental quantities characterizing the interatomic bonding and representing the anharmonic behaviors of crystals. They are related to the thermal expansion coefficient, specific heat, and Grüneisen parameter. Acoustoelasticity, at the same time, provides a unique and attractive means for nondestructive stress determination. Although many other techniques detect the surface stresses, it can determine the interior stresses of solid materials, that is, the average along the path of ultrasound. The only alternative to acoustoelasticity is neutron diffraction, which is rarely accessible and limited to laboratory studies.

Acoustoelasticity bears a close analogy with photoelasticity. The ultrasonic shear waves and light waves behave similarly. When a plate of isotropic transparent material (glasses and polymers) is loaded in a plane stress condition, the plane-polarized monochromatic light, passing through the plate thickness, splits into two polarizations in the principal stress directions. This phenomenon occurs, because the stress makes the material optically anisotropic and the two components travel with different velocities; this is the birefringent effect (double refraction). Exactly the same occurs with the ultrasonic shear waves. Acoustoelasticity, in the narrow sense, then means the shear wave birefringence. (In a broad sense, it comprises the stress-induced velocity changes of all the elastic wave modes, bulk waves, and guided waves.) In photoelasticity, the phase difference between the two components causes mutual interference, resulting in the two-dimensional pattern of light intensity, which visualizes the stress field with polarizing plates. This makes the basis of experimental stress analysis with models. Such a stress measurement is, however, impossible with acoustoelasticity. The maximum velocity anisotropy is on the order of 10^{-2} in both of them. But, there is a huge gap of frequency in use, as large as order of 10^7, and the phase difference of the polarized shear waves cannot produce a measurable interference, as has been discussed by Crecraft (1967) and Hsu (1974). The acoustoelastic stress measurements should then rely on the velocity variation itself (V) by sensing the change in either the time-of-flight ($\sim L/V$), phase ($\sim fL/V$), or resonance frequency ($\sim V/L$).

12.2 Acoustoelastic Response of Solids

Dislocation damping may associate a transient change of ultrasonic velocities when the stress varies (see Sect. 7.2 for pure copper). But, structural metals contain many obstacles to suppress dislocation movement and dislocation-induced velocity change can be ignored. The major competing effect is the texture anisotropy. In polycrystalline metals, nonrandom distribution of crystallographic orientation produces intrinsic anisotropy of elastic constants (Chap. 11). They exhibit orthorhombic symmetry in rolled plates, for example, with the principal axes along the rolling, transverse, and thickness directions. The relative magnitude is limited to order of 10^{-2}, but it well exceeds the acoustoelastic response. Separation from the texture-induced anisotropy has been a long-running issue in this field. As a potential solution, King and Fortunko (1983) proposed the SH-wave approach, which makes use of different effects of texture and stress on obliquely propagating SH waves. Thompson et al. (1984) presented a texture-independent stress measurement by exchanging the propagation and polarization directions of surface-skimming SH waves. This exchange removes the texture-induced anisotropy in velocities owing to the elastic constant symmetry (e.g., $C_{1212} = C_{2121}$), but leaves the velocity anisotropy proportional to the difference of the in-plane stresses. The rate of proportionality is given by the shear modulus, μ. However, these methods postulate homogeneous stress and texture over the measuring volume, which is not always the case. Usefulness is limited by a tight trade-off between the spatial resolution and the required accuracy. After all, birefringence acoustoelasticity has been most often used in actual stress measurements on plate-like objects, where the texture effect remains to be separated.

Traditionally, piezoelectric transducers were used to measure the time-of-flight of shear wave echoes. To cope with the low sensitivity to stress, lengthy preparation was needed to finish the surfaces in a high grade to assure the thinnest and the stablest coupling layer of highly viscous fluid, thereby achieving the required accuracy of velocity determination. A critical concern was the signal distortion caused by internal reflections in the transducers (see Fig. 1.1a). They were mixed up with the outgoing reflection signals with some delays and disturbed the time phase, from which the acoustoelastic effect was detected. This occurred at every reflection from the transducer-material interface. The influence thus accumulated in multi-reflected signals. This error source was inevitable with contact transduction and ironically increased with the transfer efficiency. Further difficulties aroused when the path length was not long enough as encountered with the thin-walled structures such as aircraft and fuel-storage vessels. The successive echoes overlapped each other even if high-frequency, broad-band signals were used. These are the main reasons why acoustoelasticity has not been accepted as a standard technique of stress measurement despite the prolonged studies and many attractive features.

Noncontact measurements with EMATs have made acoustoelastic stress measurements easily accessible, robust, and maneuverable. They allow higher accuracy and less time/labor consumption, providing enough signal intensity is attained. The EMATs are free from the problems and error sources caused by intimate contact. Incorporation into the resonance measurement (EMAR; see Chap. 5) compensates for the weak coupling efficiency and also permits one to measure stresses in thin plates.

Iwashimizu and Kubomura (1973) discussed the bulk-wave velocities in a stressed solid with a weak and uniform texture-induced anisotropy. They started the analysis on superimposed elastic waves with the equation of motion:

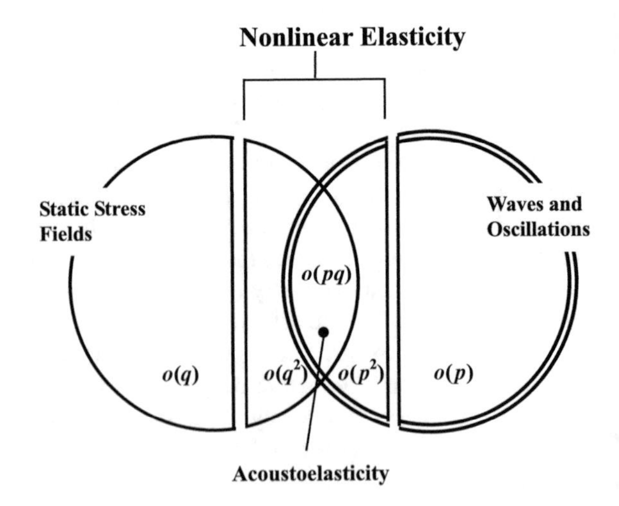

Fig. 12.1 Order of magnitude of acoustoelastic effect and surrounding fields. $o(p)$ and $o(q)$ symbolize the magnitudes of dynamic and static deformations involved, respectively

$$\rho \ddot{u}_i = \frac{\partial}{\partial x_j}\left\{ \left(T_{ijkl} + \delta_{ik}\sigma_{jl}\right)\frac{\partial u_k}{\partial x_l}\right\}. \tag{12.5}$$

Here, T_{ijkl} denotes the effective elastic constant of a stressed anisotropic solid. It has the same symmetry relations as the elastic constant tensor. $\rho = \rho_0(1 + S_{ii})$ is the density at the deformed state (ρ_0: initial density), u_i the displacement vector, σ_{ij} the stress tensor, S_{ij} the strain tensor, and δ_{ij} the Kronecker delta. The main part of T_{ijkl} consists of the generalized Hooke's law with Lamé constants λ and μ. It is completed by adding small but primary perturbations; the deviation from the isotropic elasticity $\overline{C}_{ijkl}$ and the acoustoelastic effect. The explicit form of T_{ijkl} is given as

$$\begin{aligned} T_{ijkl} &= \lambda\delta_{ij}\delta_{kl} + \mu\left(\delta_{ik}\delta_{jl} + \delta_{il}\delta_{jk}\right) + \overline{C}_{ijkl} \\ &+ \left\{\lambda\delta_{ij}\delta_{kl} + \mu\left(\delta_{ik}\delta_{jl} + \delta_{il}\delta_{jk}\right)\right\}S_{mm} + 2\lambda\left(\delta_{ij}S_{kl} + \delta_{kl}S_{ij}\right) \\ &+ 2\mu\left(\delta_{ik}S_{jl} + \delta_{jk}S_{il} + \delta_{il}S_{jk} + \delta_{jl}S_{ik}\right) + C_{ijklmn}S_{mn}. \end{aligned} \tag{12.6}$$

For the infinitesimal strain tensor S_{ij} and the third-order elastic constants of an isotropic solid, (ν_1, ν_2, ν_3),

$$\begin{aligned} C_{ijklmn} &= \nu_1\delta_{ij}\delta_{kl}\delta_{mn} + \nu_2\left\{\delta_{ij}(\delta_{km}\delta_{ln} + \delta_{kn}\delta_{lm}) + \delta_{kl}\left(\delta_{im}\delta_{jn} + \delta_{in}\delta_{jm}\right)\right. \\ &\left. + \delta_{mn}\left(\delta_{ik}\delta_{jl} + \delta_{il}\delta_{jk}\right)\right\} + \nu_3\left\{\delta_{ik}\left(\delta_{jm}\delta_{ln} + \delta_{jn}\delta_{lm}\right)\right. \\ &\left. + \delta_{jl}(\delta_{im}\delta_{kn} + \delta_{in}\delta_{km}) + \delta_{jk}\left(\delta_{im}\delta_{jn} + \delta_{in}\delta_{lm}\right)\right\}. \end{aligned} \tag{12.7}$$

Equation (11.5) provides $\overline{C}_{ijkl}$ in terms of the three orientation distribution coefficients (ODCs) for cubic polycrystalline materials.

Figure 12.1 shows the order of magnitude of acoustoelastic effect compared with the surrounding fields in the case of isotropic solids. The orders of magnitude of dynamic and static fields are represented by $o(p)$ and $o(q)$, respectively.

The basic Eq. (12.5) is linear for the static deformation (S_{ij}) and the superimposed dynamic displacement gradient $\left(\partial u_i/\partial x_j\right)$. But, it contains cross-terms between them of $o(pq)$, assuming that $o(pq)$ and $\left|\overline{C}_{ijkl}\right|/\mu$ are comparable. The nonlinear acoustic effects of $o(p^2)$ (Chap. 10) and the texture-induced anisotropy in C_{ijklmn} (Johnson 1982) belong to higher-order perturbations, which we neglect.

12.3 Birefringence Acoustoelasticity

It is straightforward to calculate the propagation velocities of plane bulk waves in a uniformly deformed solid. Substituting a harmonic solution for u_i, Eq. (12.5) reduces to the usual Christoffel equation, whose eigenvalues give the (phase) velocities and the eigenvectors give the polarization directions (Musgrave 1970;

Auld 1973). Care is taken to retain the terms in $o(p)$ and $o(pq)$, and those proportional to $\overline{C}_{ijkl}$.

We now confine the analysis to the case where the solid exhibits a weak orthotropic symmetry, like a rolled plate, in the stress-free state. Taking the Cartesian coordinate system O-$x_1x_2x_3$ oriented along the symmetry axes, we consider the bulk-wave propagating in the x_3 direction, which coincides with one of the principal stress directions. Another principal direction, accompanied by the principal stress σ_1, makes an angle θ from the x_1 axis (see Fig. 12.2). In this limited but practical situation, the bulk-wave velocities change with the applied (or residual) stress through

$$
\begin{aligned}
\rho_0 V_{\mathrm{L}}^2 &= \lambda + 2\mu + (\lambda + \nu_1 + 2\nu_2)S_{ii} + 2(2\lambda + 5\mu + 2\nu_2 + 4\nu_3)S_{33} + \overline{C}_{33}, \\
\left.\begin{array}{l}\rho_0 V_{\mathrm{S1}}^2 \\ \rho_0 V_{\mathrm{S2}}^2\end{array}\right\} &= \mu + (\lambda + \mu + \nu_2 + \nu_3)S_{ii} + (2\mu + \nu_3)S_{33} + \frac{1}{2}\left(\overline{C}_{44} + \overline{C}_{55}\right) \\
&\quad \mp |\mu + \nu_3| \sqrt{\left\{\frac{\overline{C}_{55} - \overline{C}_{44}}{2(\mu + \nu_3)} + (S_{11} - S_{22})\right\}^2 + 4S_{12}^2}.
\end{aligned}
\tag{12.8}
$$

Voigt's contracted notation is used to write $\overline{C}_{ijkl}$. These equations are the bases of determining the third-order elastic constants with uniaxial loading tests, for which the material's isotropy, $\overline{C}_{ijkl} = 0$, is assumed (see Eq. 7.5). We observe that the resultant anisotropy from the texture and stress effects splits the shear wave into two orthogonally polarized components of different velocities. Because of the texture anisotropy, the shear wave velocities are not always proportional to stresses.

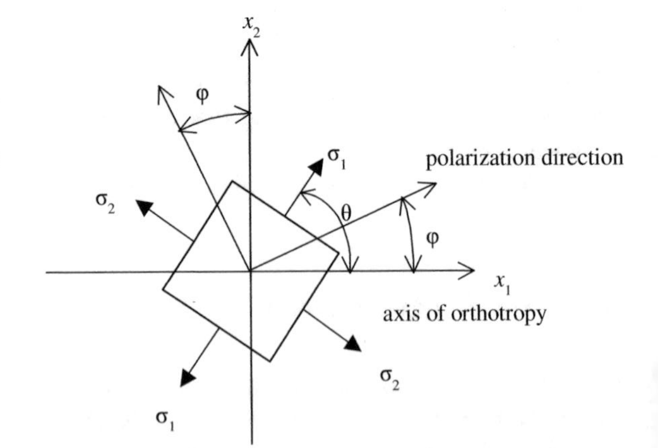

Fig. 12.2 Two-dimensional coordinate system aligned along the axes of orthotropic-texture anisotropy with the principal stress directions and the shear wave polarization

The anisotropy of shear wave velocities is normalized by $V_0 = \sqrt{\mu/\rho_0}$ to give the acoustic birefringence B

$$
\begin{aligned}
B &\equiv \frac{V_{S1} - V_{S2}}{V_0} \\
&= \sqrt{B_0^2 + 2B_0 C_A(\sigma_1 - \sigma_2)\cos 2\theta + C_A^2(\sigma_1 - \sigma_2)^2}.
\end{aligned} \tag{12.9}
$$

Here, B_0 denotes the initial anisotropy due to texture and C_A denotes the birefringence acoustoelastic constant;

$$
B_0 = \frac{\overline{C}_{55} - \overline{C}_{44}}{2\mu}, \quad C_A = \frac{1}{2\mu}\left(1 + \frac{\nu_3}{\mu}\right). \tag{12.10}
$$

The usual procedure is to replace V_0 with $(V_{S1} + V_{S2})/2$ without losing accuracy. The shear wave polarization, measured from the x_1 axis, is provided by an angle φ that satisfies

$$
\tan 2\varphi = \frac{C_A(\sigma_1 - \sigma_2)\sin 2\theta}{B_0 + C_A(\sigma_1 - \sigma_2)\cos 2\theta}. \tag{12.11}
$$

The polarization direction also depends on both of the texture-induced and stress-induced anisotropy.

The acoustic birefringence B thus depends on the initial anisotropy B_0 and the principal stress difference, $\sigma_1 - \sigma_2$. The velocity anisotropy is not directly linked to the stress effect because of the presence of the initial anisotropy. When $\sigma_1 - \sigma_2 = 0$, we have $B = B_0$ and $\varphi = 0, \pi/2$. There is then no stress effect on the birefringence and the shear wave is polarized in the principal axes of the initial anisotropy (the x_1 and x_2 directions). If $B_0 = 0$, $B = C_A(\sigma_1 - \sigma_2)$, and $\varphi = \theta$, $\theta + \pi/2$, then the shear wave is polarized in the principal stress directions. Note that B_0 measures one of the orientation distribution coefficients (ODCs) introduced in Eqs. (11.3) and (11.4) to quantitatively express the texture:

$$
B_0 = \frac{16\pi^2 c}{7\sqrt{5}\mu} W_{420}. \tag{12.12}
$$

When $\theta = 0$, that is, the principal stress directions coincide with the axes of initial orthotropy, Eq. (12.9) degenerates to

$$
B = B_0 + C_A(\sigma_1 - \sigma_2). \tag{12.13}
$$

Then, B is a simple sum of initial anisotropy and the acoustoelastic effect. At the same time, Eq. (12.11) shows $\varphi = 0, \pi/2$; that is, the maximum and minimum shear wave velocities can be found with the polarization in the principal directions. This is the *acoustoelastic birefringence formula* and the basis of many practical stress measurements. (Okada (1980) has independently derived the same formula using the analogy with the index matrix of optical refraction, whose principal values give inverse velocities. In the limit of the weak anisotropy, his result reproduces Eq. (12.13).)

Ultrasonic experiments demonstrate the linear stress dependence of B with many structural metals, supporting Eq. (12.13). Figure 12.3 plots B measured with EMAR as a function of uniaxial stress (σ_1) on a steel and on pure titanium. It is worth noting that the heat treatment on the steel totally alters the offset B_0, but not the slope C_A. Heat treatment may modify the crystallite orientation to make different texture, while the elastic constants originate in the atomic bonding and remain unchanged. Roughly, the calibrated values of C_A are -8×10^{-6} MPa^{-1} for

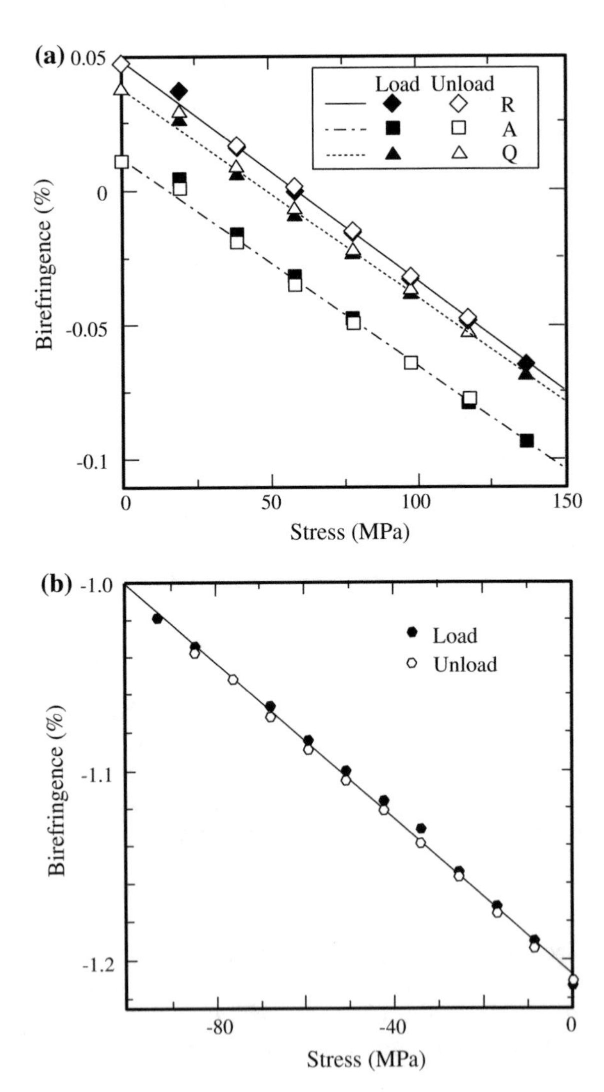

Fig. 12.3 Shear wave birefringence measurements with EMAR on **a** SCM440 steel and **b** pure titanium. *R* as-received, *A* annealed, and *Q* quenched and tempered. *Solid* and *open marks* indicate the data during loading and unloading, respectively

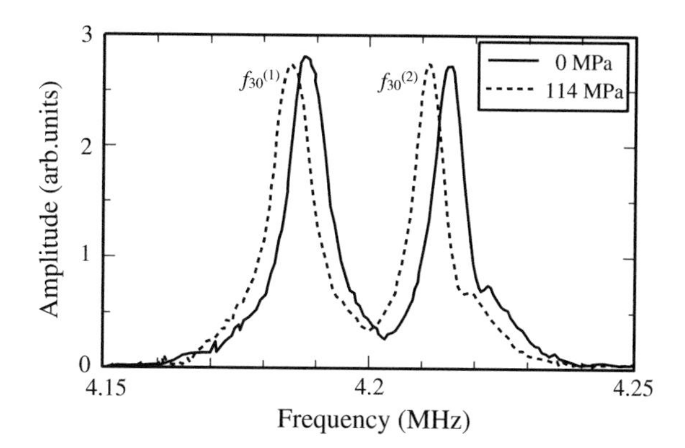

Fig. 12.4 EMAR spectra at 0 and 114 MPa tensile stress with SUS304 stainless steel. Thickness was 12 mm. A shear-wave EMAT was oriented by 45° from the stress direction to simultaneously detect the two orthotropic polarizations

ferritic steels, -4×10^{-5} to -6×10^{-5} MPa^{-1} for aluminum alloys, and -2×10^{-5} MPa^{-1} for titanium. This means that 10-MPa resolution needs an accuracy of measuring B on the order of 10^{-5} for steel products; ignorance of, say, $B_0 = 0.1$ % results in more than 100 MPa error. Austenitic stainless steels are not suitable for birefringence measurements, because V_{S1} and V_{S2} vary in the same direction as shown in Fig. 12.4, giving small C_A on the order of 10^{-7} MPa^{-1}.

Measurement of B thus leads to the determination of the principal stress difference with known C_A and B_0. Separation of them is possible by incorporating the longitudinal velocity V_L measured at the same measuring points. Within the range of approximation used so far, Eq. (12.8) yields the longitudinal/shear velocity ratio, R, for $\theta = 0$:

$$R \equiv \frac{V_L}{(V_{S1} + V_{S2})/2} = R_0 + C_R(\sigma_1 + \sigma_2), \tag{12.14}$$

with another acoustoelastic constant C_R. Combination of Eqs. (12.13) and (12.14) then permits individual determination of σ_1 and σ_2. For isotropic solids, $R_0 = \sqrt{(\lambda + 2\mu)/\mu}$. For anisotropic solids, $\overline{C}_{33}, \overline{C}_{44}, \text{and} \overline{C}_{55}$ enter R_0, which depends only on W_{400} (see Eqs. (11.7) and (11.8));

$$R_0 = \sqrt{\frac{\lambda + 2\mu}{\mu}} \left\{ 1 + \frac{\lambda + 4\mu}{\mu(\lambda + 2\mu)} \frac{8\sqrt{2}\pi^2 c}{35} W_{400} \right\}. \tag{12.15}$$

12.4 Practical Stress Measurements with EMAR

The EMAR method has realized noncontact acoustoelastic stress measurements. The following sections are devoted to describing examples of practical stress measurements, mainly with birefringence acoustoelasticity. It measures the resonance frequencies, not the velocities, and Eqs. (12.13) and (12.14) are rewritten using $f_n = nV/2d$ as

$$B = \frac{f_n^{(1)} - f_n^{(2)}}{\left(f_n^{(1)} + f_n^{(2)}\right)/2} = B_0 + C_A(\sigma_1 - \sigma_2),$$
$$R = \frac{n}{m}\frac{f_m^{(3)}}{\left(f_n^{(1)} + f_n^{(2)}\right)/2} = R_0 + C_R(\sigma_1 + \sigma_2). \tag{12.16}$$

Here, $f_n^{(1)}$ and $f_n^{(2)}$ are the nth shear-mode resonance frequencies with the polarization in the principal stresses σ_1 and σ_2 (see Fig. 12.5). $f_m^{(3)}$ is the mth longitudinal-mode resonance frequency. Measurement of higher resonance modes is advisable to magnify the sensitivity to stress and to lessen the diffraction artifacts. Use of B and R has definite merits:

(i) It takes advantage of relatively large values for C_A and C_R.
(ii) The thickness disappears, which may change with stress and from point to point.
(iii) Subtracting the two measurements in B removes the common errors arising from the temperature and liftoff variations.

Required are parallel surfaces and the orientation alignment between the orthotropic texture and the principal stress directions ($\theta = 0$).

12.4.1 Axial Stress of Railroad Rails

Installed railroad rails suffer abrupt temperature changes. They are fixed to the crossties and then temperature changes create axial stresses in the rails through the constraint to the thermal expansion/contraction. During winter nights, the thermally induced tensile stress promotes the growth of fatigue cracks and causes the weldment separation in continuously welded rails. In summer days, they are exposed to the open air and solar heat. Temperature rise can build up sufficient compressive

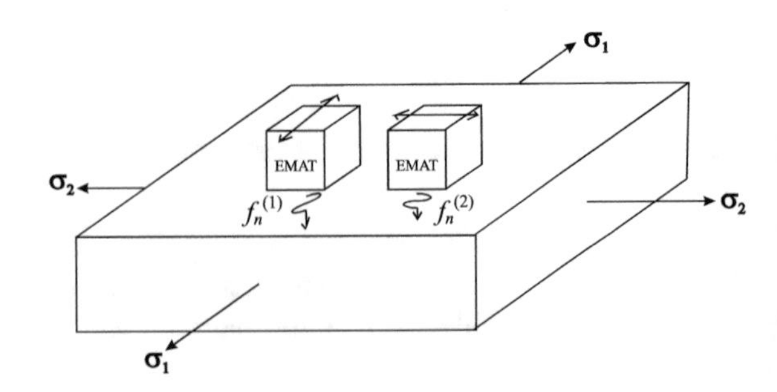

Fig. 12.5 Polarized shear waves to measure acoustic birefringence in the plane stress condition

axial stress to trigger hazardous track buckling (lateral movement of the track), which potentially produces derailment.

The axial load assessment is then important in railroad maintenance to ensure safe service. Current practice relies on human efforts in Japan and probably in other countries, too. The costs are high and it is highly time-/labor-consuming. There has been a great interest in developing a stress-detection device to be available for regular maintenance. This should be portable and robust to cope with on-site environments or it can be implemented in a test car moving at standard speeds, for which noncontact sensing is required. Other requirements are quick and easy measurement and, more importantly, enough stress resolution that is not affected by the actual measuring conditions (service period, track curvature, temperature, surface contaminations, operator, etc.). To meet these demands, EMAT techniques were developed. Alers and Manzanares (1990) devised a magnetostrictively coupling EMAT, basically the same design as Fig. 3.12, to measure the surface-skimming SH wave on the rail web. As mentioned above, the velocity can indicate the surface stress, being independent of the intrinsic anisotropy. Schramm et al. (1993) used the EMATs to measure the shear wave birefringence and also the time-of-flight of the surface-skimming longitudinal wave in the railhead. The longitudinal wave shows the highest sensitivity to normal stress in the propagation direction. Hirao et al. (1994) explored the feasibility of the EMAR for the birefringence acoustoelasticity on Shinkansen (bullet train) railroad rails. The results are shown below.

Two samples cut from the new 60-kg/m rails were tested. Both were 0.5 m long and had the cross-sectional dimensions as shown in Fig. 12.6. One was the rail whose head was slack quenched to give a deep hardened area (called DHH), which was used to study the acoustoelastic response to the compressive axial load ($\sigma_2 = 0$). The other was the standard as-rolled rail (called STD). The distribution of B at unloaded condition was measured on this sample. They were normally rusted, but surface preparation was unnecessary. A shear-wave EMAT (Sect. 3.1) of $14 \times 20\ \text{mm}^2$ effective area was placed on the web of the rails, where the thickness was 16.5 mm; the surfaces were nearly parallel to each other and supported the

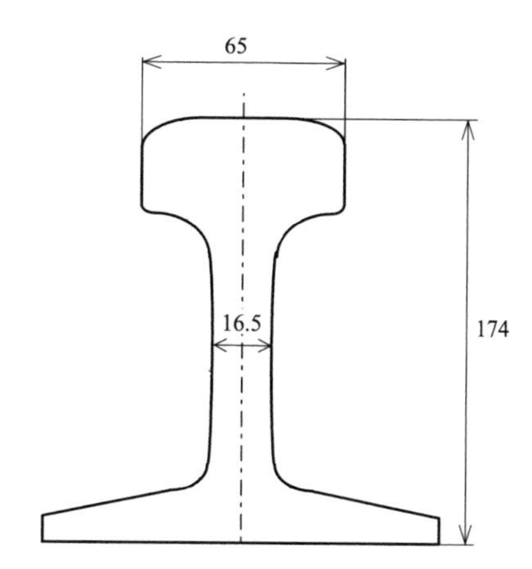

Fig. 12.6 Dimensions of JIS-60 rail cross section. (After Hirao et al. 1994.)

resonance. The frequency was swept in the range of 3.79–3.88 MHz at 600 Hz steps to detect the 40th resonance peak.

The acoustoelastic measurement was performed by applying an axial load up to 60 tons, which exceeds the buckling load of long rails. Seven strain gauges were attached around the rail to watch the balanced loading. Figure 12.7 presents the EMAR spectrum response to the polarized shear wave in the axial direction ($f_{40}^{(1)}$). The birefringence response to the stress is shown in Fig. 12.8. The good linearity concludes the stress evaluation within 3-MPa error band. The major source of scattering is the manual rotation of the EMAT to 0°/90° from the rail axis.

Measurement of B at unloaded stage on the STD rail reveals that it is considerably inhomogeneous along the length. The results are shown in Fig. 12.9, where the deviations relative to the midpoint (25 cm) measurement are divided by C_A to have equivalent values in stress. The nearly symmetric variation implies that the thermomechanical process had introduced a large compressive residual stress (~150 MPa) in the web of the original full-length rail. The residual stress is released at the end regions, but it persists in the middle part of the short section. This stress evaluation agrees with the destructive measurement using strain gauges on the same type of rail (Urashima et al. 1992). The larger cooling rate at the web, because of smaller thickness, produces a compressive residual stress there and, for balance, produces tensile stresses in the head and the base. The initial birefringence

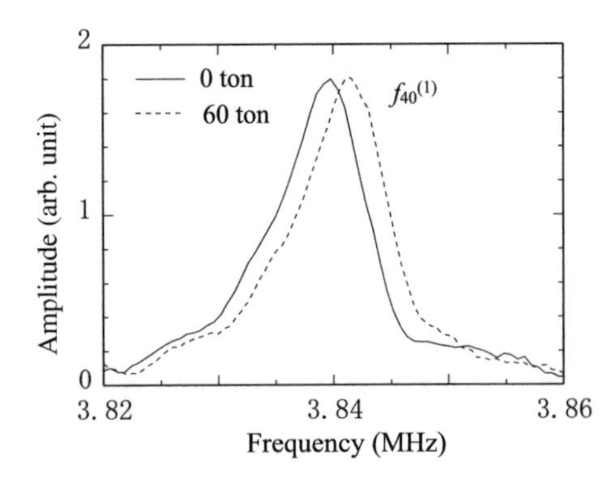

Fig. 12.7 Applied axial load and resonance frequency shift of the 40th shear mode polarized parallel to the axial load. (After Hirao et al. 1994.)

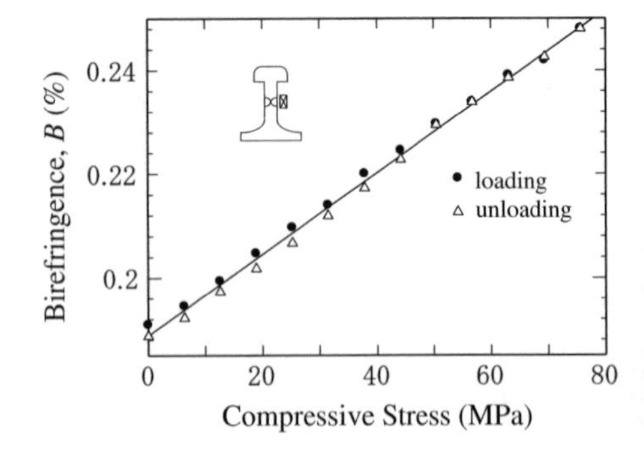

Fig. 12.8 Dependence of B on compressive axial stress at the rail web; $B_0 = 0.189$ % and $C_A = -7.85 \times 10^{-6}$ MPa^{-1}. (After Hirao et al. 1994.)

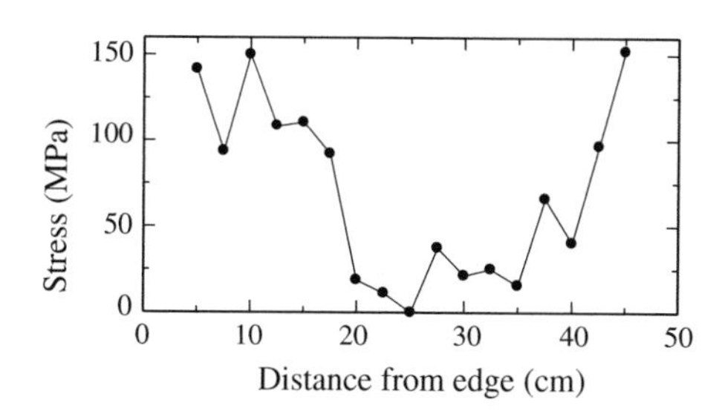

Fig. 12.9 Inhomogeneous B at the rail web converted to stress by dividing with C_A. (After Hirao et al. 1994.)

thus results from the combined effect of the residual stress and the rolling texture. The texture is considered to be uniform along the rolling lines. This observation emphasizes the necessity of measuring initial birefringence before the stress generation. In the worst possible case, it varies from point to point and should be measured at every scanning point along the rail before installation. However, the influence will not be so serious in long rails, seeing that the central part appears to be uniform in Fig. 12.9.

The liftoff effect was studied with the STD rail. The surface roughness and rusting are equivalent to adding the liftoff on EMATs. They tend to change the electrical responses. The variations occur in the same way for both polarizations, and the measurement of B is less vulnerable to these influences. However, the liftoff diminishes the signal intensity, causing more scattering. The measurement showed that the liftoff up to 1.0 mm may cause an evaluation error less than 10 MPa.

The same measurement setup allowed sensing the stress-induced phase shift of discrete reflection echoes propagated long distances (top-to-bottom echoes and transverse multiple echoes in the head of the rail). Long paths amplified the small velocity change, resulting in easily detectable phase shift. They also demonstrated the linear stress dependence as the theory predicts. Although the measuring time was 0.01–0.1 s, being much shorter than 10 s for EMAR, the phase shift detection is considered to be less suitable for transferring to the field. The EMAR is stabler and robust, since the EMAR employs a large number of echoes overlapping in phase. The web is close to the neutral axis of bending and the axial stress is isolated. It is also distant from the railhead, which undergoes wear and cold working from the wheel treads.

12.4.2 Bending Stress in Steel Pipes

The electric-resistant-welded (ERW) pipes are widely used in gas pipelines. They are usually constructed underground and often suffer from a large bending moment from earthquake and uneven settlement. Nondestructive stress measurement is therefore needed to ensure their safety. Birefringence acoustoelasticity with EMAR was applied to determine the bending stress in thin-walled ERW steel pipes (Ogi et al. 1996a). Tested pipes were 75 mm diameter and 2.3 mm thick (Pipe 1), and

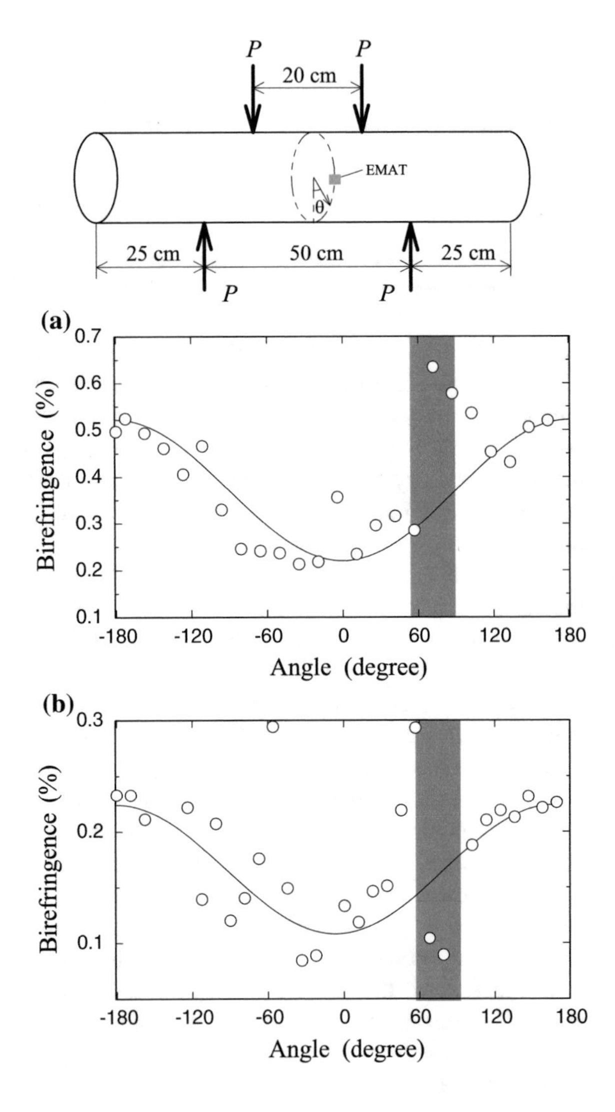

Fig. 12.10 Measurements of B around circumferences of the bent pipes and the fitting cosine curves. Applied load P was 1 ton for Pipe 1 **a** and 1.5 ton for Pipe 2 **b**. The gray bands indicate the heat-affected zones along the welding lines

101.6 mm diameter and 2.8 mm thick (Pipe 2), both being 1 m long. Figure 12.10 shows the four-point-bending configuration. A shear-wave EMAT, having a flat face of 7 × 6 mm^2 aperture, was manually moved along the center circumference and rotated at each location by right angles. The cylindrical surfaces were parallel to each other and supported the local resonance to construct resonance spectra. The acoustic birefringence B was then obtained from the resonance frequencies for the axial and circumferential polarizations. Highest possible resonance peaks, around 6.3 MHz for Pipe 1 and 5.3 MHz for Pipe 2, were detected. Strain gauges mounted at 90° intervals along the circumference, 10 mm away from the center, monitored the bending stress in the axial direction.

Figure 12.10a, b depicts the measured variation of B with θ, the angle from the vertical position. According to beam-bending theory, pure bending produces only the axial stress, σ_z, and $\sigma_z = \sigma_{max}\cos\theta$. From Eq. $(12.16)_1$, we have

$$B(\theta) = B_0 + C_A \sigma_{max} \cos\theta. \tag{12.17}$$

Here, σ_{max} (>0) is the maximum bending stress.

In the ERW pipes, B at zero applied load comes from the texture and the residual stress as in the rail of the previous section. The manufacturing process (rolling, forming, and welding) introduces residual stress and inhomogeneous texture. The initial B value distributes around the circumference. There are then two unknowns, B_0 and σ_z, at each scanning point, while only B is measurable. This troublesome situation can be overcome with the stress distribution prediction of Eq. (12.17). Fitting a cosine function to the measured $B(\theta)$ can determine σ_{max} without knowing the initial value of B. With the plausible value of $C_A = -8 \times 10^{-6}$ MPa^{-1}, the calculation gives σ_{max} = 188 MPa for Pipe 1 and σ_{max} = 73 MPa for Pipe 2 from the amplitude of the cosine curves, which reasonably compares the strain-gauge method and beam-bending theory. The strain-gauge readings give σ_{max} = 177 MPa for Pipe 1 and σ_{max} = 109 MPa for Pipe 2. The beam-bending formula gives σ_{max} = 180 MPa for Pipe 1 and σ_{max} = 120 MPa for Pipe 2. On and close to the welding lines, the multiple resonance peaks appeared at lower frequencies (Fig. 12.11). This observation is attributable to the metallurgical changes and the thickness variation, caused by welding, within the area covered by the EMAT aperture. Some peaks were very sharp, which suggests fine grains produced by heat treatment. The regression calculation excludes the birefringence data of these heat-affected regions. Use of measured $B(\theta)$ at many sites around the circumference and fitting the theoretical prediction to them smoothed out the nonuniform B_0 by minimizing the influence.

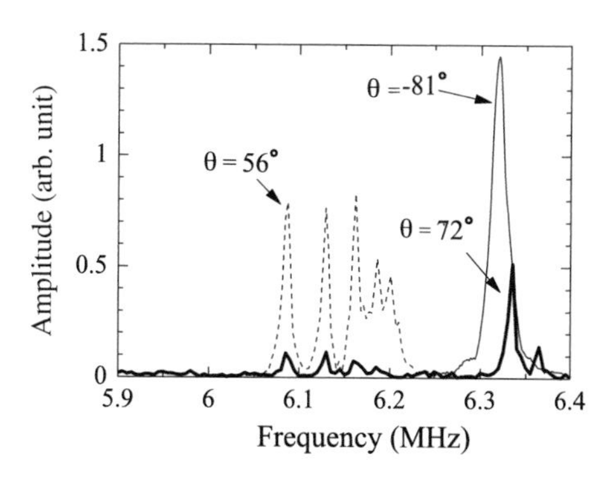

Fig. 12.11 EMAR spectra around $f_9^{(z)}$ observed at three angular positions on Pipe 1

12.4.3 Residual Stress in Shrink-Fit Disk

Shrink-fit is a conventional gripping method often used for tool/holder and wheel (or roll)/shaft systems. Usual procedure is to make the outer diameter of inner part larger than the inner diameter of the outer part. The outer part is heated (and the inner part is cooled). After assembling them and cooling, the thermal expansion/contraction builds a pressure at the boundary. It has also been a standard method to generate a two-dimensional stress distribution in circular disks for acoustoelastic studies. The stress magnitude can be controlled through the diameter misfit and the diameter ratio. Following is one of such measurements with the EMAR method.

The circular ring and plug were machined from a 15-mm-thick SCM440 rolled plate. The geometry is shown in Fig. 12.12, where a = 200 mm, b = 80 mm, and the misfit Δ was set to 0.25 mm. The ring was heated to 400–450 °C, into which the plug was inserted by making a flat face and aligning the rolling directions. If the yielding does not occur in the whole area, then the axis-symmetric stress field is given by

$$\begin{aligned} \text{plug:}\quad & \sigma_\theta = \sigma_r = -p, \\ \text{ring:}\quad & \sigma_\theta = \frac{pb^2}{a^2-b^2}\left(1+\frac{a^2}{r^2}\right), \\ & \sigma_r = \frac{pb^2}{a^2-b^2}\left(1-\frac{a^2}{r^2}\right). \end{aligned} \tag{12.18}$$

where p is the interface pressure of

$$p = \frac{E\Delta(a^2-b^2)}{4a^2b^2}, \tag{12.19}$$

with Young's modulus E.

The acoustoelastic birefringence formula (Eqs. (12.13) or $(12.16)_1$), being assisted by this theoretical prediction, can determine the residual stress field by

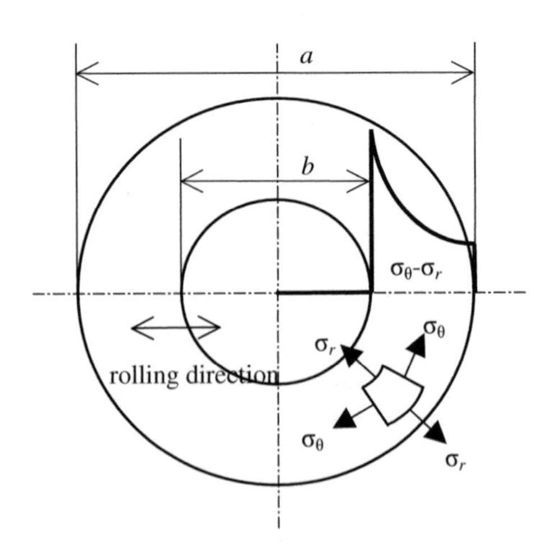

Fig. 12.12 Geometry of shrink-fit disk and the resultant residual stress field

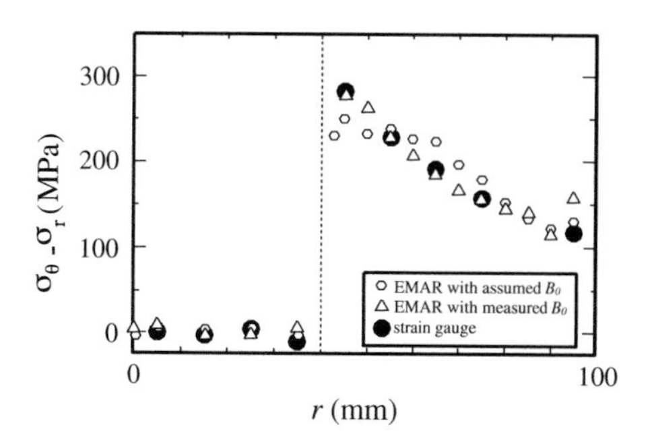

Fig. 12.13 Comparison of nondestructive stress measurement with the destructive measurement and strain-gauge reading

removing the initial anisotropy B_0 in two ways. Equation (12.18) indicates that $\sigma_\theta = \sigma_r$ and then $B = B_0$ in the plug. Measured B in the plug thus substitutes B_0 in the ring. An alternative way is to use the orthotropic symmetry of rolling texture and the axis symmetry of residual stress. Measuring B along two orthogonal diameters and subtracting one from the other at the same radius can isolate the stress-induced birefringence. Figure 12.13 shows the resultant principal stress difference with the second procedure. The EMAT's acting area was 5×6 mm^2. After this measurement, strain gauges were attached along a diameter on both faces of the specimen, which was then sectioned to relieve the stress. This process offered the chance to directly evaluate B_0 on the small pieces. Figure 12.13 also compares the *destructive* acoustoelastic birefringence with the strain-gauge method. Agreement reveals that the yielding took place in the inner part of the ring (r = 40–60 mm) at the time of insertion and B_0 had been changed from the original value.

12.4.4 Residual Stress in Weldments

Welding is a complex thermomechanical process, which involves heating, melting, solidifying, and cooling of metals. Phase transformation and recrystallization make it further complicated. In the course of cooling, thermal shrinkage occurs in the hot metals, but the surrounding materials do not follow it, since they remain at a low temperature. The hot metals then undergo tensile plastic deformation with yield stress lowered by the elevated temperature. This elongation will not be recovered at the later stage of cooling, because the yield stress increases at lower temperatures. Finally, the tensile residual stress appears along the welding line. Together with the coarsened grains, the tensile residual stress deteriorates the strength and the tolerance to fatigue of the welded components. (The plastic deformation produces the fine grains in the outer edges of heat-affected zone as suggested in Fig. 12.11.)

Reliable operation of welded structures therefore needs a stress-relieving technique and nondestructive evaluation of the residual stress. Acoustoelasticity is basically capable of measuring the residual stresses. Here are the examples of measuring residual stress profiles in welded steel plates and pipes, again using birefringence acoustoelasticity (Ogi and Hirao 1996). A plasma beam was used for cutting out the pieces, which were butt-welded to produce the specimens shown in Figs. 12.14 and 12.15. The materials are low-carbon steels. The birefringences along the broken line in Fig. 12.14 were measured to have the stress differences, $(\sigma_1 - \sigma_2)$ for the plate and along Lines 1 and 2 in Fig. 12.15 to have $(\sigma_\theta - \sigma_z)$ for the pipe, by manually moving and rotating the EMAT by right angles. Although the specimens underwent shot-blast to have rough surfaces (Ra = 6.5–7.2 μm), all measurements have been performed without any surface preparation.

A shear-wave EMAT of 5 × 3-mm^2 effective area generates and detects the shear waves propagating normal to the specimen surfaces. The intrinsic birefringence B_0 is estimated from the measurements close to the free ends, where the stress is considered to be absent. The principal stress difference calculated with $C_A = -7.8 \times 10^{-6}$ MPa^{-1} is plotted in Figs. 12.14 and 12.15, where the stress components along the welding lines, σ_1 and σ_θ, are dominant. The stress increases approaching the welding line and also the edges, especially with the welded pipes.

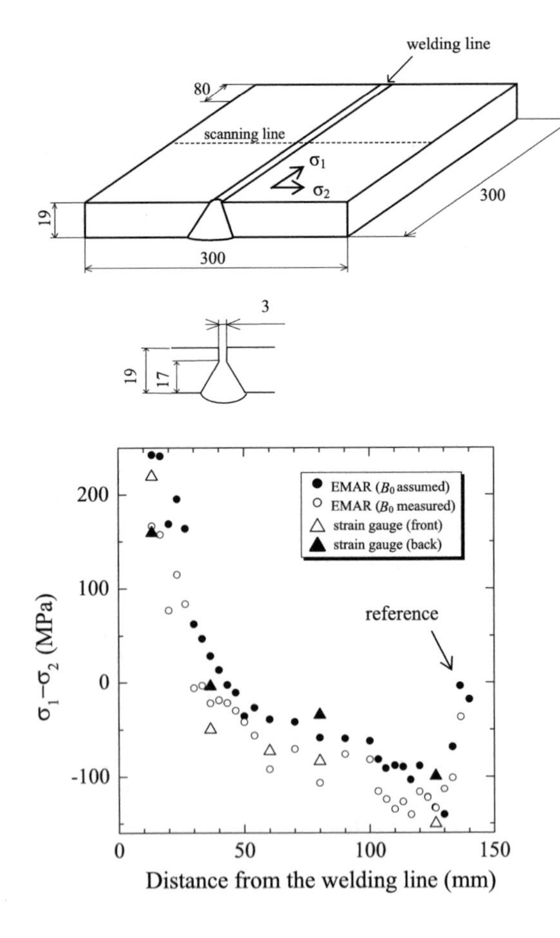

Fig. 12.14 Nondestructive and destructive residual stress measurements in butt-welded steel plates compared with the destructive measurement using strain gauges along the scanning line. Reprinted with permission from Ogi and Hirao (1996)

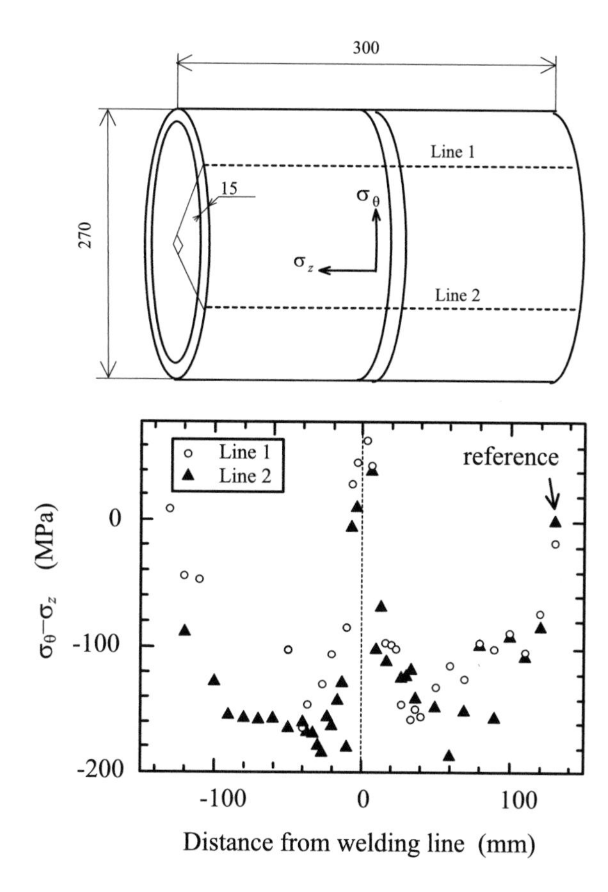

Fig. 12.15 Nondestructive residual stress measurement in butt-welded steel pipes. Reprinted with permission from Ogi and Hirao (1996)

To verify the EMAR results, strain gauges were mounted at several points along the scanning lines on the front and back surfaces of the welded plates and the residual strain was obtained by cutting it into small pieces to release the residual stresses. Figure 12.14 compares the stress distribution measured by the EMAR method and the strain-gauge method. The larger stresses in the EMAR method with the assumed B_0 are caused by unfavorable estimation of B_0. Even near the edge portion, there exist the residual stresses, which would have been introduced by the high temperature during cutting. In the figure, the recalculated stresses are also shown, which are obtained using B_0 measured at each measuring point after releasing the residual stresses. The strain-gauge results indicate the stress gradient in the thickness direction, which may cause a discrepancy with the EMAR measurement because it is sensitive to the average stress over the thickness. Near the welding line, the front surface has a stress larger than the back surface by about 60 MPa, while the relationship is reversed as the distance increases. The ultrasonic method cannot detect these gradients. Comparing the nondestructive and destructive EMAR measurements, we observe the overall difference of 30 MPa. Getting close to the welding line, it increases to 100 MPa, which is caused by the modified B_0 with heating.

As a whole, the EMAR stress evaluation with the assumed B_0 showed agreement with the strain-gauge technique within 10–40 MPa discrepancy. The comparison with the strain gauges will get worse when the stress gradient in the thickness direction is steeper.

12.4.5 Two-Dimensional Stress Field in Thin Plates

Acoustoelastic formulas (12.16) are applied for the measurement of the individual principal stresses in a thin plate. A two-dimensional stress field is generated by perforating the plate and applying a tensile load to it as shown in Fig. 12.16 (Hirao et al. 1993). The material is an aluminum alloy (Al 2017) of 1.22 mm thick, 150 mm wide, and 670 mm long. The hole is 45 mm in diameter. Nonholed strip of the same material is used for the calibration test to get the acoustoelastic constants C_A and C_R and the offset values for B_0 and R_0.

Figure 12.17 shows the EMAR spectra at two tensile stresses obtained during the calibration tests. A dual-mode EMAT (Fig. 3.3) was used, which has a circular effective area of 10 mm diameter. It detects two shear-wave resonance frequencies $f_8^{(1)}$ and $f_8^{(2)}$ and the longitudinal component $f_4^{(3)}$ at the same time and at the same probing area. The frequency was incremented for the 10.2–10.6 MHz range at 800 Hz steps. The resonance frequencies demonstrate easily detectable changes showing acoustoelastic phenomena. Two shear wave peaks for the orthogonal polarizations were superimposed on each other in the absence of stress because of very small texture anisotropy, B_0. The stress-induced anisotropy made them separate. Both B and R showed the linear dependence on stress as the theory predicts (Fig. 12.18).

Since B_0 and R_0 were nearly homogeneous over the scanning area, their representative values are used to calculate the stresses; the error is estimated to be less than 2.7 MPa. A typical measurement was done with the 500-V rf bursts of 40-μs duration, approximately 50 times the round-trip time, 34 dB for receiver gain after the preamplifier, and the 250-μs-long integrator gates. Frequency increment was 600 Hz. (See Sect. 5.2 for the details of operating EMAR.)

In scanning along the y-axis, the shear-wave EMAT (5×4 mm^2) was oriented by 45° to simultaneously detect the two shear-wave resonance frequencies, which

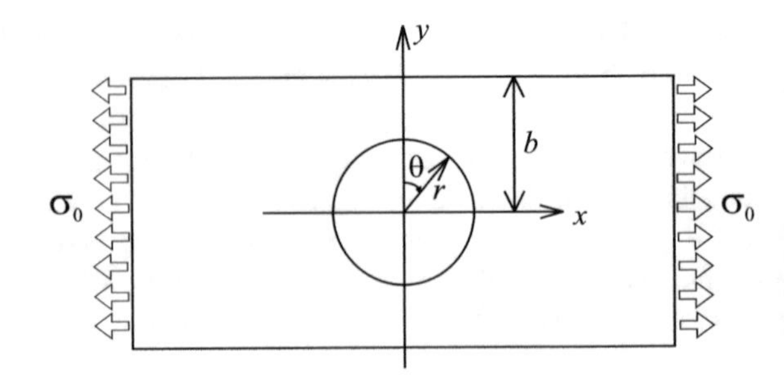

Fig. 12.16 Strip with a center hole in the plane stress condition. Reprinted with permission from Hirao et al. (1993), Copyright (1993), AIP Publishing LLC

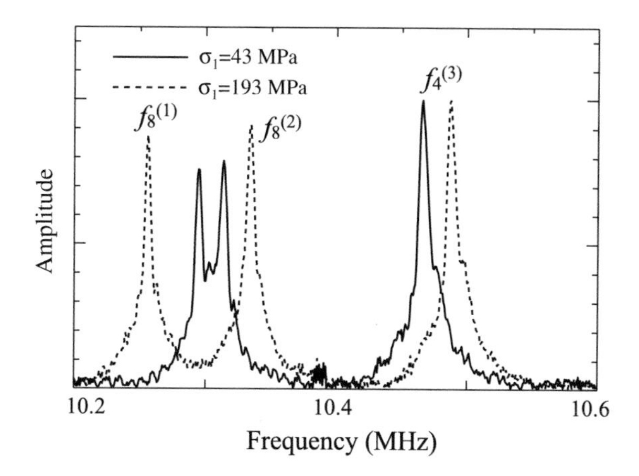

Fig. 12.17 Acoustoelastic response in EMAR spectra to uniaxial stress in aluminum. Reprinted with permission from Hirao et al. (1993), Copyright (1993), AIP Publishing LLC

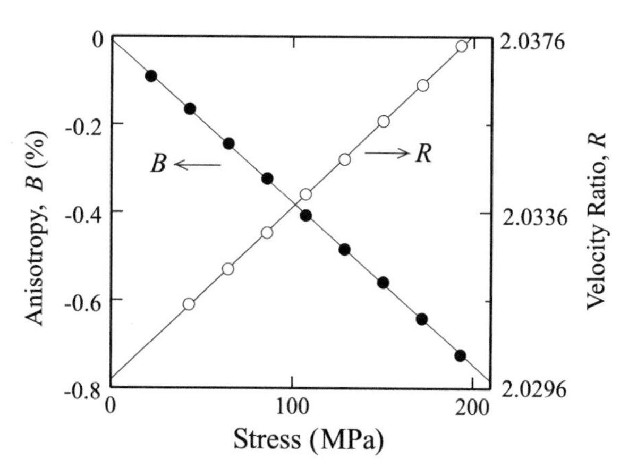

Fig. 12.18 Linear dependence of B and R on uniaxial stress in Al 2017; $B_0 = 0.006$ % and $C_A = -3.69 \times 10^{-5}$ MPa^{-1}; $R_0 = 2.0298$ and $C_R = 3.93 \times 10^{-5}$ MPa^{-1}. Reprinted with permission from Hirao et al. (1993), Copyright (1993), AIP Publishing LLC

yields B and depicts the variation of the principal stress difference ($\sigma_{xx} - \sigma_{yy}$). σ_{yy} is smaller than a tenth part of σ_{xx}. The stress difference was too small to resolve the shear wave polarization on the x-axis. The above EMAT was then rotated by 0°/90° to obtain each of $f_8^{(1)}$ and $f_8^{(2)}$. The longitudinal resonance frequency was measured using the dual-mode EMAT (10 mm diameter). Their combination results in the separate determination of σ_{xx} and σ_{yy}, where the calibrated values of C_A and C_R were used from the uniaxial tests. The acoustoelastic measurements are compared with the analytical solution (Howland 1930) in Fig. 12.19, where the averages on both sides of the hole are given. The local stress field has been averaged over the EMAT aperture. Such a correction was, however, unnecessary except for the proximity of the edges and the rim. The measurements agree with theory within the fluctuation of 5 MPa as a whole. Approaching the edges and the rim, disagreement increases close to 10 MPa. This is attributed to the abrupt variation of stresses within the EMAT's apertures, inclusion of reflections from the lateral boundaries, and the inhomogeneity of B_0 and R_0, which might be introduced when the hole was drilled.

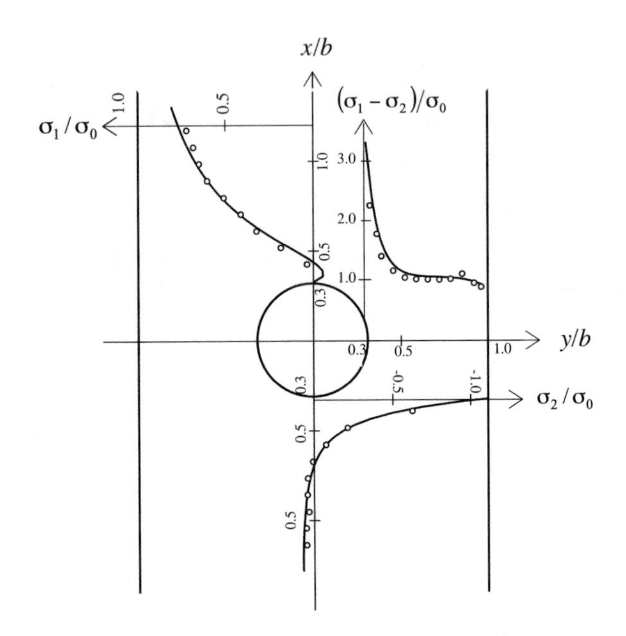

Fig. 12.19 Stress measurement along the *x*-axis and *y*-axis around the center hole of stretched strip. Howland's solution is shown in *solid lines*. Stresses are scaled by the tensile stress at infinity (σ_0 = 53.5 MPa). Reprinted with permission from Hirao et al. (1993), Copyright (1993), AIP Publishing LLC

Along the rim, the stress is not acting in the *x*- and *y*-axes and *B* is related to the principal stresses through a more general formula of Eq. (12.9) with an angle θ between the principal direction and the specimen axis. There is only one nonzero principal stress on the rim. However, the radial component is included in the calculation to account for the EMAT's finite aperture and the steep stress changes in the vicinity of the hole. Figure 12.20 compares the results of stress measurement using the above shear-wave EMAT to the theory calculated at 2 mm outside the rim, where the center of the EMAT traced. Disagreement among the quarters could be caused by the slight eccentricity and imperfect roundness of the hole. Maximum error was 15 MPa in these measurements.

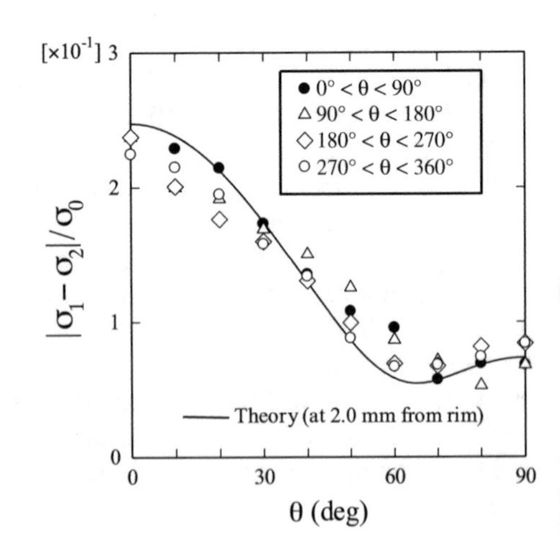

Fig. 12.20 Comparison between EMAR measurement and theory along each quarter of the rim. (σ_0 = 53.5 MPa). Reprinted with permission from Hirao et al. (1993), Copyright (1993), AIP Publishing LLC

As another issue, the tolerance to the liftoff has been tested using the same aluminum plate and the 10-mm-diameter dual-mode EMAT. Polymer sheets were inserted between the EMAT and the plate to simulate a liftoff. The measuring condition included 1500 V maximum voltage and 46 dB receiver gain. Resonance frequencies were measurable until 2.67-mm liftoff, for which both $f_8^{(1),(2)}$ and $f_4^{(3)}$ were decreased by 0.45 and 0.32 %, respectively, relative to those for the zero-liftoff measurements. Changes in B and R are usually one order less than the frequency changes because they are derived from the difference and the ratio of the resonance frequencies. The above is the result when the measuring conditions remain unchanged. Optimization of the measuring parameters enabled detecting the resonance peaks up to 5-mm liftoff in this particular specimen. Use of larger EMATs, more driving input power, and less-attenuating materials would allow us to extend the limiting liftoff.

The measurements described up to this point illustrate the difficulty of measuring the *residual stress* with acoustoelastic birefringence. Separation of B_0 remains a tough problem. Generally, the residual stresses occur in metals accompanying inhomogeneous plastic deformation, which is inevitable in thermomechanical processes and welding. Plastic deformation simultaneously changes the crystallite orientation, renewing the texture anisotropy. The same is true for heating above the recrystallization temperature.

Applied stress (principal stress difference or uniaxial stress) is easily measurable with acoustoelastic birefringence as in a rail's axial stress (Sect. 12.4.1), because the initial state is accessible for B_0. The rails are found to possess residual stress before installation, but it is not obtainable with a nondestructive method. Theoretical prediction on the induced stress field (Sect. 12.4.2) and the geometrical symmetry (Sect. 12.4.3) are useful to separate acoustoelastic and texture effects. But, these procedures are not generally applicable. Rough estimation of B_0 is available from a region where the residual stress is expected to be small enough. As shown in Sect. 12.4.4, the results are acceptable but erroneous to a degree. The principal stresses can be separated by incorporating the velocity ratio R for plates with uniform texture contribution (Sect. 12.4.5).

To summarize this section, EMAR permits practical acoustoelastic stress measurements even for thin plates, showing sufficient stress resolution. The measurement is available not only for plates with parallel surfaces but also for components with curvature such as the web portion of rail and ERW pipes. The stresses evaluated with EMAR agreed with those from theory and the strain-gauge method to the accuracy of 5–40 MPa. Although the inherent problem of the birefringence acoustoelasticity with the texture-induced anisotropy remains unsolved, the accuracy of the measurement has been improved considerably and also the difficulties with the conventional contacting methods have been removed with the noncontacting nature of the EMATs. The measurable thickness ranges from 0.1 to 50 mm, mainly depending on attenuation in the material, not on the surface quality or the surface curvature. The birefringence measurement is insensitive to temperature and the thickness, because the difference between the resonance frequencies for the two polarizations is only considered.

12.5 Monitoring Bolt Axial Stress

Bolts and nuts are important elements for the detachable assembly of components in structures and machines. They secure the working forces on the components. A bolt's axial load or preload has to be carefully controlled to ensure the safety and reliability of structures. Current practice relies on the torque-wrench technique. With this technique, however, the torque dissipates in the friction of the bolt threads and between the nut and the part, preventing accurate measurements of the axial load. This situation leads to over-designing for safety and increases both mass and cost. There has been a strong need for evaluating bolt axial stress nondestructively. Indeed, portable instruments are now commercially available, which measure transit time of longitudinal wave traveling along the bolts. This technique requires an intimate contact between the piezoelectric transducers and the bolt end, which causes inevitable errors from the stamped mark, surface conditions, and the unstable couplant property. The measurement then needs laborious surface preparation. It is apparent that further development is necessary to meet inspection demand for innumerable bolts used everywhere.

12.5.1 Shear Wave Acoustoelasticity along the Bolts

The EMAT is an appropriate choice, which makes a quick and accurate measurement of the traveling time of polarized shear waves along the bolt length. It accommodates unfavorable surface conditions including stamped marks on the heads. Requirement is only head-to-end parallelism. In what follows, the phase method and the resonance method are presented using the same instruments and shear-wave EMAT placed on the bolt heads (Hirao et al. 2001). It should be noted that Heyman (1977) has already reported a similar system using the resonance along the bolts with a piezoelectric transducer.

We consider a simple one-dimensional model to calculate the plane-wave response to the axial load (Fig. 12.21). The total length of the bolt is L_0 and the shear-wave velocity is c_0 in the stress-free condition; l_N and l_H are the lengths of the nut and the bolt head, respectively. When the tensile stress σ is applied along the length, a small segment dx in the bolt shank is elongated to $(1 + \sigma/E)\mathrm{d}x$, where E is Young's modulus. Assuming that the velocity c varies linearly with the stress, that is,

$$c = c_0(1 + C_{\mathrm{axis}}\sigma), \tag{12.20}$$

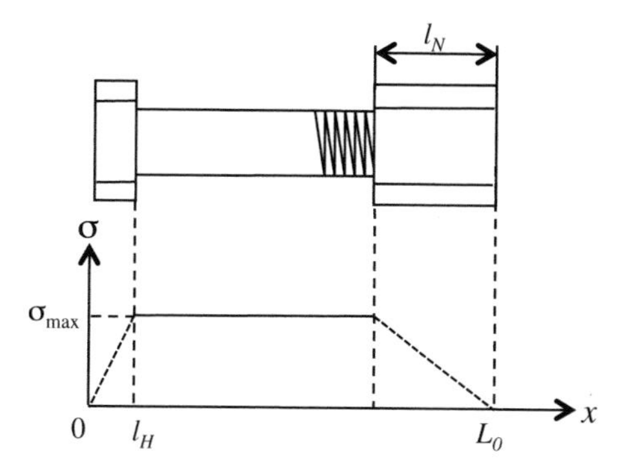

Fig. 12.21 One-dimensional axial stress model. $\sigma_{max} = P/(\pi d_0^2/4)$, where P is the load. Reprinted with permission from Hirao et al. (2001), Copyright (2001), Elsevier

with an acoustoelastic constant C_{axis}, the traveling time through this segment becomes

$$dt = \left\{1 + \left(E^{-1} - C_{axis}\right)\sigma\right\}c_0^{-1}dx. \tag{12.21}$$

We have neglected the higher-order terms in the derivation. The round-trip traveling time along the bolt is given by integration to become

$$T = T_0(1 + C_B\sigma_{ave}). \tag{12.22}$$

Here, $T_0 = 2L_0/c_0$, $C_B = E^{-1} - C_{axis}$, and $\sigma_{ave} = \beta\sigma_{max}$. The effective length ratio β is defined as

$$\beta = 1 - (l_N + l_H)/(2L_0). \tag{12.23}$$

Here, we assume that there is a linear stress variation in the bolt head and under the connection with the nut. The transit time is then proportional to the average and maximum stresses with the rates T_0C_B and βT_0C_B, respectively. Because C_{axis} takes a negative value (see Table 12.1), both the acoustoelastic effect and the elongation contribute to increase the transit time in Eq. (12.21). For short bolts, the stress distribution is somewhat different from the one in Fig. 12.21. Even in such cases, we define the average stress, to which the ultrasonic shear wave responds and provides the useful information on the acting stress.

In the phase method, the EMAT is driven with short tone bursts of a fixed frequency f to detect the individual shear wave reflections. The phase of the first echo, for example, is delayed by $\Phi = 2\pi fT$ relative to the driving signal. From Eq. (12.22), we have

$$\Phi = 2fT_0(1 + C_B\sigma_{ave}). \tag{12.24}$$

This shows a linear dependence of phase shift on σ_{ave}. The stress sensitivity is proportional to the frequency times the propagation time.

Table 12.1 Acoustoelastic response in bolts measured with shear-wave EMAT and PZT transducer up to 800 MPa (Bolt A), 920 MPa (Bolt B), 250 MPa (M24 bolt), and 210 MPa (aluminum bolt). The estimated error at a load of 5 tons is included. (L) and (S) indicate the responses shown by the longitudinal and shear waves

	Method	L_0 (mm)	β	Slope (10^4 ns/N)	C_B (βC_B) (10^{-6} MPa^{-1})	C_{axis} (10^{-6} MPa^{-1})	Error at 5 tons (10^3 N (ton))
Bolt A	Phase (EMAT)	27.5	0.68	8	6.1 (4.1)	−1.3	2.4 (0.24)
	Resonance (EMAT)			8.1	6.2 (4.2)	−1.5	2.4 (0.24)
	Transit time (PZT)			7.9	6.0 (4.1)	−1.2	2.5 (0.25)
Bolt B	Phase (EMAT)	53.7	0.83	37	7.9 (6.6)	−3.1	2.5 (0.25)
	Transit time (PZT)			35	7.5 (6.2)	−2.8	2.1 (0.21)
M24 bolt	Phase (EMAT)	155	0.84	16	7.7 (6.5)	−2.9	1.9 (0.19)
M30 bolt	Transit time (PZT)	234	0.7	–	7.5 (5.3)	−2.8	–
Aluminum bolt	Phase (EMAT)	320	0.96	317.2 (L) 396.3 (S)	56.2 (L) 29.8 (S)	−41.4 (L) −15.0 (S)	– –

The resonance frequency f_n is given by $f_n = nT^{-1}$ with an integer n defining the resonance order. Using Eq. (12.22) and neglecting the higher-order terms, we obtain

$$f_n = n(1 - C_B \sigma_{ave})/T_0. \tag{12.25}$$

Again, the resonance frequency shows a linear dependence on σ_{ave}. The slope is proportional to the resonance order, indicating enhanced sensitivity with higher-order resonances.

Measurements were done on two types of bolts used in cars, both having flanges at the heads; Bolt A (L_0 = 27.5 mm, d_0 = 10.5 mm, l_N = 10 mm) and Bolt B (L_0 = 53.7 mm, d_0 = 8.6 mm, l_N = 12 mm). Bolt A is a differential-case and ring-gear set bolt, and Bolt B is a connecting-rod bolt. They have a yield stress around 1000 MPa. Both ends were machined to be parallel to each other; the head was originally centerbored.

For the phase measurement, the EMAT (8.5 × 12 mm^2) was driven with 5-MHz tone bursts, typically at 300 V, and two echoes were selected from the received train of reflections (Fig. 12.22). An echo signal was processed with an analog superheterodyne phase-sensitive detector to obtain the in-phase and out-of-phase components into independent channels (Sect. 5.2). This signal processing extracted the component of the current operating frequency from the raw signal,

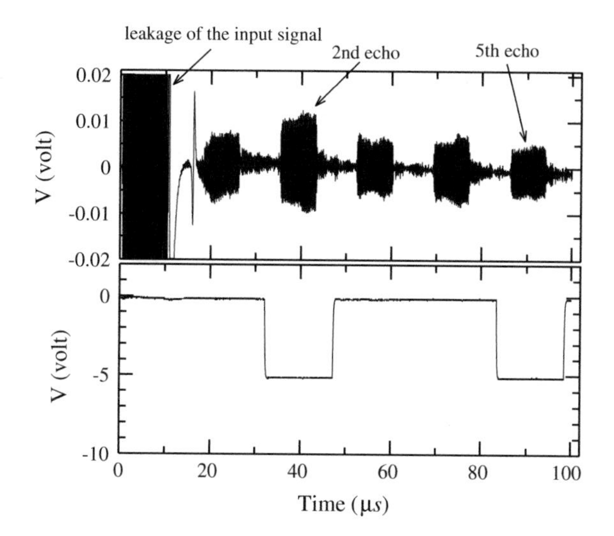

Fig. 12.22 Received echoes from Bolt A (not averaged) and integrator gates to pick up the second and fifth reflections. Reprinted with permission from Hirao et al. (2001), Copyright (2001), Elsevier

compensating for the weak transfer efficiency of the EMAT. The outputs were then digitized and their ratio provided the as-measured phase, Φ_M (Fortunko et al. 1992; Eq. (4.14)). This usually contained the phase delay Φ of Eq. (12.24) along with some artifacts caused by diffraction, transduction, and the system error plus the ambiguity of $2N\pi$ from wrapping the phase in the 0–2π range. However, subtracting the phases of the two echoes eliminated these errors, because the diffraction-induced phase shift converges to a constant value in a far-field condition (Sect. 6.4.1).

Bolts A were measured with EMAR. Long-tone bursts of 200 μs duration excited the EMAT to cause the overlapping multiple reflections in the bolts. Since the round-trip time along Bolt A was about 17 μs, more than ten echoes were superimposed at every moment. After excitation, the same EMAT received the reverberation with the analog integrator gate, from which the in-phase and out-of-phase components were extracted to calculate the amplitude from the root of the sum of their squares. We then swept the frequency to obtain the amplitude spectrum and fit to the Lorentzian function to determine the resonance frequency (Sect. 5.3). See Fig. 12.23 for a typical resonance spectrum. For monitoring the axial stress, the resonance frequency around 6.015 MHz (n = 102) was traced for Bolt A.

Measurement of the resonance frequency took approximately 10 s for stepping the frequency in a certain range, while the phase measurement finished in several milliseconds, being best suited to quick monitoring. However, from a limitation, we need to use short tone bursts to avoid signal overlapping when short bolts are measured with the phase method. This deteriorates the signal intensity and the accuracy of the phase measurement. Moreover, the stress resolution decreases for short bolts (see Eq. (12.24)). It is advisable to choose either the phase or resonance method based on the bolt length, the required accuracy, and the time allotted.

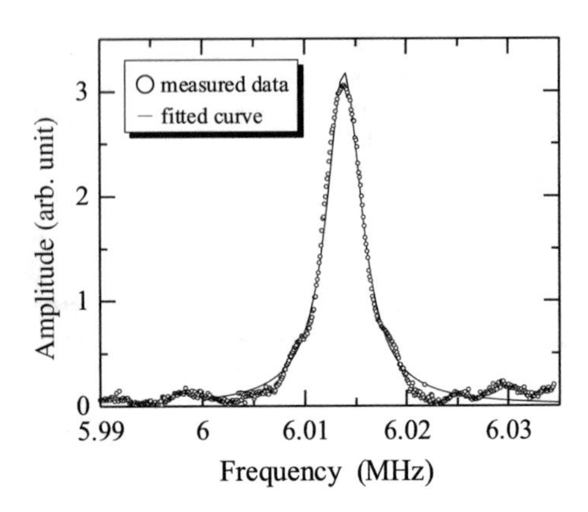

Fig. 12.23 Measured amplitude spectrum and fit to a Lorentzian curve with Bolt A. Reprinted from Hirao et al. (2001), Copyright (2001), with permission from Elsevier

Figure 12.24 plots the phase response of ten bolts of each type. Measured was the phase difference between the second and fifth echoes with Bolt A and between the first and second echoes with Bolt B. Detected phase during the loading and unloading sequences showed no difference.

Roughly, phase difference appears to be proportional to the load. Offset scattering is attributable mainly to the bolt length variation within 0.04 mm for both types. Assuming c_0 = 3240 m/s, this variation amounts to a phase variation of approximately 0.8 rad for a round-trip echo, and if converted to load, to 0.7 tons. Temperature also influences the phase measurements. Around room temperature, the phase increased at a rate of 0.23 rad/°C for Bolt A and 0.16 rad/°C for Bolt B. Figure 12.25 gives the resonance frequency shift caused by the stress application to Bolt A, where the initial measurements vary for the same reason with the above, while the slope appears to be constant.

To evaluate the measuring accuracy, the stress dependence of the phase and resonance frequency is transformed to a load/transit time relation by fitting to a linear function. Such data with over ten bolts of each type give the standard deviation of the slope. Another aspect of accuracy is seen from the data scattering of the regression line for each series of measurements. Adding up these two calculations at 5 tons provides an accuracy of 0.24 tons with Bolt A and 0.25 tons with Bolt B. For comparison, contact measurements were carried out with both types of bolts using a piezoelectric shear wave transducer (5 MHz, 6.4 mm in diameter), which shows the similar accuracy for the first-arriving echo. Moreover, the phase change was measured with M24 bolts of 155 or 203-mm total length, whose material was JIS-SCM435 (Fig. 12.26).

These results are summarized in Table 12.1, showing the slope and the acoustoelastic constants. We compare also with the reported measurements on an M30 high-tension bolt (L_0 = 234 mm, β = 0.70; SCM440). Acoustoelastic constants, C_B and C_{axis}, from Bolt B are similar to the results of the M24 and M30 bolts. (Note that the elongation effect (E^{-1}) exceeds the acoustoelastic effect C_{axis} in C_B.) But, those from Bolt A appear to be much lower. This discrepancy comes from the measurements for relatively low stresses with Bolt A. The threads close to the head

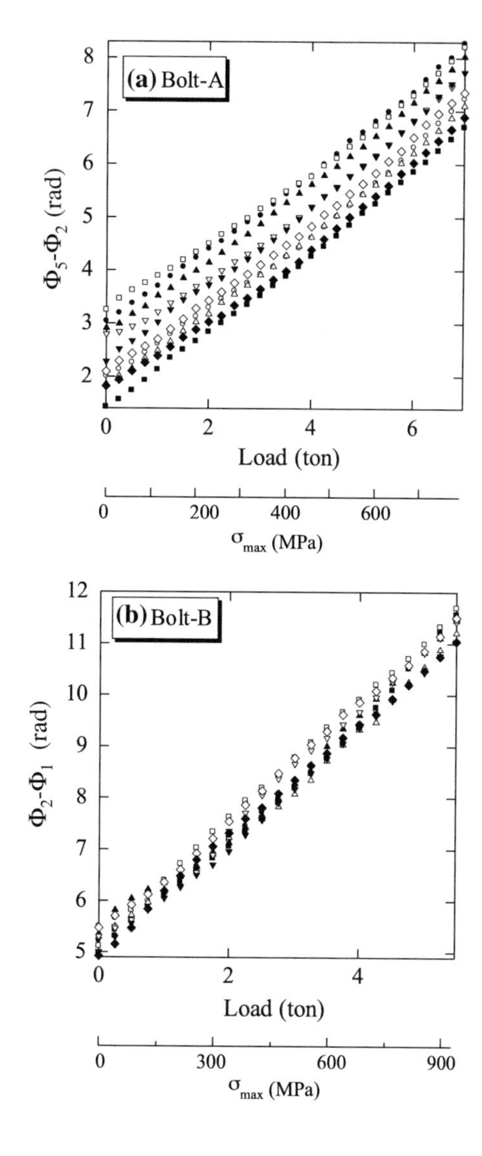

Fig. 12.24 Monitoring of the phase shift for the loading sequence. Reprinted from Hirao et al. (2001), Copyright (2001), with permission from Elsevier

support the applied load when it is small, which tends to diminish σ_{ave}. As the load increases, the nut deforms and the effective supporting threads move toward the end, increasing σ_{ave}. Deformation of the hexagonal head may also be involved. In fact, Figs. 12.24 and 12.25 depict nonlinearity, in particular for the small load range with Bolt A, indicating that the stress distribution differs from Fig. 12.21. The stress distribution under the nut may change as the load increases. For a slender bolt, this behavior exerts negligible influence on the axial stress determination, because the elongation and acoustoelastic effect in the shank are large enough. Figure 12.26 demonstrates an ideal linearity in the phase dependence on stress for longer M24

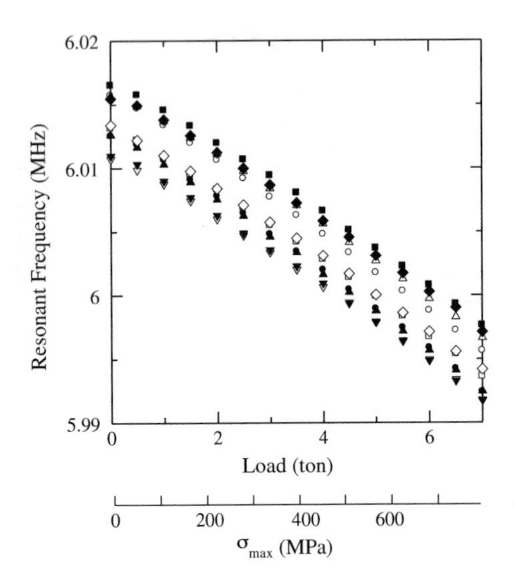

Fig. 12.25 Resonance frequency shift during the loading sequence (Bolt A). Reprinted from Hirao et al. (2001), Copyright (2001), with permission from Elsevier

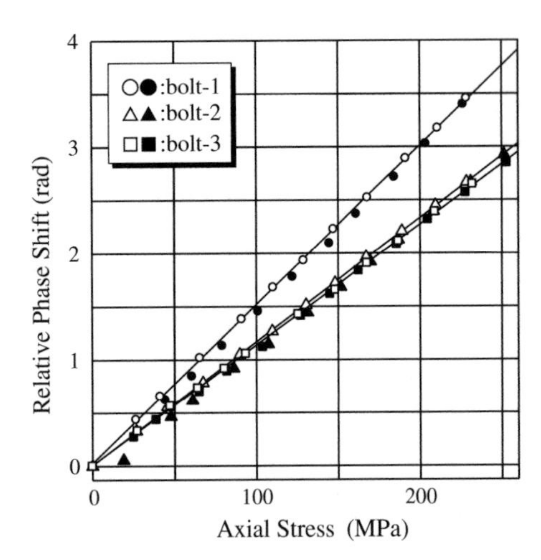

Fig. 12.26 Relative phase shift in M24 bolts at 3 MHz. Bolt-1 is 203 mm total length; others are 155 mm long. Solid and open points show the loading and unloading sequences

bolts; the accuracy seems improved relative to the short bolts, as expected. The value for C_B thus depends on the maximum applied load and the ratio β, although C_{axis} has little dependence on bolt steels. We ignored this influence in calculating σ_{ave} and then underestimated C_B and C_{axis} of Bolt A.

The phase and resonance methods using a shear-wave EMAT have shown a high accuracy (approximately 5 % at 5 ton) of measuring axial stress even for short bolts. They showed stable responses to the tensile load up to near the yield point, proving the feasibility for practical applications. Comparing the two methods, the phase detection, which takes a very short time, is more suitable for long bolts and the resonance measurement is more suitable for short bolts. The problem of stress gradient at the nut-joint portion remains, which influences the apparent acoustoelastic response for short bolts. It seems difficult to remove this influence, because both geometrical and elastic factors of the bolt, nut, and the component being fastened are involved in a complicated way. The same is true for the deformation in the bolt head. There is another fundamental problem that remains unsolved. The above methods, such as many conventional methods, need the initial state (T_0 in Eqs. (12.22), (12.24), and (12.25)), that is, the measurement before tightening the bolt. No technique is currently available to determine the axial stress of already fastened bolts. Possible solutions for this problem are addressed in the following two sections.

12.5.2 Combination of Longitudinal/Shear Velocity Ratio

If the longitudinal and shear waves can be simultaneously generated with a single EMAT, then this problem can be resolved (Ogi et al. 1996b). As discussed in Sect. 3.1, such a measurement is achievable with nonmagnetic metals. They travel the same distance, carrying different acoustoelastic effects. Their phase ratio eliminates the propagation length and extracts the acoustoelastic response. Equation (12.24) for the shear wave holds for the longitudinal wave with another T_0 and C_B. The normalized difference of the longitudinal wave and shear wave phases, $\Phi^{(L)}$ and $\Phi^{(S)}$ (or their transit times $T^{(L)}$ and $T^{(S)}$), is free from L_0:

$$R_B = \frac{\Phi^{(S)} - \Phi^{(L)}}{\Phi^{(L)}} = \frac{T^{(S)} - T^{(L)}}{T^{(L)}} \approx \frac{c_0^{(L)} - c_0^{(S)}}{c_0^{(S)}} + C_{L/S}\sigma_{ave},$$

$$C_{L/S} = \frac{c_0^{(L)}}{c_0^{(S)}}\left(C_B^{(S)} - C_B^{(L)}\right). \tag{12.26}$$

Here, $c_0^{(L)}$ and $c_0^{(S)}$ are the longitudinal and shear wave velocities at a free-stress state and $C_{L/S}$ denotes the combined acoustoelastic constant.

The shear-wave EMAT (now dual-mode EMAT, Sect. 3.1) detected the reflection signals of Fig. 12.27 with 320-mm-long bolt of aluminum alloy 7075-T771. The frequency was 5 MHz. The first-arriving echo should be composed of the longitudinal wave. This signal and another one for the shear wave around 210 μs delay were selected by two integrator gates for their phase response to stress. They are the top-to-bottom echoes, experiencing no mode conversion at the lateral surface. Their acoustoelastic characteristics are presented in Table 12.1, showing one-order-of-magnitude larger sensitivity than steels. Measured phases of these two echoes show a good linearity with the axial stress, for which Eq. (12.26) provides $C_{L/S} = -5.25 \times 10^{-5}$ MPa^{-1} (see Fig. 12.28). This is an attractive stress

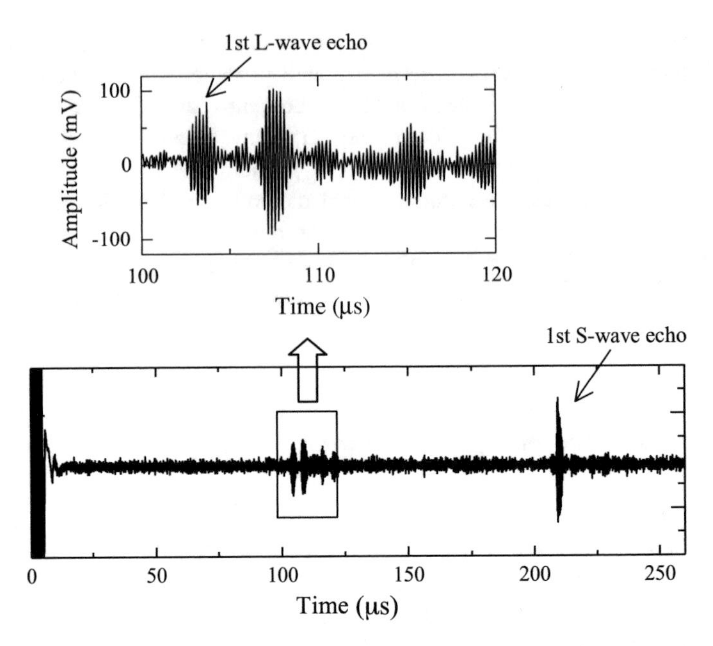

Fig. 12.27 Received echoes composed of longitudinal, shear, and mode-converted echoes. Upper trace magnifies the part containing the first arrival of longitudinal echo subjected to no mode conversion. (After Ogi et al. 1996b.)

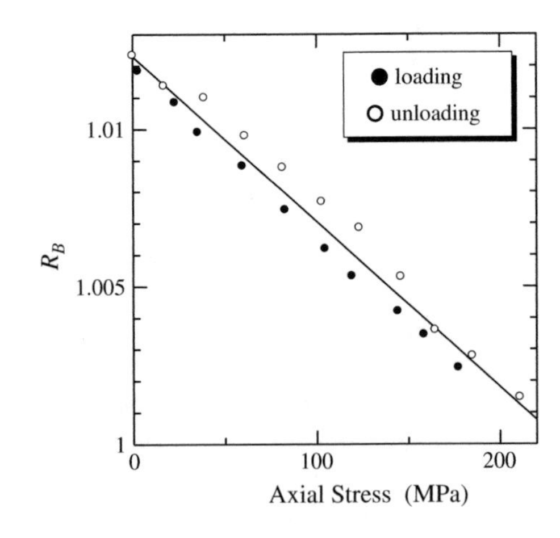

Fig. 12.28 Axial stress versus the normalized difference of the detected phases of longitudinal and shear waves in aluminum bolt. The slope of fitting line is -5.25×10^{-5} MPa^{-1}. (After Ogi et al. 1996b.)

indicator, because the offset value depends only on Poisson's ratio. (If textured, it contains W_{400}.) Unfortunately, this method is only applicable to the bolts of non-magnetic metals, while most bolts are made of ferromagnetic steels, for which the present EMAT technique cannot excite the longitudinal wave of sufficient intensity.

12.5.3 SH-Wave Resonance in the Bolt Head

The bolt head plays a role of transmitting the axial force in the shank to the objects being fastened. Its small volume is subjected to a complicated elastic deformation of tension, bending, and shearing depending on the radius and height. This stress field is axis-symmetric and develops as the tightening advances. If the axial symmetric SH-wave resonance is available as in the cylindrical rod (Sect. 3.7 and Chap. 13), the different resonance frequencies will then sense the radial gradient of stress field in the bolt head.

When the specimen is a cylindrical rod, the resonance modes are easily calculated with Eq. (3.3). At the fundamental resonance mode, the shear wave propagates near the outer surface and as the mode becomes higher order, the vibration penetrates inside. This property is essentially true for a hexagonal rod, although the resonance modes cannot be explicitly obtained with a simple frequency equation. The EMAT assembly (Fig. 3.20), which operates with the Lorentz force mechanism, detected the EMAR spectra with the M24 bolt head (JIS-S45C) and the hexagonal steel rod, which has nominally the same cross section as the bolt head (Ogi and Hirao 1998). Eight permanent magnets of 1.5 mm thick were arrayed on every side of the head. The spectra are compared in Fig. 12.29. The spectrum from the hexagonal rod is constructed purely with the axial shear wave, because the length of the rod is long enough (250 mm) and the end-reflected signals are excluded from the spectrum. But, the spectrum from the bolt head contains the axial shear modes as well as other modes caused by the comparable length and diameter.

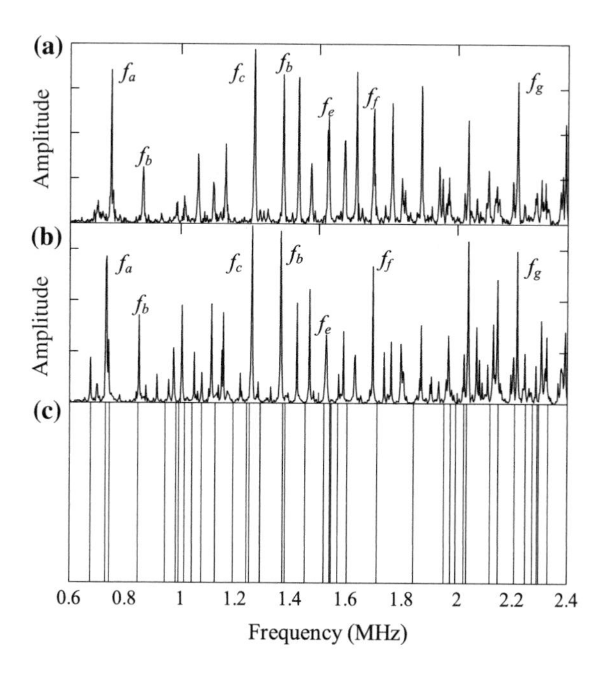

Fig. 12.29 EMAR spectra from **a** the head of M24 bolt (SCM435) and **b** hexagonal rod (25 cm long) compared with **c** the results of FEM computation. (After Ogi and Hirao 1998.) FEM calculation

A two-dimensional FEM calculation identified the axial-shear-wave resonances in the spectra. It also provided the vibration distribution for individual modes; the lower-mode oscillation is concentrated near the outer region, while the higher mode oscillates in the whole region of the head.

Now, we consider the acoustoelastic effect on these resonance frequencies by expressing the nth resonance frequency by $f_n = \eta_n c/L$. η_n is a factor specific to the mode, c the shear wave velocity, and L the characteristic length of the bolt head. The resonance frequency will shift linearly with the applied stress:

$$\frac{f_n^{(\sigma)} - f_n^{(0)}}{f_n^{(0)}} = C\sigma, \tag{12.27}$$

where $f_n^{(\sigma)}$ and $f_n^{(0)}$ are the resonance frequencies at stress σ and at the stress-free condition; C is an acoustoelastic constant. When the bolt is tightened, a stress field arises in the bolt head. We denote f_{low} as the lower resonance frequency and σ_{low} for the representative stress in the surface region, where this mode vibrates; the same for f_{high} and σ_{high} for a higher resonance mode. Taking the ratio $f_{\mathrm{low}}^{(\sigma)}/f_{\mathrm{high}}^{(\sigma)}$ and retaining the terms in the first-order smallness, we have

$$\frac{f_{\mathrm{low}}^{(\sigma)}}{f_{\mathrm{high}}^{(\sigma)}} = \frac{f_{\mathrm{low}}^{(0)}}{f_{\mathrm{high}}^{(0)}}\left(1 + C_{\mathrm{low}}\sigma_{\mathrm{low}} - C_{\mathrm{high}}\sigma_{\mathrm{high}}\right). \tag{12.28}$$

Here, we use the acoustoelastic constants of individual modes, C_{low} and C_{high}. Assuming that σ_{low} and σ_{high} are proportional to the axial stress results in

$$\frac{f_{\mathrm{low}}^{(\sigma)}}{f_{\mathrm{high}}^{(\sigma)}} = \frac{f_{\mathrm{low}}^{(0)}}{f_{\mathrm{high}}^{(0)}} + C_{\mathrm{head}}\sigma_{\mathrm{axis}}. \tag{12.29}$$

Here, C_{head} is another acoustoelastic constant related to the two modes. The first term of Eq. (12.29) is a geometrical parameter and solely depends on the factors η_n of the two modes involved, being free from the material, bolt dimension, and temperature. Actually, measurements of $f_a^{(0)}/f_g^{(0)}$ range between 0.3370 and 0.3381 for three hexagonal rods of low-carbon steel, aluminum alloy, and austenitic stainless steel.

Four M24 bolts of SCM-435 were tested by applying axial loads: two 185 mm long (L_1 and L_2) and two 155 mm long (S_1 and S_2). Strain gauges monitored the axial stress at the shank. Figure 12.30 is the resonance frequency shift of eight modes observed with a long bolt. We observe that f_a shifts at the largest rate, although not linearly, and as the mode becomes higher, the shift tends to be less. It is unclear as to which stress component or combination is responsible for the resonance frequency shift. But, all the stress components continuously and steeply vary in an axis-symmetric way. This leads to the prediction that the averaged stress

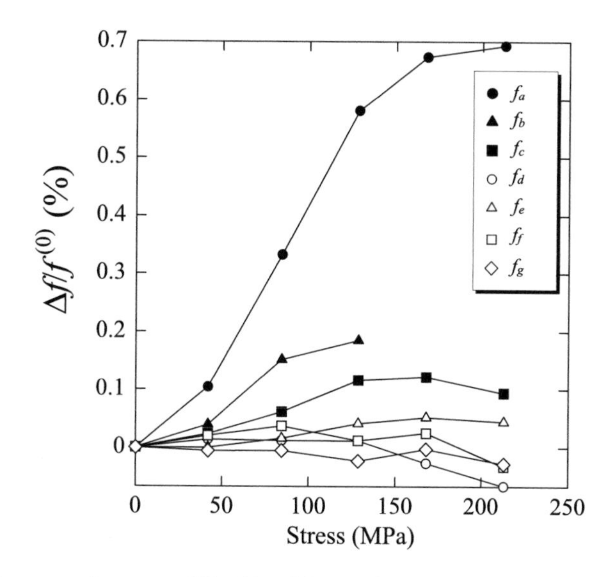

Fig. 12.30 Resonance frequency shift with axial stress in M24 bolt. (After Ogi and Hirao 1998.)

over the wide region (higher modes) is smaller than the stress acting in the limited region (lower modes).

Figure 12.31a plots f_a/f_g against the axial stress read from the strain gauges. One regression line roughly covers the measurements, being independent of the bolt length. C_{head} was calibrated to be 2.58×10^{-5} MPa^{-1}. Figure 12.31b shows the temperature effect of the measurement in the same vertical scale with Fig. 12.31a. The measurement fluctuation due to the temperature change is equivalent to 8 MPa for the range from −12 to 35 °C.

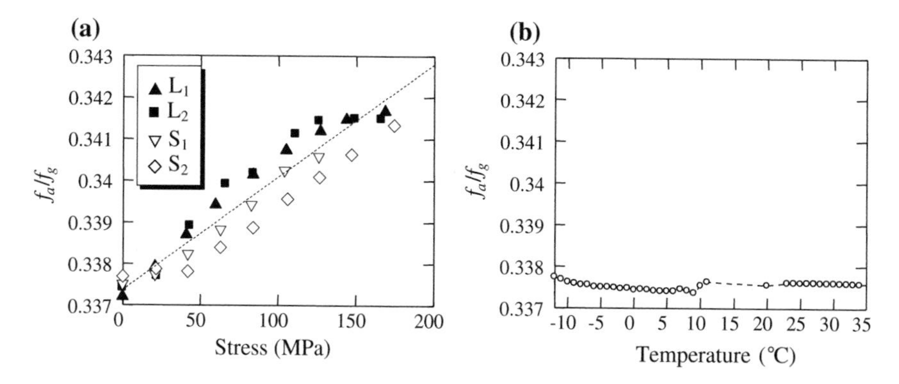

Fig. 12.31 Responses of the frequency ratio to **a** axial stress and **b** temperature without stress. (After Ogi and Hirao 1998.)

The EMAR method presented here is capable of evaluating the bolt axial stress without the information of the bolt length and the temperature. The axial-shear-wave spectrum contains a series of resonance peaks and the individual modes have the unique vibration regions in the bolt head. The axial stress changes the resonance frequencies depending on the stress profile in the each probing region. Other factors such as the temperature shift them at the same rate and their effects are minimized by taking the ratio of the two resonance frequencies. This method provided the axial stress for the M24 bolts with 25-MPa error band, and the error caused by the temperature effect is estimated as less than 8 MPa. It should be noted that this method entails another difficulty caused by the radial inhomogeneity of microstructures, which varies depending on the manufacturing process. Further study should involve investigation of the initial value for $f_a^{(0)}/f_g^{(0)}$ and the acoustoelastic constant C_{head} for a large number of bolts of different materials and different forming processes.

References

Alers, G., & Manzanares, A. (1990). Use of surface skimming SH waves to measure thermal and residual stresses in installed railroad tracks. In *Review of Progress in Quantitative Nondestructive Evaluation* (Vol. 9, pp. 1757–1764).

Auld, A. B. (1973). *Acoustic Fields and Waves in Solids*. New York: Wiley.

Brugger, K. (1964). Thermodynamic definition of higher order elastic coefficients. *Physical Review, 133*, A1611–A1612.

Crecraft, D. I. (1967). The measurement of applied and residual stresses in metals using ultrasonic waves. *Journal of Sound and Vibration, 5*, 173–192.

Eringen, A. C., & Şuhubi, E. S. (1975). *Elastodynamics*. New York: Academic Press.

Fortunko, C. M., Petersen, G. L., Chick, B. B., Renken, M. C., & Preis, A. L. (1992). Absolute measurement of elastic-wave phase and group velocities in lossy materials. *Review of Scientific Instruments, 63*, 3477–3486.

Green, R. E., Jr. (1973). *Ultrasonic Investigation of Mechanical Properties*. New York: Academic Press.

Heyman, J. S. (1977). A CW ultrasonic bolt-strain monitor. *Experimental Mechanics, 17*, 183–187.

Hirao, M., Ogi, H., & Fukuoka, H. (1993). Resonance EMAT system for acoustoelastic stress evaluation in sheet metals. *Review of Scientific Instruments, 64*, 3198–3205.

Hirao, M., Ogi, H., & Fukuoka, H. (1994). Advanced ultrasonic method for measuring rail axial stresses with electromagnetic acoustic transducer. *Research in Nondestructive Evaluation, 5*, 211–223.

Hirao, M., Ogi, H., & Yasui, H. (2001). Contactless measurement of bolt axial stress using a shear-wave EMAT. *NDT and E International, 34*, 179–183.

Howland, R. C. J. (1930). On the stresses in the neighborhood of a circular hole in a strip under tension. *Philosophical Transactions of the Royal Society of London. A, 229*, 49–86.

Hsu, N. N. (1974). Acoustical birefringence and the use of ultrasonic waves for experimental stress analysis. *Experimental Mechanics, 14*, 169–176.

Hughes, D. S., & Kelly, J. L. (1953). Second-order elastic deformation of solids. *Physical Review, 92*, 1145–1149.

Iwashimizu, Y., & Kubomura, K. (1973). Stress-induced rotation of polarization directions of elastic waves in slightly anisotropic materials. *International Journal of Solids and Structures, 9*, 99–114.

Johnson, G. C. (1982). Acoustoelastic response of polycrystalline aggregates exhibiting transverse isotropy. *Journal of Nondestructive Evaluation, 3*, 1–8.

King, R. B., & Fortunko, C. M. (1983). Determination of in-plane residual stress states in plates using horizontally polarized shear waves. *Journal of Applied Physics, 54*, 3027–3035.

Landau, L. D., & Lifshitz, E. M. (1956). *Theory of Elasticity*. Oxford: Pergamon.

Murnaghan, F. D. (1951). *Finite Deformation of an Elastic Solid*. New York: Dover.

Musgrave, M. J. P. (1970). *Crystal Acoustics*. San Francisco: Holden-Day.

Ogi, H., & Hirao, M. (1996). Noncontacting measurement of residual stresses in weldments by electromagnetic acoustic resonance. *Journal of the Japan Society of Nondestructive Inspection, 45*, 875–878 (in Japanese).

Ogi, H., & Hirao, M. (1998). Electromagnetic acoustic spectroscopy in the bolt head for evaluating the axial stress. In *Nondestructive Characterization of Materials* (Vol. 8, pp. 671–676).

Ogi, H., Hirao, M., & Fukuoka, H, (1996a), Ultrasonic measurement of bending stress in electric resistance welded pipe by electromagnetic acoustic resonance. In *Proceedings of the 14th World Conference on Non-Destructive Testing* (14th WCNDT) (pp. 875–878).

Ogi, H., Hirao, M., & Fukuoka, H. (1996b), Noncontacting ultrasonic measurement of bolt axial stress with electromagnetic acoustic transducer. In *Proceedings of the 1st US-Japan Symposium on Advances in NDT* (pp. 37–42).

Okada, K. (1980). Stress-acoustic relations for stress measurement by ultrasonic technique. *Journal of the Acoustical Society of Japan (E), 1*, 193–200.

Pao, Y. -H., Sachse, W., & Fukuoka, H. (1984). Acoustoelasticity and ultrasonic measurements of residual stresses. In *Physical Acoustics* (Vol. 17, pp. 61–143). New York: Academic Press.

Schramm, R. E., Clark, A. V., McGuire, T. J., Filla, B. J., Mitraković, D. V., & Purtscher, P. T. (1993). Noncontact ultrasonic inspection of train rails for stress. In *Rail Quality and Maintenance for Modern Railway Operation* (pp. 99–108). Dordrecht: Kluwer Academic Press.

Thompson, R. B., Lee, S. S., & Smith, J. F. (1984). Microstructure independent acoustoelastic measurement of stress. *Applied Physics Letters, 44*, 296–298.

Thurston, R. N. (1964). Wave propagation in fluids and normal solids. In *Physical Acoustics* (Vol. 1A, pp. 1–110). New York: Academic Press.

Thurston, R. N. (1974). Waves in solids. In *Encyclopedia of Physics* (Vol. VIa/4, pp. 109–308). Berlin: Springer.

Toupin, R. A., & Bernstein, B. (1961). Sound waves in deformed perfectly elastic materials, acoustoelastic effect. *The Journal of the Acoustical Society of America, 33*, 216–225.

Urashima, C., Sugino, K., & Nishida, S., (1992). Generation mechanism of residual stress in rails. In *Residual Stress-III Science and Technology* (pp. 1489–1494). Amsterdam: Elsevier.

Chapter 13
Measurement of Induction Hardening Depth

Abstract In case of inhomogeneous materials, the resonance frequency is governed by the elastic constants and density at the positions of maximum stress and amplitude, respectively, which locate in the medium, depending on the individual resonant mode. Different resonant modes provide the information on the elastic property at different positions, and the compilation of them can inversely construct their distribution. This idea is applied in the present chapter, where the axial-shear-wave resonance infers the case depth in cylindrical steel rods induced by induction hardening.

Keywords Carbon steel rod · EMAR spectrum · Phase transformation · Residual stress · Surface treatment · Vickers hardness

13.1 Sensing Modified Surface Layers

Induction hardening is one of the classical surface modification methods for steel products such as shafts and gears. The surface is heated using a high-frequency AC current up to the austenitic phase (fcc) and then quenched. Phase transformation (fcc to bct) produces a case of hardened metal and compressive residual stress, both of which enhance tolerance for fatigue, wear, and corrosion during service. There are currently two techniques for case depth testing. One is direct measurement of hardness on the cross section, but it is destructive. The other detects the ultrasonic backscattering signals from the grain-structure change in the steel parts, which is less laborious and nondestructive (Good 1984; Fujisawa and Nakanishi 1989). Because the core metal has larger grains, the core-to-case interface produces a backscattering echo, whose arrival time relative to the front-surface echo gives the case thickness in an A-scope display. It, however, requires the parts to be immersed in a water tank for acoustic coupling, making online inspection inapplicable. Another shortcoming is that the dominant front echo masks the signal from a shallow interface, and this technique fails for a thin surface layer.

M. Hirao and H. Ogi, *Electromagnetic Acoustic Transducers*,
Springer Series in Measurement Science and Technology,
DOI 10.1007/978-4-431-56036-4_13

The surface treatment also changes the elastic constants and density from their initial values. This effect is often neglected, but it is of great value from a standpoint of nondestructive evaluation of surface hardening. Although the magnitude is only on the order of 10^{-2}, sensitive ultrasonic velocity measurements can easily detect and characterize the modified surface layer. For the cylindrical workpieces, radial variation of the elastic wave velocities occurs, accompanying the metallurgical modification. The best choice in such conditions is to use the axial shear wave (Sect. 3.7), which travels along the circumference of a cylindrical surface with the axial polarization and penetrates from the surface depending on the individual modes. EMAR spectroscopy with an axial shear wave has been proven to be an alternative to the destructive and immersion testing.

13.2 Axial-Shear-Wave Resonance

An axial shear wave is a pure shear wave trapped at the free surfaces of cylindrical rods (and also pipes) and is only available with EMATs. In the cylindrical coordinate system r, θ, and z, the axial shear wave carries only the axial displacement w and travels in the circumferential direction. Displacement w is independent of z, that is, $w = w(r, \theta)$. The governing equation becomes:

$$\rho \ddot{w} = \frac{1}{r}\frac{\partial}{\partial r}\left(\mu r \frac{\partial w}{\partial r}\right) + \frac{1}{r}\frac{\partial}{\partial \theta}\left(\frac{\mu}{r}\frac{\partial w}{\partial \theta}\right), \tag{13.1}$$

where ρ is the mass density and μ the shear modulus. We assume a long cylinder of an isotropic material so that the end-reflected waves are neglected. The solution should satisfy the boundary condition that the outer surface be free from any traction, that is, $\partial w / \partial r = 0$ at $r = R$ for radius R of the solid cylinder. To find the resonance frequencies for homogeneous ρ and μ, we substitute a harmonic solution $w = W(r)\exp[i(\omega t - n\theta)]$, in which an index n defines the periodicity around the circumference (see Eq. (3.4)). Equation (13.1) yields $W(r) = AJ_n(kr)$ with an amplitude A and wavenumber k. Application to the boundary condition leads to the resonance condition $J'_n(kR) = 0$ or $nJ_n(kR) - kRJ_{n+1}(kR) = 0$ (Eq. (3.3)), whose roots correspond to the resonance frequencies. For large n, the minimum root is asymptote to $kR \approx n + 0.809n^{1/3}$.

From the character of the Bessel function, $|W(r)|$ shows m peaks for a resonance at $f_m^{(n)}$, of which the innermost one has the maximum height, and it decays approaching the center of the cylinder (see Fig. 13.1). The oscillation is concentrated to the surface area for the fundamental mode ($m = 1$). The degree of concentration intensifies with n. Overtones vibrate in deeper regions as m increases. Considering that the individual resonance frequency is determined by a convolution integral of the eigenfunction $J_n(kr)$ with the radial variation of ρ and μ, the fundamental mode provides a measure of the material modification in the surface skin,

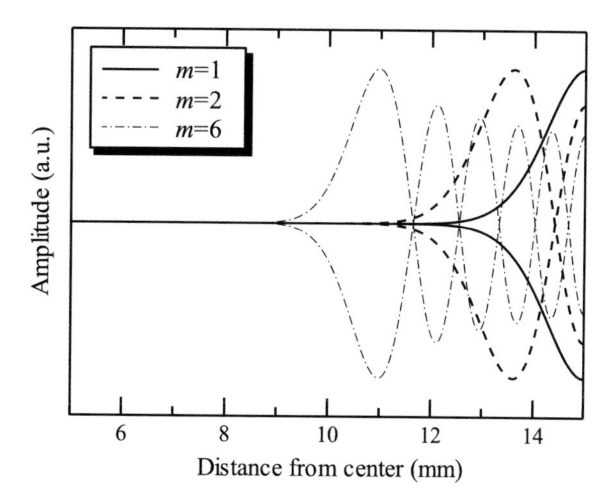

Fig. 13.1 Amplitude distribution of three axial shear resonance modes (R = 15 mm, n = 73)

and the higher modes probe the inner region. Slight radial inhomogeneity of elastic properties will then cause different frequency shifts, which provide the basis of inverse evaluation of the subsurface gradient.

This unique nature of an axial shear wave was first explored by Johnson et al. (1994), who demonstrated the principle of systematically selecting the penetration into the material. They designed an EMAT (n = 12), relying on the Lorentz force mechanism (Fig. 3.18); they detected the resonance frequency shifts of the lowest ten modes with case-hardened automobile shafts of 23.8 mm diameter. But, the hardness profile was not reconstructed from the measurements.

Hirao et al. (2001) dealt with the same problem using a magnetostrictively coupling EMAT (Fig. 3.17) of n = 73 and 104. Measured spectral response allows reconstructing the surface layer profile in conjunction with a linear perturbation scheme and a single-layer model for the effect of the material's radial inhomogeneity. Inferred case depth is favorably compared with the destructive measurement of hardness profile within 0.1 mm error. The analytical approach and results are described in the following sections.

The advantage over the conventional immersion technique includes contactless measurement and the capability of evaluating thin surface layers. Moreover, this ultrasonic method applies to other surface modification such as carburizing, nitriding, and cold working, because these treatments also cause small fractional changes of elastic properties (Hirao et al. 1983) and the basic formula is equally valid. These aspects facilitate the installation of this less consuming, more accurate, and robust inspection technique for steel cylinders in an actual production environment. It should be noted at the same time that the proposed EMAT technique requires a *smooth* cylindrical surface and axisymmetric surface layers. Automobile shafts are often machined to have spiral surface shapes. Unless the groove is very shallow, the present approach is inapplicable. For an eccentric layer, the frequency change will give a circumferential average of microstructural effects. In the following, we neglect the effect of compressive residual stress, which occurs with the volume expansion accompanied by the martensitic transformation and contributes to change the resonance frequencies as well. But, referring to a reported magnitude of 10^{-6} MPa^{-1} for the acoustoelastic constants of steels (Chap. 12), we estimate the magnitude to be on the order of 10^{-3}, one order less than that caused by the material modification itself. For more accurate determination, the acoustoelastic effect should be accounted for in the calibration and analysis.

13.3 Linear Perturbation Scheme

For interpreting the measured shift of resonance frequencies and inferring the case depth, we derive the resonance frequency shift following the perturbation theory developed by Auld (1973). The theory is based on the reciprocal theorem of linear elasticity and the first-order approximation in terms of small deviations.

Consider two cylinders of different ρ and μ, but subjected to the same n, R, and boundary condition. Their density and elastic modulus, (ρ_0, μ_0) and (ρ_1, μ_1), may change in an axisymmetric way. The axial shear wave accompanies the displacements w_0 and w_1 in them, which are the solutions of $L(w_0; \rho_0, \mu_0) = 0$ and $L(w_1; \rho_1, \mu_1) = 0$, respectively:

$$L(w; \rho, \mu) \equiv \rho\, \ddot{w} - \frac{1}{r}\frac{\partial}{\partial r}\left(\mu r \frac{\partial w}{\partial r}\right) - \frac{1}{r}\frac{\partial}{\partial \theta}\left(\frac{\mu}{r}\frac{\partial w}{\partial \theta}\right). \tag{13.2}$$

Calculating the identical equation

$$\int_S L(w_0; \rho_0, \mu_0) w_1 \mathrm{d}S - \int_S L(w_1; \rho_1, \mu_1) w_0 \mathrm{d}S = 0, \tag{13.3}$$

with Eq. (13.2), assuming harmonic solutions of angular frequencies ω_0 and ω_1, partial integrating, and using the stress-free boundary condition at $r = R$, we find

$$\begin{aligned} 2\frac{\Delta\omega}{\omega_0} \int_S \rho_0 w_0 w_1 \mathrm{d}S + \int_S \Delta\rho\, w_0 w_1 \mathrm{d}S \\ - \frac{1}{\omega_0^2} \int_S \Delta\mu \left(\frac{\partial w_0}{\partial r} \cdot \frac{\partial w_1}{\partial r} + \frac{1}{r}\frac{\partial w_0}{\partial \theta} \cdot \frac{1}{r}\frac{\partial w_1}{\partial \theta} \right) \mathrm{d}S = 0, \end{aligned} \tag{13.4}$$

in which $\Delta\rho(r) = \rho_1 - \rho_0$, $\Delta\mu(r) = \mu_1 - \mu_0$, and $\Delta\omega/\omega_0 = (\omega_1 - \omega_0)/\omega_0$ are the normalized difference between the resonance frequencies observed in the two cylinders. Integrations are done over the whole cross section, S. This is an exact expression for w_0 and w_1 and is valid for any combination of (ρ_0, μ_0) and (ρ_1, μ_1).

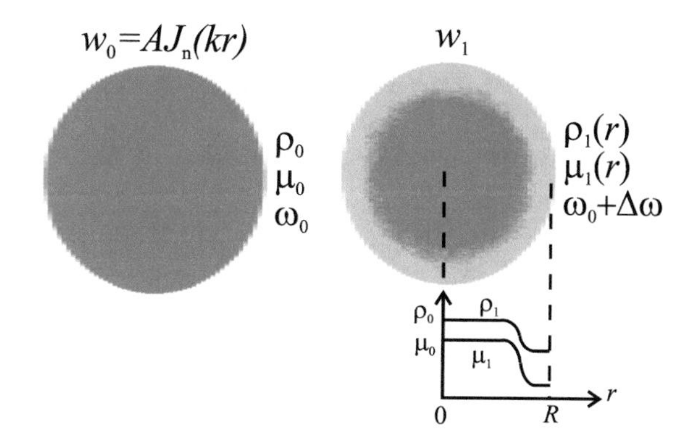

Fig. 13.2 Model for axial-shear-wave resonance in a case-hardened cylinder. Reprinted from Hirao et al. (2001), Copyright (2001), with permission from Elsevier

Now, suppose that (ρ_0, μ_0) and (ρ_1, μ_1) are the properties before and after the heat treatment, respectively, and that (ρ_0, μ_0) is constant throughout the cylinder (Fig. 13.2). Because the density and elastic moduli are insensitive to the microstructural changes in common metals, being limited to the relative order of 10^{-2} or less, we can restrict the problem to the weak perturbation of $|\Delta\rho|/\rho_0 \ll 1$ and $|\Delta\mu|/\mu_0 \ll 1$. We already had the homogeneous solution of w_0, but not for w_1. Their relative difference must be in the same order of magnitude as $|\Delta\rho|/\rho_0$ and $|\Delta\mu|/\mu$. By neglecting the second- and higher-order terms in the small quantities in Eq. (13.4), which implies making $w_0 = w_1$ in the integrands, and substituting $w_0 = AJ_n(kr)\cos n\theta$, we obtain the relation between the frequency shift and the material perturbation:

$$\left.\frac{\Delta\omega}{\omega_0}\right|_n = \frac{\int_0^R r\left\{-\left(\frac{\Delta\rho}{\rho_0}\right)J_n^2(kr) + \left(\frac{\Delta\mu}{2\mu_0}\right)\left[J_{n-1}^2(kr) + J_{n+1}^2(kr)\right]\right\}\mathrm{d}r}{2\int_0^R rJ_n^2(kr)\mathrm{d}r}. \tag{13.5}$$

The first term in the integrand occurs with the perturbation in the kinetic energy, and the second term occurs with the perturbation in the elastic strain energy. They are normalized by the unperturbed power flow to give rise to the frequency shift. The changes in material's properties contribute to different weights and directions (increase or decrease). Density increase causes frequency decrease, and modulus increase causes frequency increase. It is reasonable because ω is proportional to the phase velocity for fixed k and the phase velocity is basically given by (modulus/density)$^{1/2}$.

It is straightforward to calculate $\Delta\omega/\omega_0$ from the known functions of r for $\Delta\rho/\rho_0$ and $\Delta\mu/\mu_0$. However, the present situation is reversed; we have to infer the case depth from the measured $\Delta\omega/\omega_0$. We solve this inverse problem with a supplementary model for the radial inhomogeneity. In the case of induction hardening of steels, the martensitic transformation saturates within a surface layer and then a step function is an appropriate model, where the matrix (core) is overlaid with a layer of hardened metal (case) of a constant thickness h and a uniform perturbation, $(\Delta\rho, \Delta\mu)$:

$$\begin{aligned} &\Delta\rho = \Delta\mu = 0 && (0 \le r \le R - h) \\ &\Delta\rho = \text{const. and } \Delta\mu = \text{const.} && (R - h \le r \le R) \end{aligned} \quad (13.6)$$

With the above model, Eq. (13.5) simplifies to

$$\left.\frac{\Delta\omega}{\omega_0}\right|_{n,m} = A(h/R; n, m)\left(\frac{\Delta\rho}{\rho_0}\right) + B(h/R; n, m)\left(\frac{\Delta\mu}{\mu_0}\right), \quad (13.7)$$

where

$$\begin{aligned} A(h/R; n, m) &= -\frac{\int_{R-h}^{R} rJ_n^2(kr)\mathrm{d}r}{2\int_0^R rJ_n^2(kr)\mathrm{d}r}, \\ B(h/R; n, m) &= \frac{\int_{R-h}^{R} r\left[J_{n-1}^2(kr) + J_{n+1}^2(kr)\right]\mathrm{d}r}{4\int_0^R rJ_n^2(kr)\mathrm{d}r}. \end{aligned} \quad (13.8)$$

We see that $\Delta\omega/\omega_0$ of individual resonance mode depends on *h/R* as well as *n* and *m*. Values for $\Delta\rho/\rho_0$ and $\Delta\mu/\mu_0$ should be calibrated with the test steel.

Auld's perturbation theory (1973) has been extensively studied to account for and utilize the Rayleigh wave dispersion caused by the surface treatments on steels (Szabo 1975; Richardson 1977; Tittmann et al. 1987; Hirao et al. 1990). The theoretical formulation was much more complicated than the present case because the Rayleigh wave propagation is a mixture of dilatational and shear deformations, being associated with two displacement components. For this, it was difficult to obtain the desired results from the measurements of the frequency-dependent phase velocity. Mathematical simplicity of axial shear resonance modes is a big advantage.

13.4 Inverse Evaluation of Case Depth

Six specimens, being 30 mm diameter and 200 mm long, from a drawn carbon steel rod (C: 0.47 mass%) were prepared. Their central parts of 100 mm long were given a series of induction hardening (not tempered) to introduce the surface layers; the thickness ranged from 0.70 to 3.06 mm. Figure 13.3 presents the nearly stepwise distribution of Vickers hardness on the cross sections.

The measurement setup was the same as that shown in Fig. 3.17. The solenoid coil supplied a bias field around 0.03 T along the length. A meander-line coil of 20 mm wide was used, which was printed onto a flexible polymer film at a fixed spacing of $\delta = 2\pi R/n \approx 1.29$ mm with $n = 73$. The specimen was loosely wrapped with the meander-line coil, providing a noncontact acoustic coupling. The meander-line coil was driven by rf bursts, typically of 100 V_{pp} and 150 μs duration, to induce the alternating dynamic field in the circumferential direction on the specimen surface,

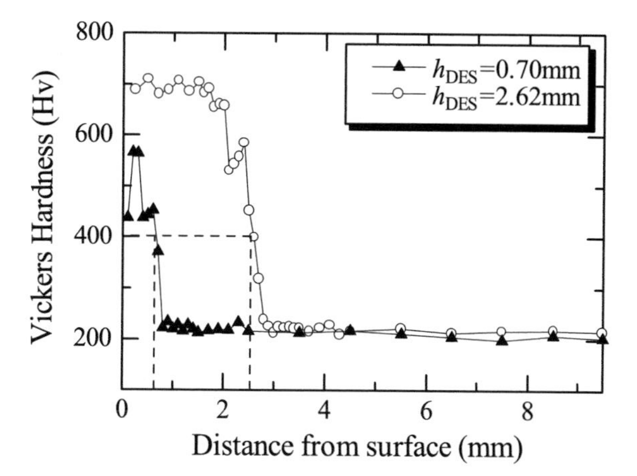

Fig. 13.3 Vickers hardness distribution measured by indentation at 2.0-N load. Reprinted from Hirao et al. (2001), Copyright (2001), with permission from Elsevier

which was superimposed on the static axial field. Then, the surface elements beneath the coil underwent shear deformation responding to the obliquely oriented total fields, which generated the shear waves of the axial polarization. Reversed magnetostrictive mechanism altered the magnetic field in the specimen surface, which was picked up by the same meander-line to form the received signals.

Swept frequency operation measured the resonance spectrum of each specimen; the details of superheterodyne analog circuitry and operations appear in Chap. 5. A single frequency scan typically took 10 s, depending on the frequency range and the stepping intervals. In Fig. 13.4, the spectra are compared for before and after induction hardening. Agreement with the frequency equation (Eq. (3.3)) was confirmed with the untreated specimens. We observe negative shifts for all the modes, reflecting a decrease of shear wave velocity in the surface layer. Going into details with the thin surface layer, a larger shift occurs for the fundamental mode (m = 1) and smaller shifts for the higher modes. For the thick layer, apparently equal shifts take place in all the detected modes. This contrast illustrates the essential feature of the axial-shear-wave resonance for measuring the radial gradient of elastic properties.

A Lorentzian function fits the spectral data around the maximum amplitude for reducing the resonance frequency from the center axis. The shifts are normalized with the original measurements and are summarized in Fig. 13.5 up to the sixth mode. As h increases, $\Delta\omega/\omega_0$ of all the modes monotonically approaches the saturation value of −1.1 %. The decrease starts to occur first with the fundamental mode, the second mode follows, then the third mode, and so on. When the penetration is entirely within the surface layer, (ρ, μ) of the hardened steel determines the resonance frequency and the core metal has nothing to do with the observed resonance. When the penetration is moderate, encompassing both core and case, the resonance frequency is given by ρ and μ of the two slightly different metals. Equation (13.7) provides the balance of influence of the two media, which depends on h and the amplitude pattern shown in Fig. 13.1.

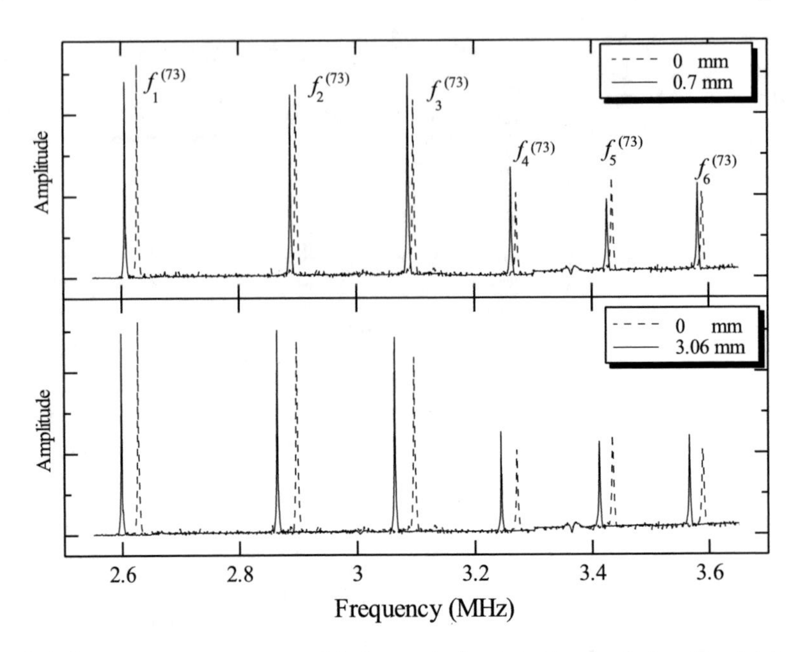

Fig. 13.4 Resonance spectra measured before and after induction hardening. Reprinted from Hirao et al. (2001), Copyright (2001), with permission from Elsevier

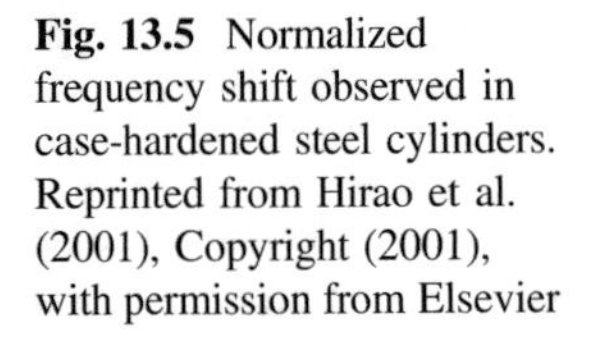

Fig. 13.5 Normalized frequency shift observed in case-hardened steel cylinders. Reprinted from Hirao et al. (2001), Copyright (2001), with permission from Elsevier

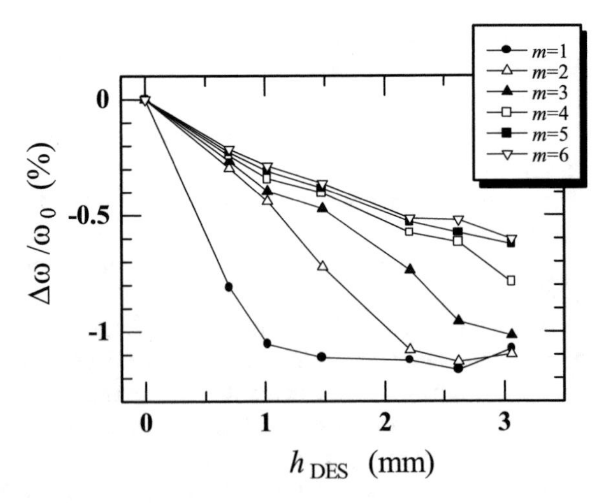

The Archimedes method gave $\Delta\rho/\rho_0 = -0.71$ % with a coupon specimen cut from the thickest surface layer. It is substantially equivalent to the estimate of $\Delta\rho/\rho_0 = -0.79$ % for the same carbon content (Lement 1956). Then, $\Delta\mu/\mu_0 = -2.88$ % results using Eq. (3.3) along with $\Delta\rho/\rho_0 = -0.71$ % and the saturation value of $\Delta\omega/\omega_0 = -1.1$ %. The martensitic transformation decreases both ρ and μ, but with more influence on μ, causing the decrease of axial-shear-wave resonance frequencies.

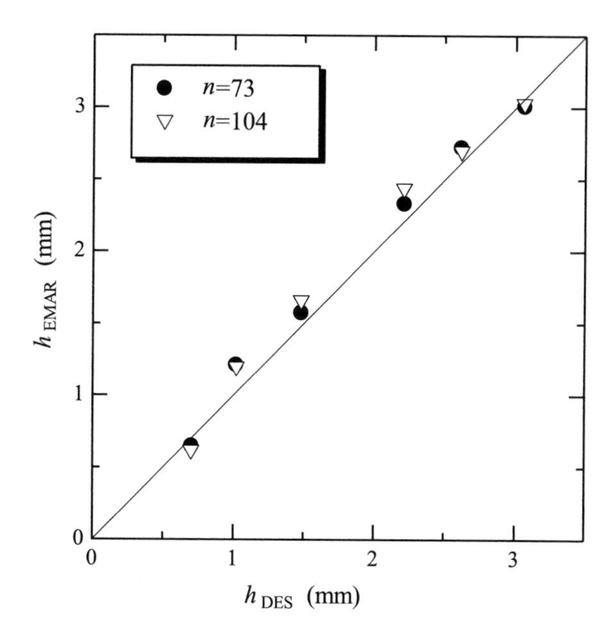

Fig. 13.6 Comparison between the case depth measured with acoustic resonance and Vickers hardness. Reprinted from Hirao et al. (2001), Copyright (2001), with permission from Elsevier

The transformation also causes the expansion of specimens. But, the magnitude is estimated to be on the order of 10^{-4}, and the dimensional change can be ignored in interpreting the measured spectral response. The remaining unknown variable is therefore h, which was determined so as to best satisfy Eq. (13.7) with the measured $\Delta\omega/\omega_0$.

Figure 13.6 compares the best fitting h against the case depth determined by the Vickers hardness measurement referring to Hv = 400. Excellent agreement is observed between them. Mode selection via n and m is important to deduce the correct result. The resonance modes are selected such that they exhibit the transitional shift for each specimen, $m = 1$ for $h = 0.70$ mm and $m = 2$ for $h = 1.48$ mm, for example. Otherwise, the resonance frequency shift leads to an inaccurate value or was unable to yield an answer.

Another meander-line coil of $n = 104$ ($\delta = 0.42$ mm) was tested. This coil was short and covered less than a half of the circumference. Despite the decreased signal strength, it was also capable of measuring the resonance spectra and resulting in essentially the same evaluation of the case depth, which is included in Fig. 13.6.

References

Auld, A. B. (1973). *Acoustic Fields and Waves in Solids*. New York: Wiley.

Fujisawa, K., & Nakanishi, A. (1989). Hardening depth measurement by using ultrasonic backscattering. *Nondestructive Testing Communications, 4*, 131–136.

Good, M. S. (1984). Effective case depth measurement by an ultrasonic backscatter technique. *Metals/materials technology series #8408-003, ASM Metals Congress*, pp. 1–6.

Hirao, M., Fukuoka, H., Toda, H., Sotani, Y., & Suzuki, S. (1983). Non-destructive evaluation of hardening depth using surface-wave dispersion patterns. *Journal of Mechanical Working Technology, 8*, 171–179.

Hirao, M., Ogi, H., & Minami, Y. (2001). Contactless measurement of induction-hardening depth by an axial-shear-wave EMAT. In *Nondestructive Characterization of Materials* (Vol. 10, pp. 379–386).

Hirao, M., Tanimoto, N., & Fukuoka, H. (1990). Nondestructive evaluation of hardening depth on cylindrical surface using dispersive Rayleigh wave. In *Review of Progress in Quantitative Nondestructive Evaluation* (Vol. 9, pp. 1603–1609).

Johnson, W., Auld, B. A., & Alers, G. A. (1994). Application of resonant modes of cylinders to case depth measurement. In *Review of Progress in Quantitative Nondestructive Evaluation* (Vol. 13, pp. 1603–1610).

Lement, B. S. (1956). *Distortion in Tool Steels*. USA: ASM.

Richardson, J. M. (1977). Estimation of surface layer structure from Rayleigh wave dispersion: Dense data case. *Journal of Applied Physics, 48*, 498–512.

Szabo, T. L. (1975). Obtaining subsurface profiles from surface-acoustic-wave velocity dispersion. *Journal of Applied Physics, 46*, 1448–1454.

Tittmann, B. R., Ahlberg, L. A., Richardson, J. M., & Thompson, R. B. (1987). Determination of physical property gradients from measured surface wave dispersion. *IEEE Transactions on Ultrasonics, Ferroelectrics and Frequency Control*, *UFFC-34*, 500–507.

Chapter 14
Detection of Flaw and Corrosion

Abstract Flaw detection and sizing on industrial materials are always of the first priority of nondestructive testing. EMAT techniques offer unique flaw detection methods, two of which are picked up in this chapter (see Chap. 18 for practical inspection cases). One is to use SH plate modes in plates or torsional modes in pipes. Group velocity measurement, at the frequency and wavenumber pinpointed on the dispersion curves, and the mode conversion that occurs at the defect edges, provide the information on the remaining thickness of the waveguides. The other is the point-focusing EMAT, which enables detecting stress corrosion cracking located close to the welded joint of austenitic stainless steel. A meander-line coil is designed to focus the generated SV waves at an intended point.

Keywords Austenitic stainless steel · Cutoff thickness · Group velocity dispersion · Guided SH modes · Mode conversion · PPM EMAT · Stress corrosion cracking (SCC) · SH_0/SH_1 modes · SV wave · Wall thinning

14.1 Gas Pipeline Inspection

Polyethylene pipes are now adopted to construct gas transmission lines and to replace damaged steel pipes. They are lightweight yet resistant to corrosion and cracking. However, steel pipes are still commonly used in many countries, and corrosion remains a major problem. Despite protective coating, the corrosion forms on the outer surfaces of buried pipes. There has been great interest to develop a reliable and easy-to-use inspection technique to ensure structural integrity of steel pipelines. One wishes to inspect underground pipelines from the inside without excavation work and service shutdown. To meet this objective, the whole apparatus should be compact to be fitted in the pipe and be mounted on a *pig* moving platform. The system should stand alone or operate under a remote control unit and consume minimum electricity. Quick inspection is a top priority, since the pipelines extend over kilometers. Detailed defect characterization is, however, not always necessary for screening purposes.

M. Hirao and H. Ogi, *Electromagnetic Acoustic Transducers*,
Springer Series in Measurement Science and Technology,
DOI 10.1007/978-4-431-56036-4_14

A possible solution is to use low-frequency guided waves excited by EMATs. They can propagate axially or circumferentially over a long distance and return with pertinent information to the remote transducer. This is feasible because of reduced propagation loss and relatively high transfer efficiency at lower frequencies as well as limited space of waveguide. The noncontact nature of EMATs is the key to establish a robust implementation, accommodating unfavorable surface conditions. The allowable liftoff is limited to a few millimeters even for frequencies less than 1 MHz. This will not matter in practice, because pipes of fixed dimensions are to be inspected. A number of EMAT techniques have been investigated and tested for on-site applications (Thompson et al. 1972; McIntire 1991; Thompson 1997). Most of them used the fundamental Lamb wave (S_0 mode) propagating in the axial or circumferential direction. They were, however, often impractical for coated steel pipes, since the polymer coating layer absorbed much wave energy. Alers and Kerr (1998) circumferentially moved a pair of EMATs to pitch/catch SH waves propagating in the axial direction of the pipe. This technique is suitable for weldment inspection, but not so for the corrosion detection over a long distance.

14.1.1 Linear Scanning with PPM EMAT

More appropriate is a periodic-permanent-magnet (PPM) EMAT (Sect. 3.3), which is moved along the steel pipe to generate and receive the SH guided waves that propagate in the circumferential direction. Results of a preliminary study demonstrate the high-speed operation, the sensitivity to small artificial corrosion, and the potential to evaluate remaining wall thickness (Hirao and Ogi 1999). There are several benefits in using SH guided elastic waves. (i) The simple dispersion character leads to easier interpretation of measurements. (ii) The wavenumber is fixed by the period of the magnet array, and the individual guided modes can be selected through the driving frequency. (iii) The SH waves tolerate damping from the protective coating of lossy polymers, partly because they are independent of the stress continuity boundary condition at the interface.

A PPM EMAT was connected to the superheterodyne phase-sensitive detector (Chap. 5). The active area was 20 mm wide and 40 mm long. It contained twelve slices of Nd-Fe-B magnets magnetized in opposite directions, being normal to the surface (Fig. 14.1). The period of the magnet array was $\delta = 7.53$ mm, which equaled the wavelength of the excited SH waves. An elongated spiral coil of 28 turns was wound with enameled copper wire. The EMAT was placed inside the pipe to generate SH waves, which propagate circumferentially with the axial polarization. The same transducer detected a series of round-trip signals. The first arriving signal was gated out to obtain the amplitude and phase at the operating frequency.

Feasibility tests were done with short sections (70 cm long) of natural gas transmission steel piping of 216 mm outer diameter and 5.4 mm (or 6.3 mm) wall thickness. They were manufactured by bending the rolled plates to the cylinder and then electric resistance welding; the welding debris was removed, and no reflection

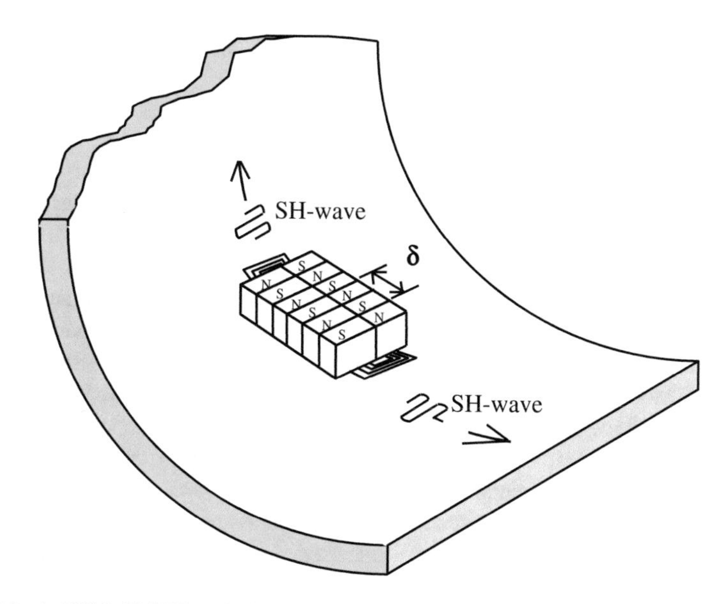

Fig. 14.1 A PPM EMAT and the generation of guided SH waves traveling along the circumference of a pipe. Reprinted from Hirao and Ogi (1999), Copyright (1999), with permission from Elsevier

signal from the weldment was observed. The outer surfaces were painted, and the inner surfaces were left as rusted.

Wheels were attached to the EMAT for laboratory tests to facilitate the linear movement in the axial direction, while continuously measuring the amplitude and phase. The gap from the EMAT case surface to the pipe surface was around 0.2 mm. The EMAT was put near one end of the pipe section and was manually pulled toward the other end in a smooth motion, taking nearly 10 s. The pipe can be vertical or horizontal. During this time, the system acquired and stored approximately five hundred sets of amplitude and phase of a selected SH mode. The amplitude represents the transmission efficiency, and the phase lag represents the transit time at an individual cross section. An axial resolution of 1 mm was possible by averaging over two or three measurements.

14.1.2 Dispersion Relation of SH Modes

Measurements are interpreted with the dispersion relation of the guided SH waves in the pipe wall. The equation of motion in cylindrical coordinates is solved for only the axial component of displacement, satisfying the boundary condition at the free surfaces (Ogi et al. 1997). Numerical calculation revealed that a large diameter–

thickness ratio reduces it to the dispersion relation in a flat plate (Meeker and Meitzler 1964; Auld 1973);

$$k^2 = (\omega/C_S)^2 - (n\pi/t)^2. \quad (14.1)$$

This equation provides the basis of interpretation. Here, ω is the angular frequency, k the wavenumber, t the thickness, and C_S the shear wave speed. The integer n characterizes the dispersion nature and is used to label the mode as SH_n.

Figure 14.2 shows the dispersion curves of SH modes in a plate, where a symmetric group of even n is drawn with solid lines and an antisymmetric group of odd n with broken lines. Except for the fundamental SH_0 mode, all are dispersive. The PPM EMAT restricts the wavelength to δ, and then, the wavenumber is fixed to $k = 2\pi/\delta$ for all the modes. This situation allows selecting the probing SH mode through the frequency to drive the EMAT. The SH_0 and SH_1 modes were measured in the following experiments. For $t = 5.4$ mm, they were excited at 0.425 and 0.52 MHz. These operating points on a flawless pipe are marked in Fig. 14.2. The resultant acoustic data are only of the intended mode and frequency, because the superheterodyne spectrometer and the fixed wavelength serve to reject other components in the raw signals from noise or from mode conversion.

In the SH_0 mode, the shear deformation occurs uniformly through the thickness, while in the SH_1 mode, it is antisymmetric around the midplane with maxima on both surfaces. Because of different deformation profiles, the two modes provide unique information on the defects, including the high sensitivity of SH_1 mode to surface anomalies.

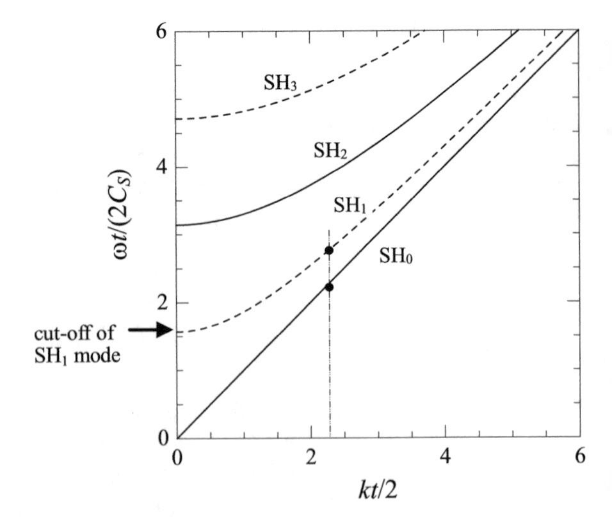

Fig. 14.2 Dispersion curves of guided SH waves for $t = 5.4$ mm. There are cutoff thicknesses for the dispersive modes of SH_n ($n > 0$). Reprinted from Hirao and Ogi (1999), Copyright (1999), with permission from Elsevier

14.1.3 *Interpretation of Measurements*

Two types of artificial defects were introduced on the pipe's outer surfaces by grinding and machining. Figure 14.3 shows results of a scanning test that focused on the SH_1 mode together with the thickness profile of the dish-shaped defects, in which the depth gradually changes in the axial and circumferential directions. The EMAT was located at a quarter circumference from the defects. Hereafter, phase data are plotted relative to the measurements at flawless parts. Both the amplitude and phase of the SH_1 mode are shifted from the normal levels and, in this sense, successfully indicate the presence of defects somewhere along the circumference. We observe a decrease in amplitude at the defects, but the magnitude does not directly indicate the defect depth. We also observe a decrease in phase for the shallow defect but an increase for the deeper defects. In practice, there is a complicated intermode energy transfer at the positions of thickness change, including nonpropagating (imaginary) modes whose amplitudes decay exponentially from the discontinuities (Meeker and Meitzler 1964; Auld 1973). Qualitative explanation is possible considering three mechanisms: mode conversion, the group velocity dispersion, and wave diffraction around the corroded area. Exact explanation will only be available by solving this three-dimensional, time-dependent elastic problem.

The defect behaves as an obstacle for the propagating SH wave. When it impinges upon the wall thinning, transmission, reflection, and mode conversion take place (see a simplified illustration in Fig. 14.4) and only a part of incident

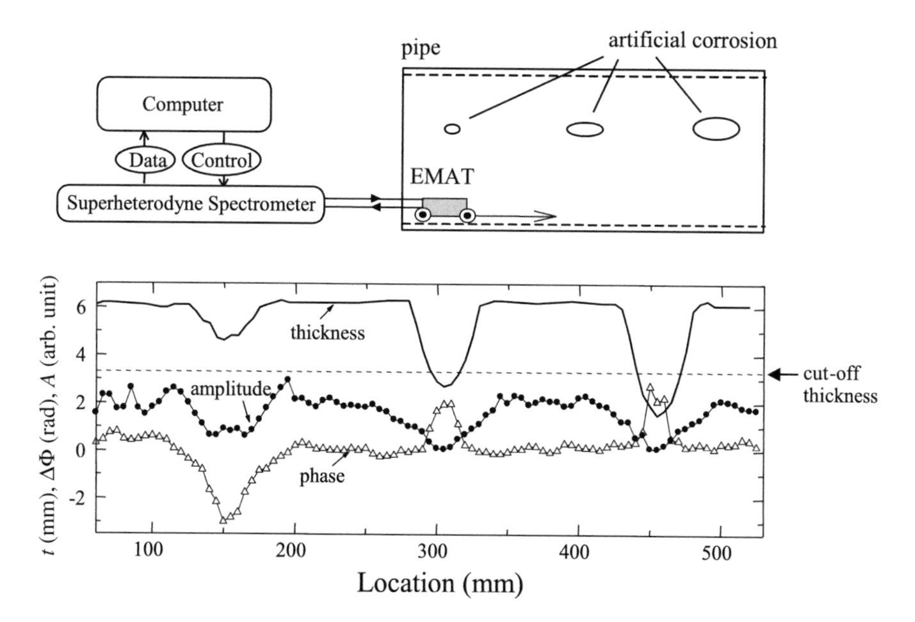

Fig. 14.3 Axial scanning data of the SH_1-mode amplitude and phase shift for dish-shaped defects (t = 6.3 mm). The defect profiles are determined with a dial gauge reading. Reprinted from Hirao and Ogi (1999), Copyright (1999), with permission from Elsevier

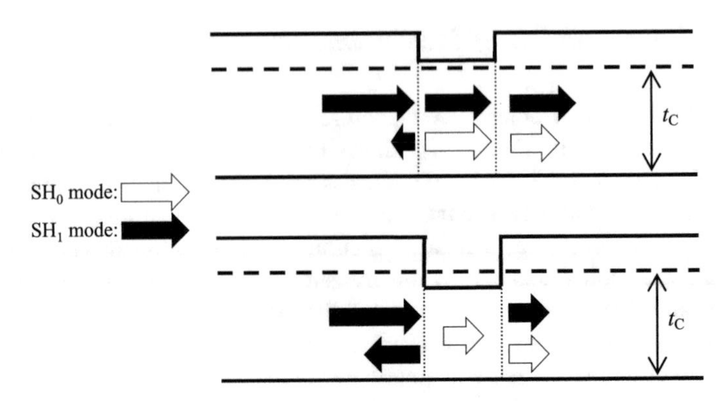

Fig. 14.4 A schematic for the mode conversion at discontinuous thickness changes. The SH$_1$ mode is incident from the right-hand side and is detected on the left-hand side. The transmitted amplitude is larger for $t_{min} > t_C$ than for $t_{min} < t_C$. Reprinted from Hirao and Ogi (1999), Copyright (1999), with permission from Elsevier

energy is eventually allotted to the original propagation mode, which returns to the transducer with amplitude smaller than that of the flawless parts. At a fixed frequency, the wavelength is adjusted in the thinner part so as to match the dispersion relation there. This occurs in a straightforward manner, and the velocity remains unchanged with the SH$_0$ mode, because t modifies the normalized frequency and wavenumber at the same rate. But, it is not so for the SH$_1$ mode. The group velocity diminishes with thinning, and this mode cannot propagate for thickness less than the *cutoff thickness* defined by $t_C = n\pi C_S/\omega$, where ω satisfies the dispersion relation, Eq. (14.1). This cutoff thickness t_C takes the value of 3.1 mm for the normal thickness of $t = 5.4$ and 3.3 mm for $t = 6.3$ mm. When the remaining thickness is larger than t_C, the defect supports the propagation both in the SH$_0$ and in the SH$_1$ modes, which are mode-converted back to the SH$_1$ mode on the far side, returning to the EMAT. Involvement of the SH$_0$ mode of larger group velocity results in the earlier arrival and the phase drop of the SH$_1$ mode relative to the flawless circumference. Because the thickness varies in the axial direction and the group velocity decreases with the thickness, the dish-shaped defect acts as a lens and focusing occurs (Thompson et al. 1972). This effect causes a small amplitude rise at the center of the shallowest defect in Fig. 14.3. When the remaining thickness is smaller than t_C, the SH$_1$ mode cannot transmit the energy straight across the defect and the most incident energy is reflected back to the near side. The SH$_1$ wave then detours around the defect, causing the delayed arrival and the phase jump. The mode-converted SH$_0$ wave seems to contribute little.

A further experiment was made using a rectangular defect with a constant axial width. The defect was machined deeper at many steps, keeping the flat face of the 50 mm axial length and the edges parallel to the axial and circumferential directions of the pipe. The amplitude and phase of the two modes are measured at the center of the defect for every step of machining. The EMAT was placed opposite the defect. Figure 14.5 shows the results. We observe that the amplitude and phase of the SH$_0$

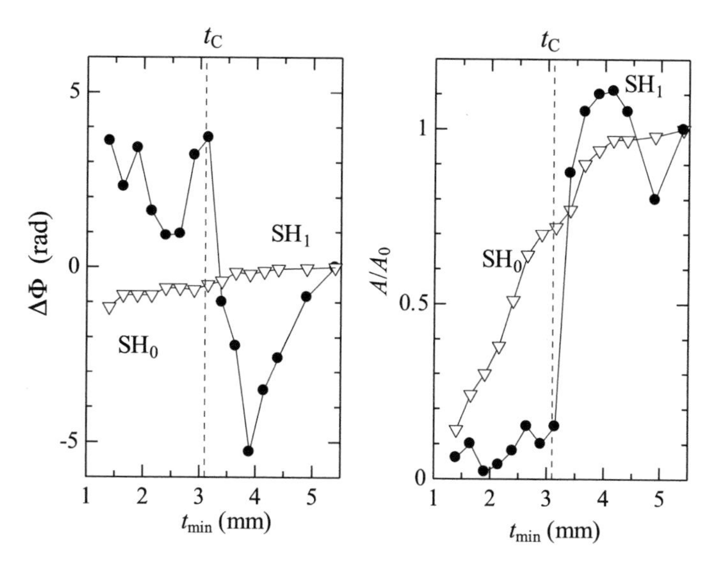

Fig. 14.5 Dependence of phase and amplitude on t_{min} of the rectangular defect (t = 5.4 mm). The broken lines indicate the cutoff thickness of the SH_1 mode. Reprinted from Hirao and Ogi (1999), Copyright (1999), with permission from Elsevier

mode monotonically decrease with the minimum thickness, t_{min}. The decrease in phase can be explained by a reduction of the effective propagation distance caused by the wall thinning from the outside. The amplitude ratio is proportional to t_{min} in the range of $t_{min} < t_C$, which is the consequence of the uniform amplitude across the thickness and no mode conversion to the SH_1 mode.

The SH_1 mode shows a drastic response to the wall thinning. The phase changes in a complicated way, but we always see an increase in phase and a very small amplitude for $t_{min} < t_C$, where the received signal consists of mode-converted (via the SH_0 mode through the defect) and diffracted SH_1 waves. There is interference among them, and the phase delay indicates that the diffraction makes a larger contribution. For $t_{min} > t_C$, the through-transmitted SH_1 mode also participates to make the further complicated interference. We observe a clear difference of amplitude ratio between above and below t_C. The amplitude ratio appears to be larger than 1 for $t_{min} \sim 4$ mm, which could be understood by a positive interference among the constituents.

Also important is the remarkable difference of the axial variation between the two thickness regions. Figure 14.6 shows the typical scanning results. Approaching the defect edges for $t_{min} > t_C$, the amplitude first increases and then decreases around them. It rises considerably and often exceeds the normal values at the midpoint because of the focusing effect. The phase is shifted down there. We see no amplitude rise for $t_{min} < t_C$, where the amplitude at the median portion is almost at the noise level. The phase increases because of diffraction of the SH_1 mode.

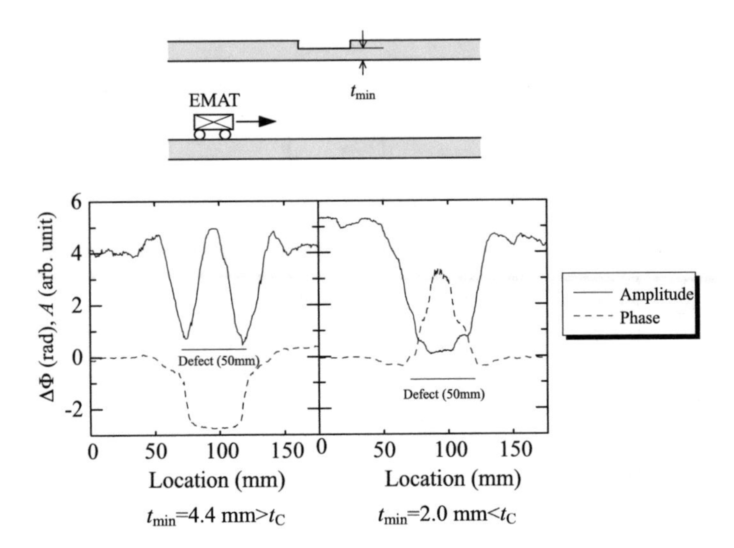

Fig. 14.6 Axial scanning with the SH_1 mode and the response to a rectangular defect at two t_{min}. Reprinted from Hirao and Ogi (1999), Copyright (1999), with permission from Elsevier

Figure 14.7 demonstrates the tolerance of the PPM EMAT technique for a coated pipeline, although the round-trip amplitude of the SH_1 wave becomes weaker by 15 dB compared with the uncoated flawless pipe of the same dimension. The standard triple-layer coating was provided, which consists of the adhesive undercoat, the anticorrosive polyethylene coating, and then the protective layer of polyolefin. A 10-mm-diameter hole of a v-shaped end was drilled through the total coating thickness of 2.5 mm to give a minimum thickness of 4.0 mm of the steel wall. Because $t_{min} > t_C$, both amplitude and phase decrease (see Fig. 14.3).

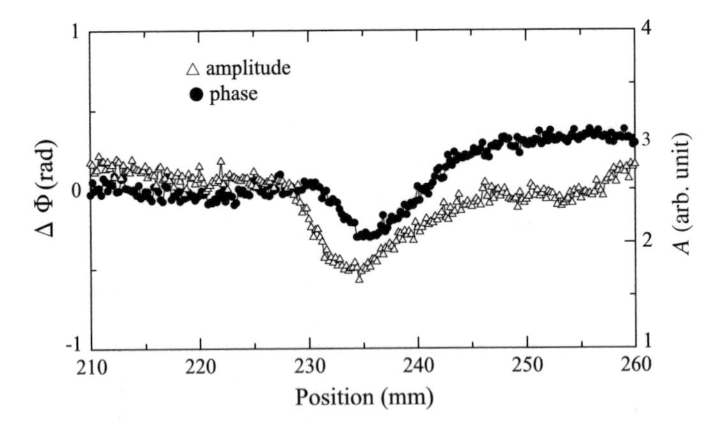

Fig. 14.7 Detection of a drilled defect in a coated steel pipe (t = 5.4 mm) using the SH_1 mode. The minimum thickness at the defect center was 4.0 mm. Reprinted from Hirao and Ogi (1999), Copyright (1999), with permission from Elsevier

This EMAT technique has shown a potential for gas pipeline inspection. The round-trip signals of SH_0 and SH_1 modes were proven to respond uniquely to the surface defects. The amplitude and phase shift of the SH_1 mode were more sensitive to the presence of the defects than those of the SH_0 mode, but they did not show a correlation with wall thinning. The phase shift of the SH_0 mode showed a good correlation with the remaining thickness. It is advisable to measure the amplitude and phase of both modes for corrosion monitoring. If one detects a phase increase and a significant amplitude decrease of SH_1 mode, it is a warning that the wall thinning has progressed below the cutoff thickness. Then, the phase delay of the SH_0 mode provides the thickness information. Further research is necessary to determine the multiple guided modes to be used, their frequencies related to the magnet spacing δ, and the size of EMAT for optimum performance.

14.2 Total Reflection of SH Modes at Tapered Edges

The reflection and transmission behaviors of SH plate modes at defects near the cutoff thickness were investigated in detail using PPM EMAT (Nurmalia et al. 2012). Used EMAT has an active area of 30 mm wide and 24 mm long, and it was constructed with 18 pieces of Nd-Fe-B magnets of 10 mm height. The magnetic periodicity was 5.22 mm, yielding the operating frequencies 0.546 MHz for SH_0 mode and 0.727 MHz for SH_1 mode. The specimens were aluminum alloy plates with 1000 mm length, 200 mm width, and 3 mm thickness. A square defect of 100 mm by 100 mm with 1.7 mm depth was machined at the center of each plate: The remaining thickness in the defected region was smaller than the cutoff thickness of SH_1 mode. The defect edge was stepped or tapered by 5° slope. A pitch-catch experimental setup was used as shown in Fig. 14.8, where the SH modes generated by the PPM EMAT were detected by a needle-type piezoelectric transducer (pinducer); the pinducer was moved in the propagation direction to investigate the mode conversion across the defect edge.

Two-dimensional representation of wave propagation was constructed with color-scaled amplitude in Fig. 14.9a, b for SH_0-mode and SH_1-mode incidences, for the stepped edge defect, where SH modes propagated from left to right. Only the first arriving signal should be considered, since the multiple reflections occur within the pinducer, leading to the prolonged waveforms. The slope in Fig. 14.9 is inversely proportional to the group velocity, from which modes are identified. When the SH_0 mode enters the defect area, it remains inside the defect. But, it mainly converts to the SH_1 mode at the far end of the defect (Fig. 14.9a). In the case of the SH_1 mode, it converts to the SH_0 mode at the near end and reconverts to the SH_1 mode at the far end (Fig. 14.9b). These unique propagation behaviors can be explained by noting the different displacement distributions across the thickness. In the SH_0-to-SH_1 mode conversion, the uniform displacement of the SH_0 mode

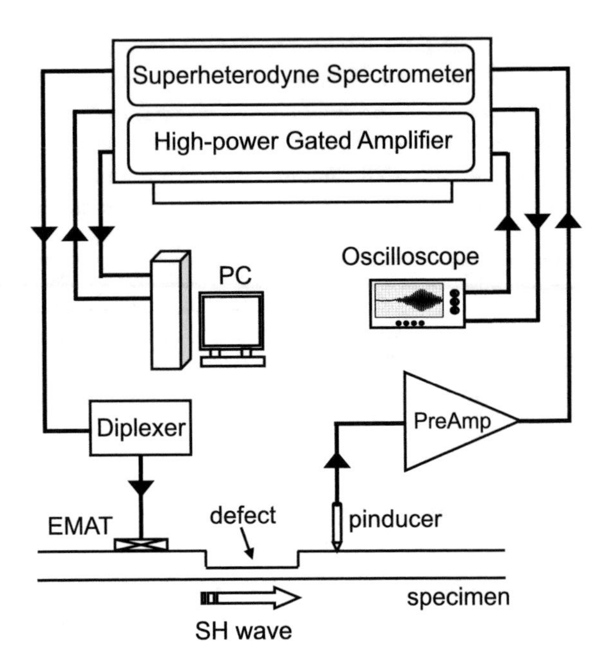

Fig. 14.8 Experimental setup. Reprinted from Nurmalia et al. (2012), Copyright (2012), with permission from Elsevier

partially matches to that of the SH_1 mode, where the asymmetric displacement distribution occurs in the thickness direction, exciting the SH_1 mode in the thicker region (Fig. 14.10a). On the other hand, when the SH_1 mode enters the thinned region, it produces the unidirectional displacement at the boundary, giving rise to the SH_0 mode there (Fig. 14.10b), which is then converted into the SH_1 mode at the far end.

A noticeable propagation behavior was found for the SH_1 mode in the tapered region as shown in Fig. 14.11: The SH_1 mode is totally reflected during propagation in the tapered region without any boundaries, and no transmitted and converted waves occur at the reflection. This total reflection occurs at the thickness of 2.17 mm, which is close to the theoretical t_C of the SH_1 mode (t_C = 2.13 mm). For clarifying the boundary-free total reflection mechanism, a two-dimensional numerical simulation was carried out for a shallower tapered region (1° slope) using the image-based elastodynamic finite integration technique (Nakahata et al. 2009). Figure 14.12 shows snapshots at three elapsed times during propagation of the SH_1 mode in the tapered region. The arrows indicate the propagation directions. As the thickness decreases, the wavenumber decreases to minimum, and then, the total reflection occurs. Shortly after this event, the wave travels in the opposite direction, gradually recovering the original wavenumber. At the minimum wavenumber, the wave energy is highly condensed and the total reflection occurs as if there were a boundary. The calculated group velocity decreases along the slope and becomes zero at the total reflection point. The thickness of this particular point is 2.12 mm, confirming the total reflection at the cutoff thickness (2.13 mm).

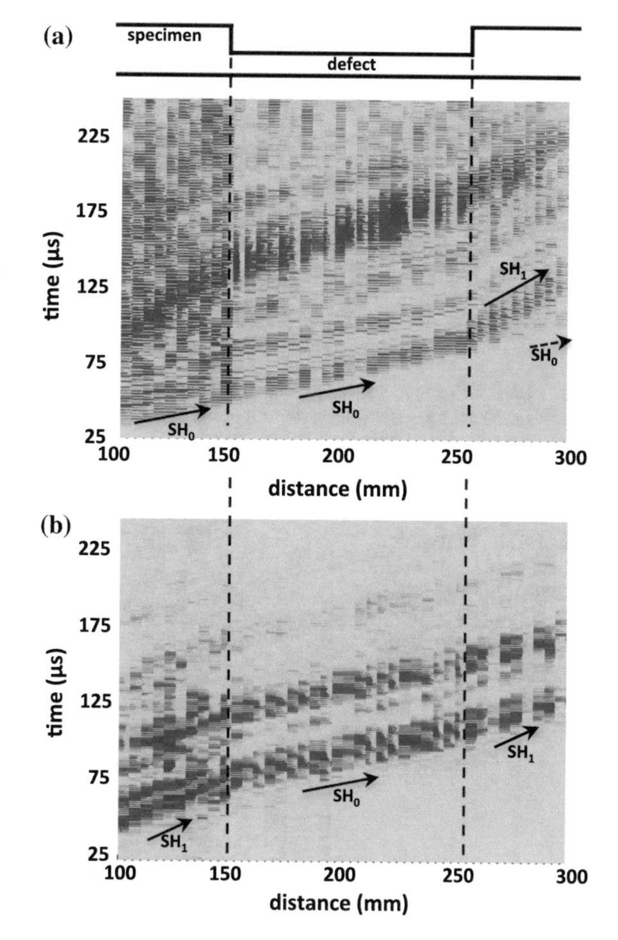

Fig. 14.9 Propagation behaviors of SH modes toward the defect with the step edges. $t_{\min} = 1.3$ mm, which is less than t_C of the SH_1 mode. Incidence waves are **a** SH_0 and **b** SH_1 modes. Reprinted from Nurmalia et al. (2012), Copyright (2012), with permission from Elsevier

14.3 Flaw Detection by Torsional Modes along Pipes

The mode conversion caused at the defect described in Sect. 14.1 and the total reflection at the tapered region in Sect. 14.2 occur for torsional waves as well. Taking advantage of the simple dispersion characteristic, a wall-thinning inspection method was proposed based on the group velocity change of the torsional wave induced by the mode conversion (Nurmalia et al. 2013). Infinite numbers of modes of torsional waves are designated as T(0, *m*), where *m* defines the radial/thickness mode parameter. T(0,1) is the nondispersive fundamental mode, and T(0,2) is the first higher dispersive mode. (They correspond to the SH_0 and SH_1 plate modes, respectively.) The dispersion equation of a torsional wave propagating in a pipe with inner and outer radii of *a* and *b*, respectively, is given by

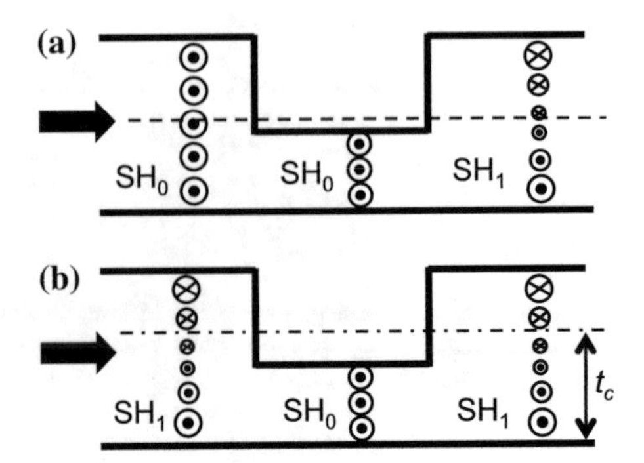

Fig. 14.10 Schematic explanations for mode conversions of **a** SH_0 to SH_1 and **b** SH_1 to SH_0. *Arrows* indicate the propagation direction, the *dashed line* shows the half-thickness, and t_C denotes the cutoff thickness of the SH_1 mode. Reprinted from Nurmalia et al. (2012), Copyright (2012), with permission from Elsevier

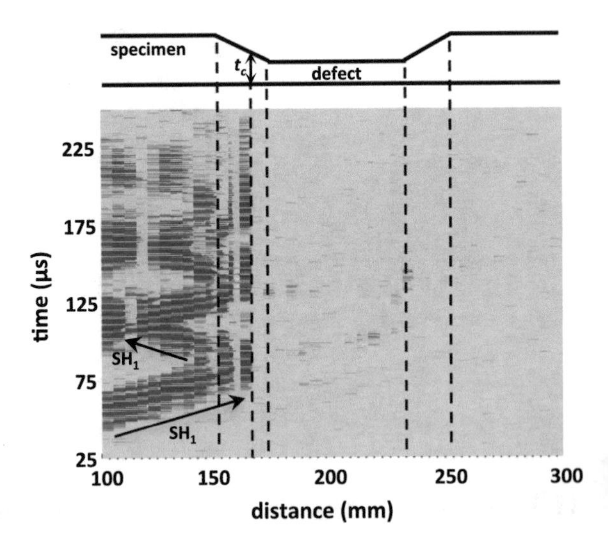

Fig. 14.11 Propagation behavior of SH_1 modes for the tapered edge. Reprinted from Nurmalia et al. (2012), Copyright (2012), with permission from Elsevier

$$\begin{vmatrix} J_1'(\kappa a) - J_1(\kappa a)/\kappa a & Y_1'(\kappa a) - Y_1(\kappa a)/\kappa a \\ J_1'(\kappa b) - J_1(\kappa b)/\kappa b & Y_1'(\kappa b) - Y_1(\kappa b)/\kappa b \end{vmatrix} = 0, \tag{14.2}$$

Here, J_1, Y_1, J'_1, and Y'_1 are Bessel functions of the first and second kinds, and their derivatives, respectively. κ denotes the wavenumber in the radial direction given by

$$\kappa = \sqrt{\left(\frac{\omega}{C_s}\right)^2 - \gamma^2}, \tag{14.3}$$

with the wavenumber in the axial direction, γ.

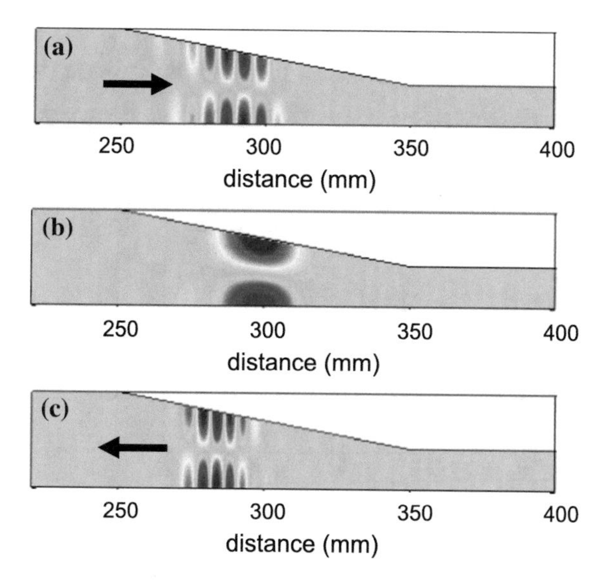

Fig. 14.12 Cross-sectional views of SH_1 mode propagating in tapered part of 1° slope at **a** 90 μs, **b** 105 μs, and **c** 125 μs after excitation. Thickness is enlarged for better observation. *Arrows* indicate the propagation directions. Reprinted from Nurmalia et al. (2012), Copyright (2012), with permission from Elsevier

Aluminum alloy pipes of a = 9.5 mm and b = 12.5 mm were used as specimens. A torsional-wave EMAT (Fig. 3.11) was developed to generate a specific torsional wave mode. The permanent magnet period was 5.22 mm, yielding the T(0,1) and T(0,2) modes at 0.615 and 0.77 MHz, respectively. The pinducer was again used for detecting the torsional waves. Figure 14.13 presents typical detected waveforms of the two modes, showing smaller group velocity of the T(0,2) mode. Figure 14.14 displays the mode conversion behavior on the dispersion curves of the torsional modes, which is similar to the results in Figs. 14.9 and 14.10 for the SH

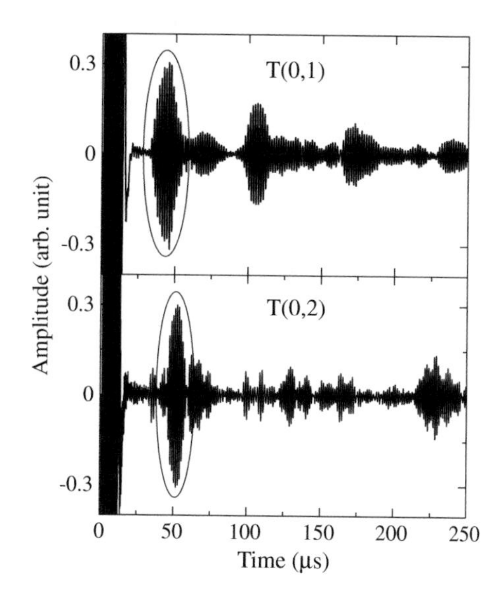

Fig. 14.13 Typical received waveforms of two modes propagating in the aluminum alloy pipe (after Nurmalia et al. 2013)

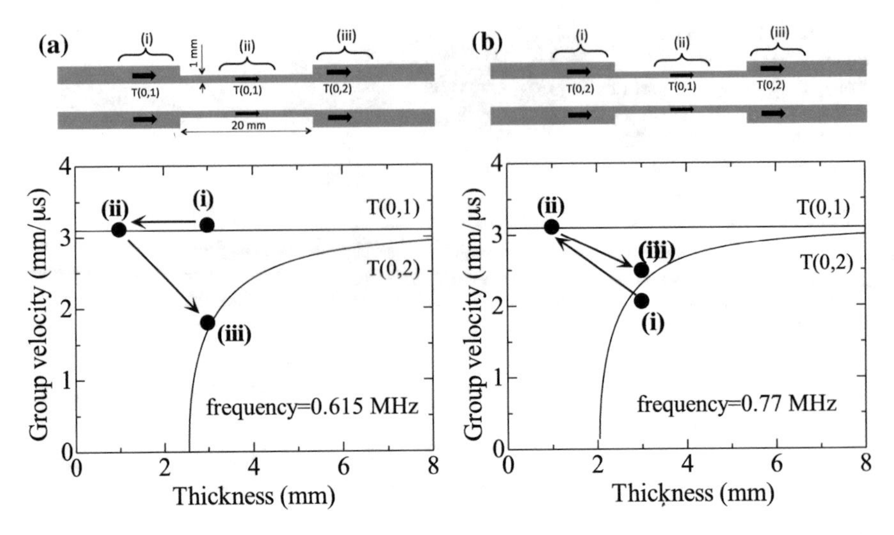

Fig. 14.14 Change in the group velocity during propagation through the defect region for **a** T(0,1)-mode incidence (0.615 MHz) and **b** T(0,2)-mode incidence (0.770 MHz). (*i*), (*ii*), and (*iii*) indicate the measurements before, in, and after the defect region (after Nurmalia et al. 2013)

modes. When the T(0,1) mode enters the thinned region through the step defect, it does not cause the mode conversion and keeps propagating through the defect region. However, it converts to the T(0,2) mode at the far end (Fig. 14.14a). When the T(0,2) mode enters the defective region, it converts to the T(0,1) mode and reconverts to the T(0,2) mode at the far end. These behaviors are also explained by the displacement distribution in the radial direction (Fig. 14.14b), being analogous to Fig. 14.10.

The boundary-free total reflection was also observed for the T(0,2) mode as shown in Fig. 14.15. The defect edge was tapered by 5° slope. The entire wave is reflected back, and there is no transmission at all through the defect. The vertical

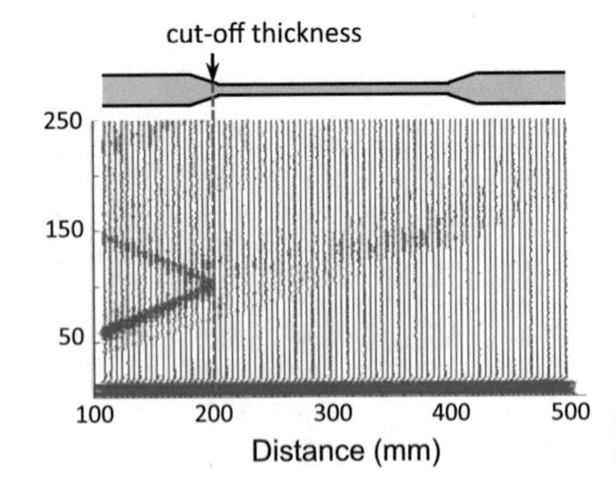

Fig. 14.15 Total reflection of the T(0,2) mode at the tapered region (after Nurmalia et al. 2013)

broken line in Fig. 14.15 indicates the location where the total reflection occurs, which corresponds to a thickness of 1.87 mm, while the theoretical cutoff thickness is 2.1 mm.

14.4 Line- and Point-Focusing EMATs for Back Surface Defects

14.4.1 Line-Focusing EMAT for Detection of Slit Defects

The principle to fabricate the line-focusing (LF) EMAT is discussed in Sect. 3.10. It utilizes the radiation directivity of an SV wave generated by a single line source oscillating parallel to the surface and normal to itself on the elastic half-space. High amplitude occurs around an angle 30° from the normal direction. Focusing is then made possible by aligning these line sources so that they radiate the SV wave rays in the directions centered around 30° to the focal line. These rays should arrive at the focal line in phase for constructive interference.

Following the calculation, two meander-line coils were designed for operating at 4 MHz in steels, but for different focal depths, LFE-S1 and LFE-S2 (Ogi et al. 1999). Table 14.1 provides their specifications. A circuit-printing technique permits one to fabricate variable-spacing coils within 1 μm accuracy. The coil segments are 110 μm wide and 18 μm thick. Polyimide sheets of 25-μm thick sandwich the copper wiring for the protection and insulation purposes.

Figure 14.16d shows the measured directivity patterns of LFE-S2, showing the sharp focusing at the intended position. The pattern is identical for generation and reception. The LF EMAT moves along the upper surface of a steel block, being 30 mm thick, 50 mm wide, and 400 mm long. The focal line is located on the opposite surface, where a bulk-wave (BW) EMAT (Sect. 3.1) receives or generates the SV wave (Fig. 14.16c). To produce good spatial resolution, the face of the BW EMAT is masked with a copper foil of 50 μm thick, leaving a slit opening of 1 mm. Generation directivity is provided by driving the LF EMAT and receiving the SV waves with the BW EMAT. Their roles are exchanged to measure the receiving directivity. In Fig. 14.16d, the directivity measurements also reveal an unwanted peak on the opposite side of the focal line ($x > 0$), which appears much broader with less intensity than the one at the focusing point.

Table 14.1 Specifications of the meander-line coils for line-focusing EMATs. n denotes the number of line segments; L and D dimensions; (x_F, z_F) coordinate of the focal line; θ angle from the normal direction (Fig. 14.16a, b). © (1999) IEEE. Reprinted, with permission, from Ogi et al. (1999)

EMAT	L (mm)	D (mm)	n	x_F (mm)	z_F (mm)	θ (°)
LSE-S1	20	24.43	30	25.18	−20	2.1–51
LSE-S2	20	28.18	31	30.28	−30	4.0–45

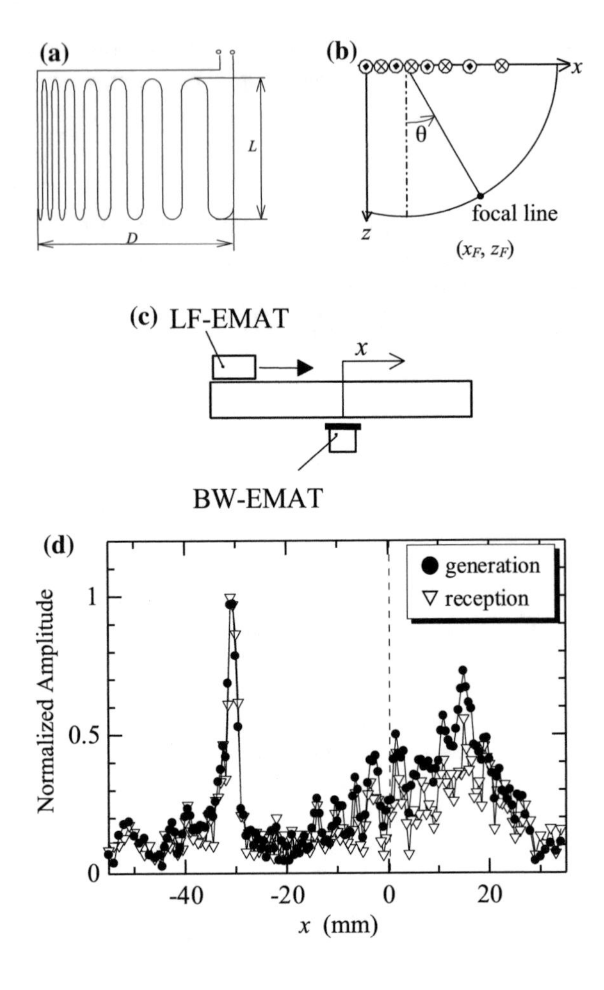

Fig. 14.16 Generation and detection directivity patterns of LFE-S2. **a** Schematic of the meander-line coil for LF EMAT. **b** Definition of the focal line. **c** Measurement setup. **d** Directivity measurement. © (1999) IEEE. Reprinted, with permission, from Ogi et al. (1999)

Two test blocks are prepared for LFE-S1 and LFE-S2, being 50 mm wide and 400 mm long. The thicknesses are 20 and 30 mm for LFE-S1 and LFE-S2, respectively. The focal lines are on the bottom surfaces. The slits of 20 mm long and 1.0 mm wide are introduced by machining; their depths are between 0.05 and 0.7 mm. The LF EMATs are manually moved on them while measuring the amplitude of the scattered signals from the defects. The duration of the driving burst signals is 12 μs and the driving voltage is typically 800 V_{pp}.

Figure 14.17 presents the result of detecting slit defects. When the focal line is far from the defects, the obliquely radiated and then reflected SV wave fails to return to the transducer, and the measured amplitude is as small as the noise level. When it hits the defect, the amplitude jumps, reaches a peak, and then drops to the noise level. The half-width of the peak is typically 2.5 mm. Figure 14.18 presents the raw signals when the focal line is far away from the defects and when it is just on the 0.3-mm-deep defect. The dead time for the equipment to recover from the high

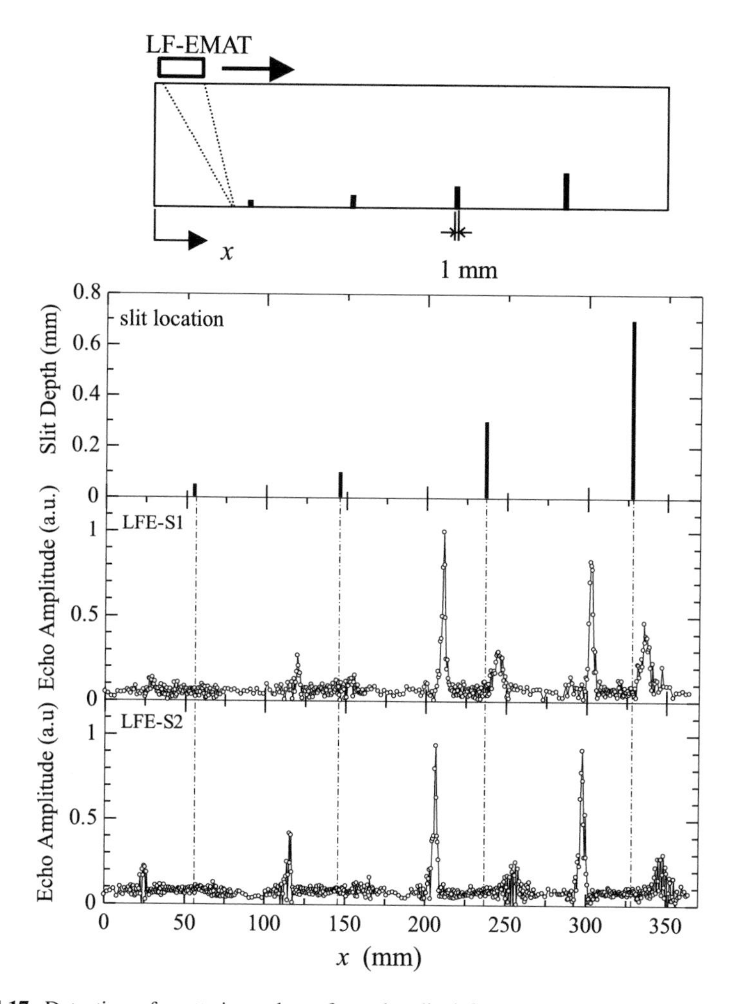

Fig. 14.17 Detection of scattering echoes from the slit defects. The frequency is 4 MHz. The defect location and depth are shown as thick bars. © (1999) IEEE. Reprinted, with permission, from Ogi et al. (1999)

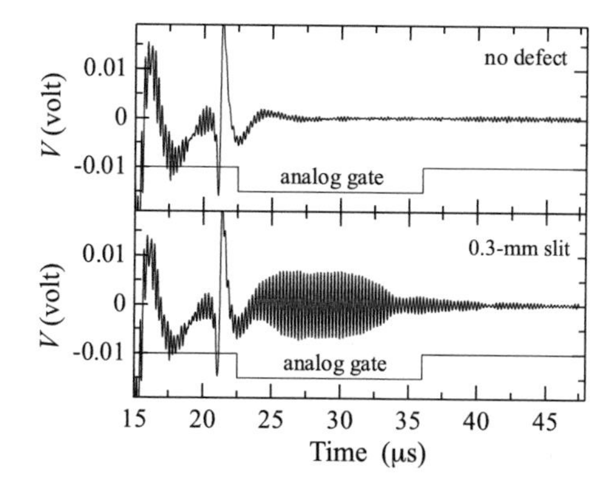

Fig. 14.18 Scattering signals generated and received by LFE-S2 (30-mm-thick steel block). Upper trace is the signal when the focal line is away from the defects. Bottom one is the scattering signal from 0.3-mm-deep slit defect. © (1999) IEEE. Reprinted, with permission, from Ogi et al. (1999)

power driving extends to before the scattering signal arrival. But, the final outputs are hardly affected owing to the superheterodyne processing on gated signals.

We observe good correlations between the defect locations and the amplitude peaks. They agree with x_F in Table 14.1 within a 0.5-mm error band. The LF EMATs can detect defects deeper than or equal to 0.05 mm. The amplitude increases with the slit depth, but it saturates beyond 0.3 mm deep. Transit times of the scattering signals at the amplitude peaks are independent of depth. These observations indicate that the SV wave is scattered at the slit mouth and returns to the transducer as sketched in Fig. 14.19. The scattered signal is detected while scanning from Fig. 14.19a–c, and it vanishes at d.

There are undesirable amplitude peaks in Fig. 14.17 besides the sharp peaks. They are caused by the SV waves radiated on the opposite side of the focal line and reflected from the defects of 0.3 mm and 0.7 mm deep. This is predictable from the directivity measurement of Fig. 14.16d. Their amplitudes are comparable with those from 0.05- and 0.1-mm-deep defects at the focusing conditions. These false peaks could be discriminated by observing the wider amplitude peaks.

The sharp directivity and focusing mechanism function twice, in generation and reception, to make high sensitivity for defect detection. The focused SV waves impinge on the slit mouth in phase and scattered signals occur, which spread like a cylindrical wave in the medium and reach the incident surface. They are picked up one after another by the same EMAT. Because the coil segments are positioned at every half-wavelength from the focal line (now at the slit corner), the scattered SV waves are received in phase, giving large amplitude after being summed up. This mechanism operates only at the designed frequency (4 MHz). Thus, the LF EMATs act as frequency filters by themselves and remove the scattered longitudinal waves and the other components of the SV waves.

Surface roughness and oxide effectively increase the liftoff in the case of using EMATs. To study the tolerance to the liftoff, polymer films of 0.1 mm thick were

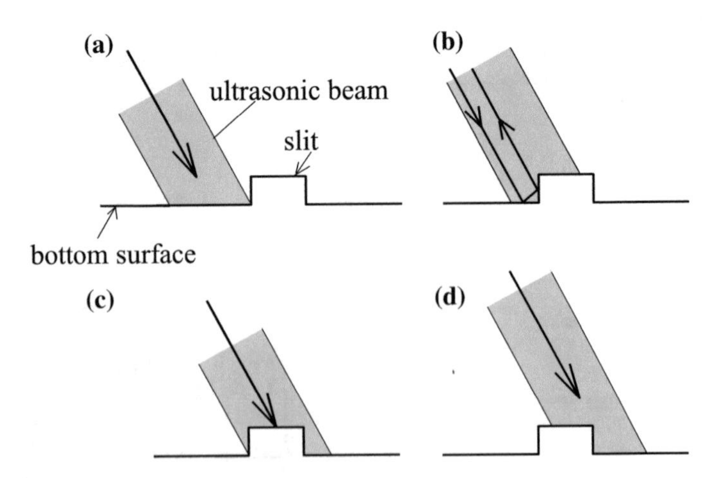

Fig. 14.19 Schematic image of the focused wave scattering and reflection from the slit. © (1999) IEEE. Reprinted, with permission, from Ogi et al. (1999)

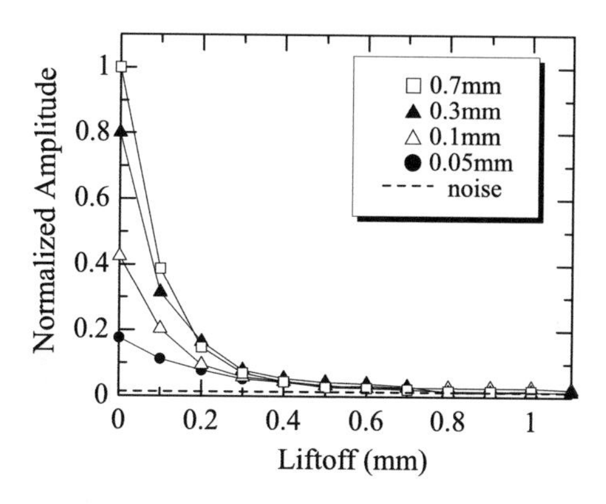

Fig. 14.20 Liftoff effect on the scattering amplitude from the defects. LFE-S2 is used. © (1999) IEEE. Reprinted, with permission, from Ogi et al. (1999)

inserted between the EMAT and the metal surface. Figure 14.20 shows the influence of liftoff on the received amplitude. When the coil has a constant spacing l, the liftoff h tends to lower the efficiency by the factor of $\exp^{-h/l}$ (Eq. 2.29). The amplitude decrease in Fig. 14.20 appears to be stronger than the exponential dependence. This discrepancy is attributable to the changing spaces of the meander-line coil. As the liftoff increases, the narrow-spaced region becomes selectively inactive, which deteriorates the concentration of the SV wave rays on the focal line and also the coherency in reception. The LF EMAT is therefore less tolerable to the liftoff than the EMATs of constant spacing. The present focusing EMAT allows liftoff up to 0.6 mm in the detection of the slit defects.

14.4.2 Point-Focusing EMAT for Detecting Back Surface Cracks

The wave energy enhancement at defect is further promoted by focusing SV waves at a point inside the specimen, that is, the point-focusing EMAT (PF EMAT). The fundamental concept of PF EMAT is the same as that of the LF EMAT. It is composed of a pair of permanent magnets and two concentric meander-line coils for transmitting and receiving SV waves (Fig. 3.26). The coils are located on the surface of a metallic specimen, and the permanent magnets are placed on them. The traction forces for SV waves are generated by the Lorentz force mechanism, and the receiving coil detects the SV waves scattered from defects through the reversed mechanism. The PF EMAT, which operates at 2 MHz, was developed for detecting stress corrosion cracks (SCCs) at welded parts in stainless steels (Takishita et al. 2015a). The coil shape was designed so that radiation angle θ from each source takes values between 13.4° and 37.5°.

Slit defects with various depths, of 10 mm length and 0.05 mm width, were machined on the back surface of a 20-mm-thick stainless steel (SUS304) block, imitating SCCs. The PF EMAT was moved on the top surface, monitoring the scattered SV wave signal by the slit defects. Figures 14.21 shows typical waveforms obtained from the flawless area and from the slit defects of 0.05 and 0.15 mm depths. The signals scattered from the defects are clearly visible in the waveforms. Figure 14.22 shows the amplitude profile of the signal scattered by the

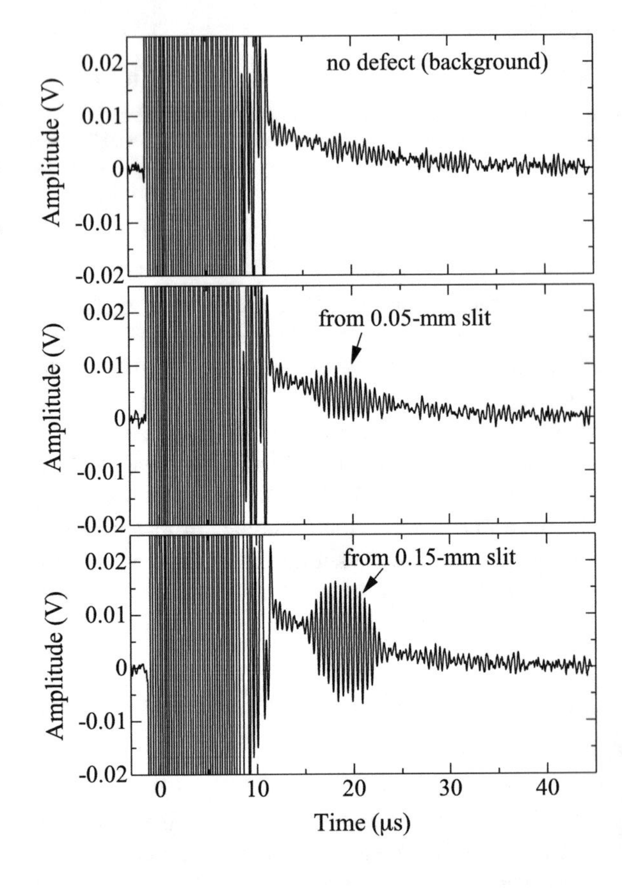

Fig. 14.21 Measured waveforms when the focal point of the PF EMAT was set to the slit defect. After the large signal until 10 μs caused by the excitation tone bursts, the scattered signals from the slit defects are observed (after Takishita et al. 2015a)

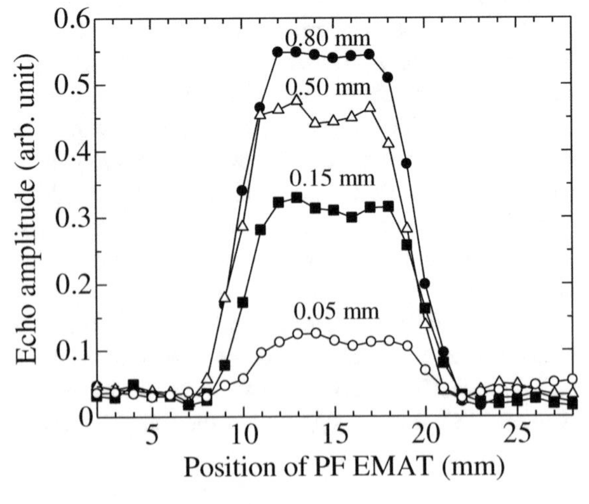

Fig. 14.22 Signal amplitude profiles when the slit defects on the back surface are scanned by the PF EMAT from the top surface. The numbers show the slit depth (after Takishita et al. 2015a)

defects obtained when the PF EMAT was moved on the top surface parallel to the slit defects. In this figure, the slit defects are located between 10 and 20 mm. The PF EMAT successfully detects all of them, including very shallow defect (0.05 mm depth), demonstrating its high capability for detection of SCCs.

A laboratory-prepared SCC specimen was used for displaying the detectability of the PF EMAT (Takishita et al. 2015b). The specimen was cut from an SUS316 pipe of 35 mm thickness and 609.6 mm diameter. There was a welded region in the circumferential direction. Two SCC cracks (Crack 1 and Crack 2) were artificially introduced using chemicals under a bending stress as shown in Fig. 14.23a. They were located at about 8 and 2 mm from the edge of the weld metal on the inner surface, and their lengths were about 19 and 12 mm in the circumferential direction, respectively.

Figure 14.23b, c shows the typical waveforms obtained from flawless area and Crack 1, respectively, and Fig. 14.23d–f shows the amplitude profiles. The horizontal axis, y, denotes the position of the focal point in the axial direction on the inner surface, and the weld metal is located at y = 299–306 mm. Signals reflected

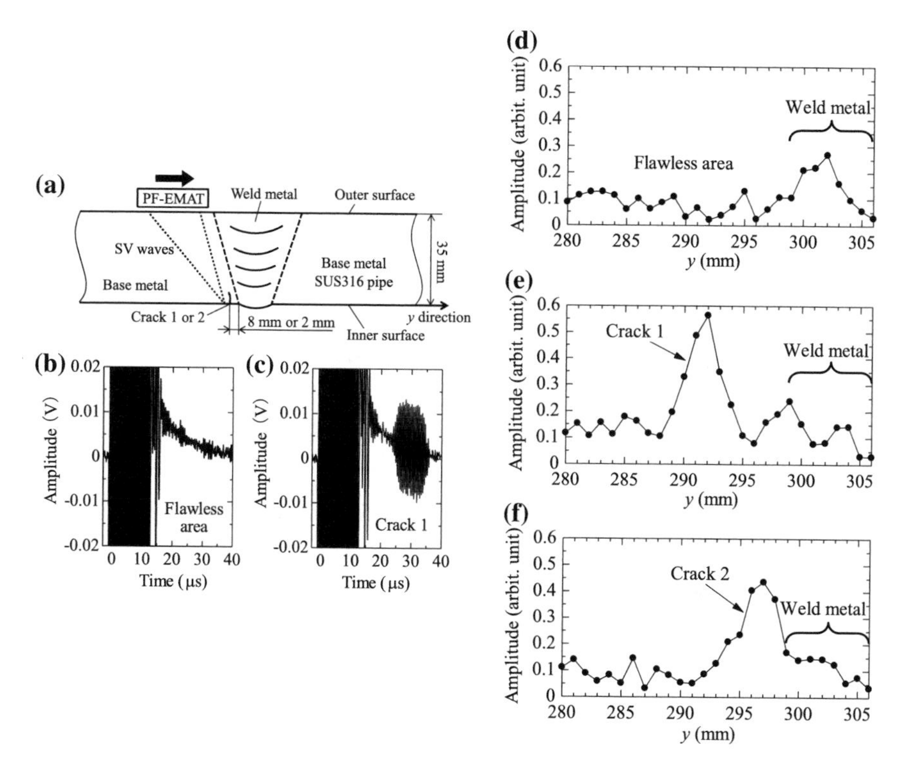

Fig. 14.23 **a** Cross-sectional schematic for the experimental setup. **b**, **c** Typical waveforms obtained from flawless area and Crack 1, respectively. **d**–**f** Amplitude profiles measured at y = 280–306 mm for flawless area, Crack 1, and Crack 2. Reprinted from Takishita et al. (2015b), with permission from Symposium on UltraSonic Electronics

from the welded region were observed at y = 299–305 mm, but notable signals were not observed in the flawless area (Fig. 14.23d). The SCC cracks (Cracks 1 and 2) are clearly detected in Fig. 14.23e, f, respectively, confirming that the SV wave PF EMAT is capable of detecting SCC cracks even near the welded region in SUS316 pipes.

References

Alers, R. B., & Kerr, D. (1998). Girth weld inspection of buried gas pipeline using EMATs. In *Proceedings of the 9th APCNDT/1998 ASNT's Spring Conference/7th Annual Research Symposium* (pp. 326–329).

Auld, A. B. (1973). *Acoustic Fields and Waves in Solids*. New York: Wiley.

Hirao, M., & Ogi, H. (1999). An SH-wave EMAT technique for gas pipeline inspection. *NDT and E International, 32*, 127–132.

McIntire, P. (Ed.) (1991). *Ultrasonic Testing: Nondestructive Testing Handbook* (2nd ed., Vol. 7) ASNT.

Meeker, T. R., & Meitzler, A. H. (1964). Guided wave propagation in elongated cylinder and plates. In *Physical Acoustics*, (Vol. 1A, pp. 111–167). New York: Academic Press.

Nakahata, K., Hirose, S., Schubert, F., & Khler, B. (2009). Image based EFIT simulation for nondestructive ultrasonic testing of austenitic steel. *Journal of Solid Mechanical Engineering, 3*, 1256–1262.

Nurmalia, Nakamura, N., Ogi, H., Hirao, M., & Nakahata, K. (2012). Mode conversion behavior of SH guided wave in a tapered plate. *NDT and E International, 45*, 156–161.

Nurmalia, Nakamura, N., Ogi, H., & Hirao, M. (2013). Mode conversion and total reflection of torsional waves for pipe inspection. *Japanese Journal of Applied Physics, 52*, 07HC14.

Ogi, H., Hirao, M., & Minoura, K. (1997). Noncontact measurement of ultrasonic attenuation during rotating fatigue test of steel. *Journal of Applied Physics, 81*, 3677–3684.

Ogi, H., Hirao, M., & Ohtani, T. (1999). Line-focusing electromagnetic acoustic transducers for detection of slit defects. *IEEE Transactions on Ultrasonics. Ferroelectrics Frequency Control, UFFC-46*, 341–346.

Takishita, T., Ashida, K., Nakamura, N., Ogi, H., & Hirao, M. (2015a). Development of shear-vertical-wave point-focusing electromagnetic acoustic transducer. *Japanese Journal of Applied Physics, 54*, 07HC04.

Takishita, T., Ashida, K., Nakamura, N., Ogi, H., & Hirao, M. (2015b). Detection of cracking near welding of SUS316 pipe by shear-vertical-wave point-focusing-electromagnetic acoustic transducer. In *Proceedings of Symposium Ultrasonics Electronics,* (Vol. 36).

Thompson, R. B., Alers, G. A., & Tennison, M. A. (1972). Application of direct electromagnetic Lamb wave generation to gas pipeline inspection. In *Proceedings of IEEE Ultrasonic Symposium* (pp. 91–93).

Thompson, R. B. (1997). Experiences in the use of guided ultrasonic waves to scan structures. In *Review of Progress in Quantitative Nondestructive Evaluation* (Vol. 16, pp. 121–128).

Chapter 15
Average Grain Size of Steels

Abstract This chapter reviews the EMAR application to determine the average grain size of carbon steels (Ogi et al. 1995) on the basis of the fourth-power frequency dependence of attenuation. The final results are favorably compared with the average of three-dimensional distribution from metallographic observations for various grades.

Keywords Absorption · Attenuation · Carbon steel · Crystallographic orientation · Diffraction correction · Rayleigh scattering · Resonance spectrum · Surface roughness

15.1 Scattering of Ultrasonic Waves by Grains

Grain size of a polycrystalline metal is one of the important factors that govern mechanical properties such as yield stress and fracture toughness (Hall 1970). Many techniques have been studied for nondestructive evaluation of grain size, including ultrasonic attenuation (scattering), X-ray diffraction, and magnetic tests. Among them, ultrasonic attenuation is the most promising, because it provides a through-thickness average of grain structure, on which mechanical properties depend.

Ultrasonic waves are scattered by grains while propagating in a polycrystalline metal because of different crystallographic orientations at their boundaries, causing an amplitude loss. Scattering does not dissipate the wave energy by itself, but it generates out-of-phase wavelets, causing apparent attenuation. Seeking a relationship between grain size and attenuation has been a principal topic (Rayleigh 1894; Mason 1958; Papadakis 1963, 1965, 1966; Klinman and Stephenson 1981; Smith 1987; Nicoletti et al. 1992). These classical studies showed a fourth-power dependence of attenuation on frequency in the region where the grain size is much smaller than the ultrasonic wavelength (Rayleigh scattering). Many studies evaluated grain size of metals using this formula, but reached unfavorable results. One of the difficulties comes from the grain size distribution in real metals, for which the

M. Hirao and H. Ogi, *Electromagnetic Acoustic Transducers*,
Springer Series in Measurement Science and Technology,
DOI 10.1007/978-4-431-56036-4_15

power law relationship needs to be modified (Papadakis 1963; Nicoletti et al. 1992). The principal problem, however, lies in the fact that *pure* attenuation is hard to measure. This basic but essential concern is solved by replacing contact transducers with noncontact EMATs as shown in Chap. 6.

15.2 Fourth-Power Law

Ultrasonic attenuation in a polycrystalline metal originates from absorption and scattering in general. Absorption, or unrecoverable energy transformed to heat, is related to dislocation's anelastic motion, thermoelastic effects, magnetic structure, interaction with electrons and phonons, and nuclear spin as summarized in the monographs (e.g., Truell et al. 1969). Among them, the interactions with electrons, phonons, and nuclear spins are negligible at room temperature and with megahertz frequencies. Furthermore, use of only shear waves almost excludes the thermoelastic effect, because no volume change occurs in the shear deformation. Consequently, dislocation damping and the magnetoelastic effect are mainly responsible for absorption. The relationship between the dislocation and the frequency-dependent attenuation shows a second power behavior at low frequencies as shown in Eq. (7.1). However, measurement in a narrow frequency range (1–10 MHz) allows the simplification to linear frequency dependence. The magnetoelastic effect is also a linear function of frequency (Levy and Truell 1953). Attenuation caused by absorption is, therefore, approximately linear with frequency.

On the other hand, attenuation caused by grain scattering in the Rayleigh regime shows a fourth power dependence on frequency (Mason 1958; Bhatia 1959; Truell et al. 1969). The frequency dependence of the shear wave attenuation is then expressed as follows:

$$\alpha = af + SD^3 f^4. \tag{15.1}$$

Here, a is a coefficient for the absorption attenuation, S the scattering factor, and D the average grain size. The different dependences of α on f permit one to separate the absorption and the scattering contributions. Bhatia (1959) derived the scattering factors for longitudinal and shear waves and demonstrated that the shear waves undergo more intensive grain scattering than the longitudinal waves. This indicates an advantage of using shear waves for grain size evaluation. As mentioned above, the longitudinal waves lose energy from the thermoelastic effect more than the shear waves. Thus, the shear wave attenuation provides a sensitive means of determining the grain size; the extraction of the scattering part from the as-measured attenuation should be much easier with shear waves, resulting in more accurate grain size evaluation. This situation well suits the use of EMATs that excite and detect shear waves with higher efficiency (see Chap. 2).

15.3 Steel Specimens and Grain Size Distribution

Thirty-six steel plates were prepared to study various influences on the grain size. They are classified as follows:

(i) Carbon content: thirteen specimens with three different carbon contents of 0.007 mass%, 0.062 mass%, and 0.147 mass%.
(ii) Pearlite fraction: four specimens with 2, 61, 80, and 98 % pearlite volume fractions.
(iii) Martensite fraction: twelve specimens with four different martensite volume fractions of 15.3, 20.5, 30.9, and 39.4 %.
(iv) Surface roughness: seven specimens with different average roughness ranging from 0.14 to 8.7 μm.
(v) Texture: three specimens with 1.58, 3.00, and 5.19 % shear wave birefringence (see Chap. 11).

All specimens were plates measuring 6 × 100 × 100 mm^3. Figure 15.1 shows some microstructures of the specimens from group (i).

There is naturally a distribution of grain size. Microscopic examination provides only the two-dimensional (2D) distribution on exposed cross sections. The determined average size is smaller than that of the three-dimensional (3D) distribution, which determines mechanical properties of the material. The difference is significant for nonequiaxial grains. It is possible to inversely calculate the 3D distribution from the measured 2D distribution using a model that the grain shape is a regular

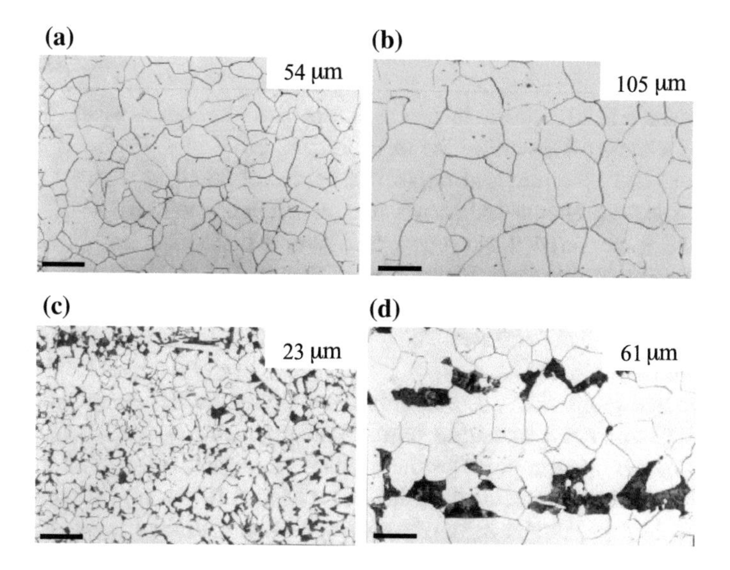

Fig. 15.1 Grain structures of carbon steels. Carbon contents are 0.007 mass% for **a** and **b** and 0.147 mass% for **c** and **d**. The *bars* indicate 100 μm. The average grain size calculated from the 3D grain size distribution is shown

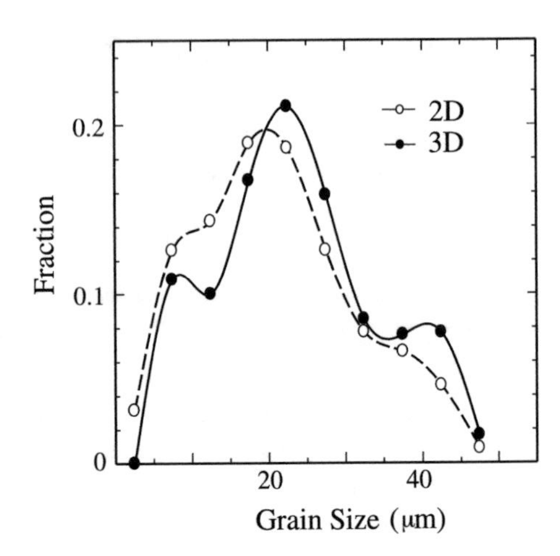

Fig. 15.2 Two-dimensional (2D) and three-dimensional (3D) grain size distributions in the specimen of Fig. 15.1c. 2D values are from microscopic examination. Reprinted with permission from Ogi et al. 1995). Copyright 1995, Acoustical Society of America

polyhedron or semiregular one whose number of faces depends only on the grain size, and each grain has a unique diameter distribution on the exposed cross section depending on the number of the faces (Matsuura and Itoh 1991). Figure 15.2 shows an example of 2D-to-3D conversion. The average grain size in the 3D distribution (23 μm) became larger by ten percent than that of the 2D distribution (21 μm). The fraction of small grains in the metallographic observation also decreases in the 3D distribution, giving a narrower distribution.

15.4 Grain Size Evaluation

A shear-wave EMAT shown in Fig. 3.1 was used to measure the shear wave attenuation. Chapter 6 describes the measurement procedure. Figure 15.3 shows a typical resonance spectrum and ring-down curves at three resonance frequencies. Figure 15.4 shows attenuation vs resonance frequency on two specimens. We plotted both as-measured α and α corrected for the diffraction loss (see Sect. 6.4). The corrected attenuation provides a better fitting to Eq. (15.1), especially for low frequencies, where the diffraction loss is prominent. The α vs f curves tend to be linear as the grain size decreases, showing the dominant absorption term over the scattering term in Eq. (15.1).

The average grain size was determined by fitting Eq. (15.1) to the measured $\alpha(f)$ and separating the scattering term. With a suitable scattering factor S, the coefficient of the f^4 term provides the average grain size. Although the scattering factor can be derived theoretically (Bhatia and Moore 1959), we calibrated S to be $S = 2.25 \times 10^{10}$ $\mu s^3/\mu m^3$ using a reference specimen with the narrowest grain size distribution. Papadakis (1961) pointed out that D^3 in Eq. (15.1) should be replaced by $\langle D^6\rangle/\langle D^3\rangle$, where $\langle\ \rangle$ denotes an average over the volume. This formula was applied to the EMAR measurements with the inferred 3D grain size distribution and

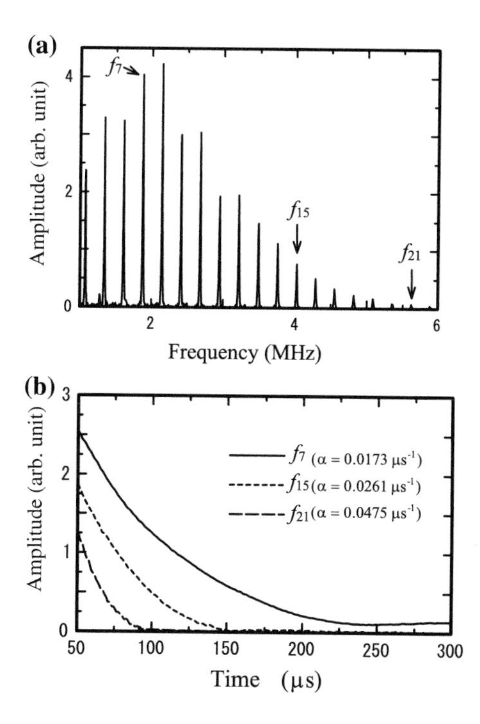

Fig. 15.3 **a** Resonance spectrum of thickness resonance of shear wave and **b** ring-down curves at three resonance frequencies. Reprinted with permission from Ogi et al. 1995). Copyright 1995, Acoustical Society of America

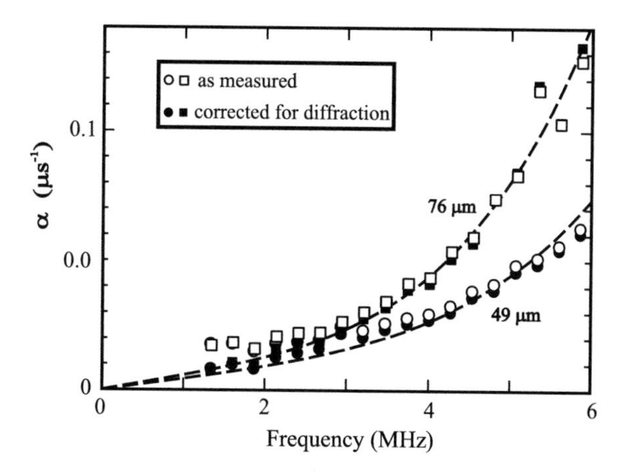

Fig. 15.4 Frequency dependence of the shear wave attenuation for carbon steels with 3D-averaged grain sizes of 76 and 49 μm. Reprinted with permission from Ogi et al. 1995). Copyright 1995, Acoustical Society of America

the recalibrated S. However, this much underestimated the grain size. It was found that the use of $\langle D \rangle^3$ for D^3 in Eq. (15.1) leads to the best agreement with the metallographic examination.

The grain size thus evaluated by the EMAR method (D_{EMAR}) is compared with those by the microscopic observations (D_{Photo}) as shown in Fig. 15.5. For almost all the specimens, they agree well within a band of 6 μm. But, D_{EMAR} exceeds D_{Photo} for several specimens. The unfavorable results for the specimens containing

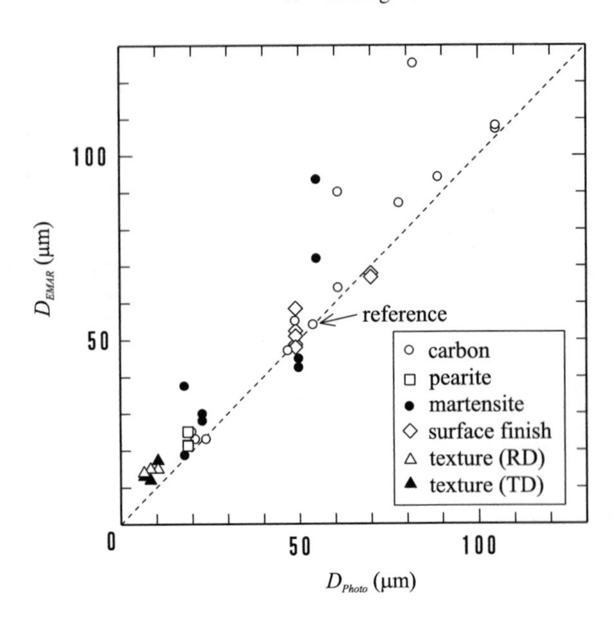

Fig. 15.5 Grain size evaluated by the EMAR method (D_{EMAR}) versus that by the 3D grain size distribution obtained by microscopic examination. A reference specimen was used to calibrate the scattering factor S. Reprinted with permission from Ogi et al. 1995). Copyright 1995, Acoustical Society of America

martensitic phase arise from two reasons: first, complexity of grain structure. These specimens contain low carbon content (<0.1 %). Microstructure of such a low-carbon martensitic phase includes many boundaries: (i) old austenite grain boundaries, (ii) boundaries of packets which consist of blocks, (iii) boundaries of blocks which consist of an oriented laths, and (iv) the lath boundaries. Because the ultrasonic scattering occurs from the different orientations of individual grains from their surrounding, it recognizes a block as a grain. On the other hand, the microscopic examination counts all the observable interfaces to be grain boundaries, leading to the overestimation with the EMAR. Second, the dual-phase system of martensitic and ferritic phases is a problem. The scattering occurs from elastic mismatch not only across grain boundaries but also across the interphase boundaries (Ying and Truell 1956). The fourth-power law in Eq. (15.1) was derived for a single-phase polycrystalline material, and it may not apply because the scattering factor will differ from that of the single phase.

There were two more outliers showing $D_{EMAR} \gg D_{Photo}$ in specimens containing various carbon contents. Both of them showed broad distributions of grain size, involving plural peaks (not shown here). Since Eq. (15.1) strictly applies to an ideal polycrystalline metal made up of monosize grains, the formula becomes progressively inadequate as the distribution broadens.

Surface roughness can also cause loss. The scattering attenuation caused by a surface unevenness has a second power dependence on frequency (Nagy and Rose 1993). Generally, separation of the attenuation caused by surface roughness and that by grain scattering is difficult; both cases take similar dependence on frequencies (f^2 and f^4). The effect is likely to be included in the scattering term in Eq. (15.1) and overestimate the grain size. But, the D_{EMAR} is larger than D_{Photo} by only 8 μm even for the roughest surfaces of Ra = 8.7 μm. The effect of surface roughness is then insignificant for the present set of specimens.

For textured specimens, two shear waves polarized in the rolling and transverse directions are independently used for the measurements. The results reveal that the effect of texture on the grain size evaluation is not significant.

The EMAR has hence demonstrated its capability of determining grain size of steels in a noncontact way. It is, however, restricted to plate-shaped specimens. In a thick plate, the resonance peaks appear at narrow frequency intervals, and they will eventually overlap, making the attenuation measurement impossible. On the other hand, for a too thin plate, the resonance peaks appear sparsely, decreasing the number of measurable modes. Suitable thickness is between 0.3 and 60 mm, covering standard metals. But, this range highly depends on the metal's attenuation to be determined.

Dubois et al. (2000) have explored a laser ultrasound technique, another noncontact ultrasound method, to determine austenite grain size at elevated temperatures. Also in the Rayleigh scattering regime with 5–20 MHz band, they claim the usefulness of measuring the grain sizes smaller than 100 μm. Besides, the truly remote sensing, high frequency, and thin specimens (1 mm) allow neglecting the diffraction correction in the spectrum difference between two consecutive longitudinal echoes in the received trace.

References

Bhatia, A. B. (1959). Scattering of high-frequency sound waves in polycrystalline materials. *The Journal of the Acoustical Society of America, 31*, 16–23.

Bhatia, A. B., & Moore, R. A. (1959). Scattering of high frequency sound waves in polycrystalline materials. II. *The Journal of the Acoustical Society of America, 31*, 1140–1141.

Dubois, M., Militzer, M., Moreau, A., & Bussière, J. F. (2000). A new technique for the quantitative real-time monitoring of austenite grain growth in steel. *Scripta Materialia, 42*, 867–874.

Hall, E. O. (1970). *Yield Point Phenomena in Metals and Alloys*. New York: Plenum Press.

Klinman, R., & Stephenson, E. T. (1981). Ultrasonic prediction of grain-size and mechanical-properties in plain carbon-steel. *Materials Evaluation, 39*, 116–1120.

Levy, S., & Truell, R. (1953). Ultrasonic attenuation in magnetic single crystals. *Reviews of Modern Physics, 25*, 140–145.

Mason, W. P. (1958). *Physical Acoustics and Properties of Solids*. Princeton: Van Nostrand.

Matsuura, K., & Itoh, Y. (1991). Estimation of three-dimensional grain size distribution in polycrystalline material. *Materials Transactions JIM, 32*, 1042–1047.

Nagy, P. B., & Rose, J. (1993). Surface roughness and the ultrasonic detection of subsurface scatterers. *Journal of Applied Physics, 73*, 566–580.

Nicoletti, D., Bilgutay, N., & Onaral, B. (1992). Power-law relationship between the dependence of ultrasonic attenuation on wavelength and the grain size distribution. *The Journal of the Acoustical Society of America, 91*, 3278–3284.

Ogi, H., Hirao, M., & Honda, T. (1995). Ultrasonic attenuation and grain size evaluation using electromagnetic acoustic resonance. *The Journal of the Acoustical Society of America, 98*, 458–464.

Papadakis, E. P. (1961). Grain-size distribution in metals and its influence on ultrasonic attenuation measurements. *The Journal of the Acoustical Society of America, 33*, 1616–1621.

Papadakis, E. P. (1963). From micrograph to grain-size distribution with ultrasonic applications. *The Journal of Acoustical Society of America, 35*, 1586–1594.

Papadakis, E. P. (1965). Ultrasonic attenuation caused by scattering in polycrystalline metals. *The Journal of Acoustical Society of America, 37*, 711–717.

Papadakis, E. P. (1966). Ultrasonic diffraction loss and phase change in anisotropic materials. *The Journal of Acoustical Society of America, 40*, 863–876.

Rayleigh, L. (1894). *The Theory of Sound*. London: Macmillan.

Smith, L. (1987). Ultrasonic materials characterization. *NDT&E International, 20*, 43–48.

Truell, R., Elbaum, C., & Chick, B. B. (1969). *Ultrasonic Methods in Solid State Physics*. New York: Academic Press.

Ying, C. F., & Truell, R. (1956). Scattering of a plane longitudinal wave by a spherical obstacle in an isotropically elastic solid. *Journal of Applied Physics, 27*, 1086–1097.

Chapter 16
Remaining-Life Assessment of Fatigued Metals

Abstract This chapter discusses the discovery of the attenuation peak in copper and carbon steels subjected to cyclic tension load and also the nonlinear peaks in carbon steels subjected to rotating-bending fatigue. These clear indications are caused by the dislocation mobility and restructuring due to cyclic loading and occur at the specific fractions to the fatigue life, being independent of the stress amplitude. Continuous observation with EMAR can only detect such precursors and tell the remaining lifetime of components being fatigued.

Keywords Acoustic nonlinearity · Attenuation peak · Axial-shear-wave resonance · Continuous monitoring · Copper · Dislocation damping · Granato-Lücke's theory · Low-carbon steel · Shear wave · Small cracks · Slip bands · TEM

16.1 Fatigue and Ultrasonic Measurements

Numerous studies on fatigue of metals have been published. As summarized in some monographs (Klesnil and Lukáš 1980; Suresh 1998), they include observations of microstructures in the incubation period (small cracks, slip bands, dislocations, etc.); nondestructive evaluation of microstructure change using ultrasonics, X-ray, and magnetic methods; and fracture mechanics to predict the remaining life. These efforts produced an essential view of fatigue and its underlying physics.

As a part of them, intensive studies appeared on ultrasonic characteristics (velocity, attenuation, nonlinearly, etc.) during fatigue tests for the purpose of remaining-life assessment (Truell and Hikata 1957; Schenck et al. 1960; Bratina and Mills 1962; Bratina 1966; Pawlowski 1964; Fei and Zhu 1993; Zhu and Fei 1994). Main attention was directed to the attenuation evolution because it shows high sensitivity to dislocations in metals (Chap. 6), which control the fatigue process via interactions with grain boundaries, twins, point defects, stacking faults, and other dislocations. Schenck et al. (1960) measured attenuation at several stages of fatigue life using a bar resonance method on a 0.22 mass% C steel exposed to rotating-bending fatigue. They varied the bending stress between 0.56 and 0.88 of

M. Hirao and H. Ogi, *Electromagnetic Acoustic Transducers*,
Springer Series in Measurement Science and Technology,
DOI 10.1007/978-4-431-56036-4_16

the yield strength (fatigue limit/yield strength = 0.66). For all cases, attenuation increased slightly at the beginning and, after a long stable period, it increased remarkably just before failure. Bratina and Mills (1962) studied attenuation at several stages of cyclically stressed steels and indicated that it is caused by diffusion of interstitial point defects to dislocations. Pawlowski (1964) showed a correlation between the slope of the attenuation increase in the early stages and remaining life. Many other studies showed similar results as reported by Schenck et al. (1960). The initial increase was attributed to the increase of the dislocation density (see Eq. (7.1)) and the final increase was attributed to the occurrence of cracks and their subsequent growth. However, no critical indication was found for remaining-life prediction; nor has in situ monitoring been realized.

The EMAR method was then applied to continuously monitor the ultrasonic characteristics throughout the fatigue lives of low-carbon steels (Ogi et al. 1997, 2001, 2002; Ohtani et al. 2000), aluminum alloy (Ogi et al. 2000), and copper (Hirao et al. 2000). The high sensitivity owing to noncontact nature had been expected to find new phenomena pertaining to the damaging process. Indeed, a remarkable and important observation was made: the attenuation peak, which appears at the same ratio of the cycle number to the failure cycle number, is independent of the applied stress amplitude. The peak can show the remaining material life. It was proven that this peak is caused by drastic changes of dislocation structure and crack growth behavior (Hirao et al. 2000; Ogi et al. 2002). This evident acoustic indication and synchronous microstructure changes were never observed before the EMAR measurements despite the long history of such studies. There are two main reasons for this achievement. First, as a nature of fatigue, the fatigue life can vary widely even for the same loading condition and material. The number of elapsed cycles does not indicate the current deterioration degree and the remaining life. Repeatable observation focused at a prescribed fractional life to fracture has been unfeasible. Second, there was no method of continuously measuring the evolution of interior microstructure change through the fatigue life. It should be nondestructive and have enough sensitivity to the dislocations below surfaces but no influence on the fatigue process. Only EMAR can do this task.

16.2 Zero-to-Tension Fatigue of Copper

An EMAT forms acoustic sources directly on the metal surface, requiring no intimate contact with the surface. This nature allows monitoring the dislocation mobility in a material subjected to a repeated stress. Hirao et al. (2000) applied the EMAR method to continuous monitoring of the fatigue-damage accumulation process in copper. Sinusoidal loading and signal sampling were synchronized to accomplish the measurements without intermission throughout the fatigue life.

The material used was 99.99 mass% polycrystalline copper, which was annealed at 200 °C for 1 h before machining. Grains appeared to be equiaxial with an average diameter of 135 μm. Proof stress was 183 MPa. Figure 16.1 shows the specimen

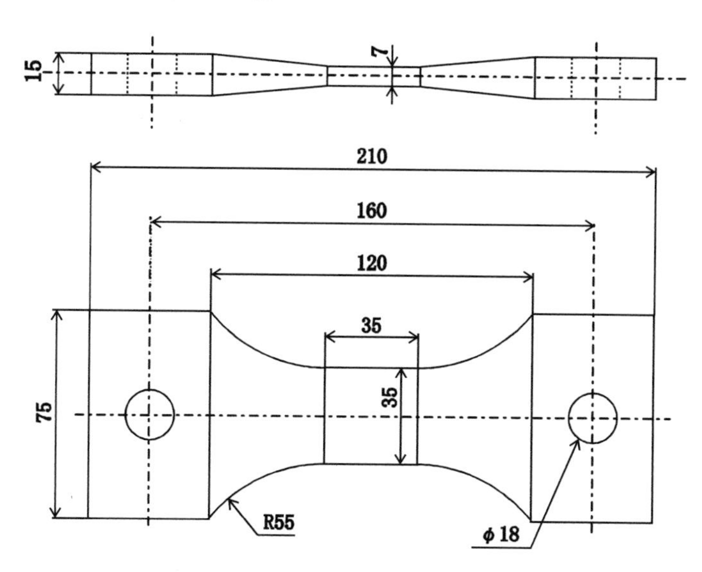

Fig. 16.1 Plate specimen and dimensions in millimeters. Reprinted from Hirao et al. (2000), Copyright (2000), with permission from Elsevier

dimensions. The center portion of the specimen was electropolished for replication. TEM foils were obtained by interrupting the fatigue tests after a certain number of cycles. The shear wave velocity was found to vary within 0.1 % when the polarization direction was rotated in the gauge plane, showing a weak texture anisotropy.

Figure 16.2 shows the experimental setup. The shear-wave EMAT (Fig. 3.1) with an active area of 10 × 10 mm^2 was placed near the gauge section of the

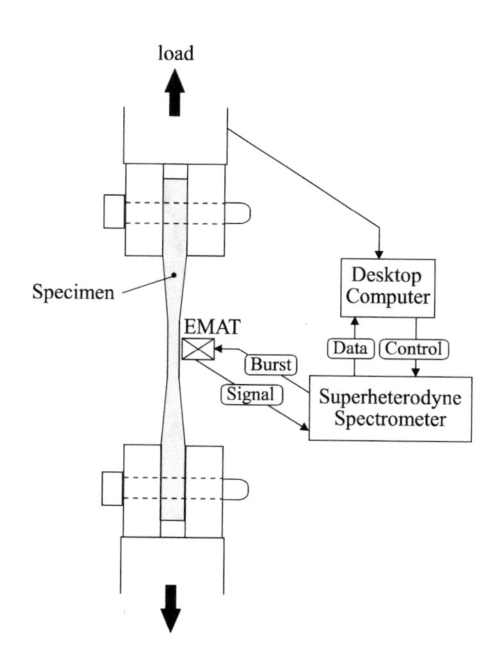

Fig. 16.2 EMAR system for monitoring the resonance frequency and attenuation coefficient of a polarized shear wave during a zero-to-tension loading process. Reprinted from Hirao et al. (2000), Copyright (2000), with permission from Elsevier

specimen. The measurement system automatically acquires the nth resonance frequency, f_n, and the attenuation coefficient, α, for one of the shear wave polarized either parallel or perpendicular to the cyclic load. Measurement parameters, such as the scanning frequency range and receiver gain (Sect. 5.2), are automatically adjusted to make in situ observation possible. Zero-to-tension sinusoidal load was applied ($\sigma_{min}/\sigma_{max} = 0$) to the specimen at a frequency of 10 Hz in laboratory air. Measurements of the resonance frequency and attenuation coefficient required a number of signal samplings while scanning the frequency and then acquiring the ring-down curve. For continuous monitoring of the fatigue process, the cyclic loading and measurements were synchronized with each other so that signal sampling, each taking approximately 5 ms, was made at minimum stress (0 MPa), during which the stress varied by 2.5 % relative to σ_{max}. The typical numbers of cycle needed for measuring f_n and α was 100, which was small enough compared with the cycle to fracture N_F of the order of 10^4–10^5.

Figure 16.3 shows the typical evolution of the resonance frequency and attenuation. The mode traced for this particular test was the 14th thickness resonance around 2.3 MHz of the shear wave polarized parallel to the load. The stress amplitude was 95 MPa ($\sigma_{max} = 190$ MPa) and $N_F = 722{,}000$. From the start to about 10 % of the lifetime, both α and f_n remain nearly unchanged. For the period of 10–40 % lifetime, α shows a large peak and f_n increases at the largest rate. The attenuation peak is ten times as large as the changes observed during the quasistatic unidirectional deformation into the plastic range for a 99.99 %-pure polycrystalline copper (Fig. 7.3). Abrupt increase of f_n indicates the thinning concentrated on this period. No remarkable indications occur after 40 % of lifetime. Occasionally, α jumps at the point of fracture. Small cracks were visible even to the naked eyes in the final stages. Among them, a few cracks grew, connected to each other, and eventually traversed the cross section.

The same experiments were repeated by varying the stress amplitude in the 90–115 MPa range (Fig. 16.4). Fatigue test was interrupted many times to make optical

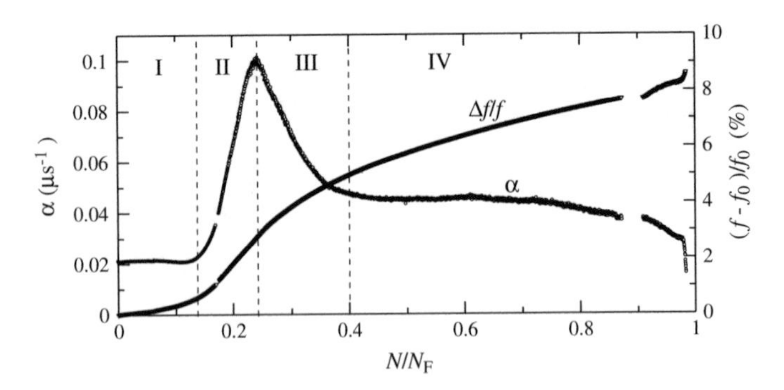

Fig. 16.3 Continuous measurements of the resonance frequency f and attenuation coefficient α. f_0 denotes the resonance frequency before fatiguing. The shear wave polarization was parallel to the stress. Stress amplitude was 95 MPa. $N_F = 722{,}000$. Reprinted from Hirao et al. (2000), Copyright (2000), with permission from Elsevier

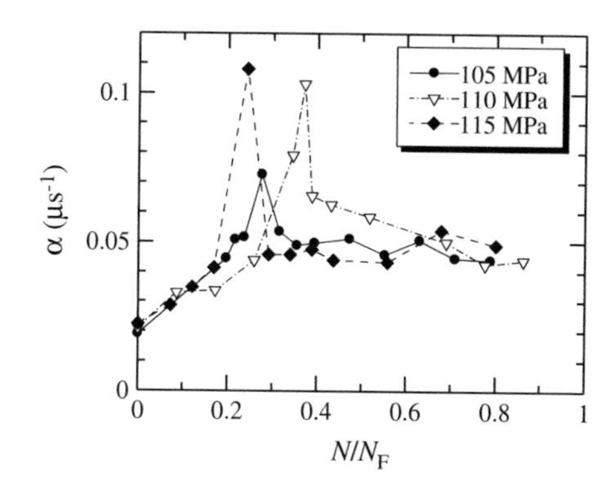

Fig. 16.4 Evolution of attenuation coefficient at the 28th resonance of the shear wave with the polarization parallel to the stress. $N_F = 253{,}800$ for the stress amplitude of 105 MPa, $N_F = 116{,}000$ for 110 MPa, and $N_F = 41{,}250$ for 115 MPa. Reprinted from Hirao et al. (2000), Copyright (2000), with permission from Elsevier

examination of the specimen surface and thickness measurement. The attenuation peak always appeared between 20 and 40 % of cycles to fracture, but there is no clear dependence on the stress amplitude. Comparing with Fig. 16.3, continuous and interrupted measurements make no noticeable difference in the acoustic behavior. Also, changing the polarization direction gave similar results.

Thickness measurements yield the shear wave phase velocity with the resonance frequency. In Fig. 16.5, the (elastic) softening takes place simultaneously with the attenuation peaks, around $N/N_F = 0.25$ in this case, for three resonance frequencies. The higher the resonance mode is, the larger the attenuation peak appears. While α levels off in the latter half of the specimen lifetime, the velocity v continues to decrease.

When cycled at the lower stress amplitude of 65 MPa, the specimen did not fracture until 1.5×10^6 cycles. For the first 5000 cycles, α increased by 0.003 μs^{-1} and f_n decreased by 0.01 MHz. Beyond this number of cycles, both changed little,

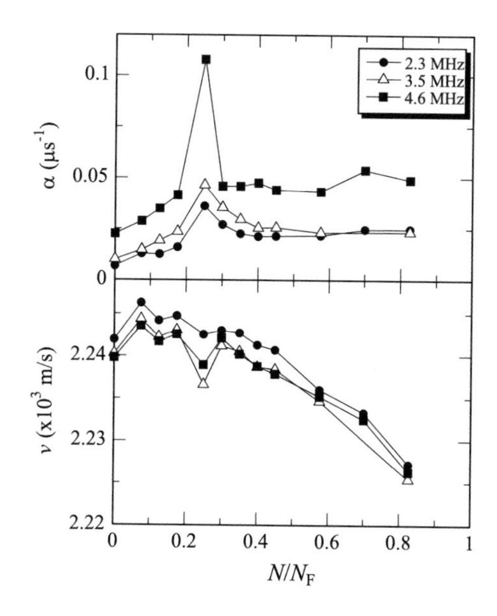

Fig. 16.5 Evolution of attenuation coefficient α and shear wave velocity v at three resonances. Stress amplitude was 115 MPa. Reprinted from Hirao et al. (2000), Copyright (2000), with permission from Elsevier

providing unremarkable indications. The specimen thickness showed no variation. This observation implies that the evolution illustrated in Figs. 16.3, 16.4 and 16.5 reflects some relevant metallurgical events leading to the eventual failure.

Replicas were obtained for optical examination of the damaged surfaces. Figure 16.6 plots the slip band density and α in the course of fatigue. Slip bands, being oriented preferentially perpendicular to the stress axis, were first observed after 5×10^4 cycles and grew in number at the highest rate, which was immediately followed by the peak in α. This indicates a close relationship between attenuation behavior and slip band development.

To further study the microstructure, dislocation structure was observed by transmission electron microscopy (TEM) at five representative phases in the response of α (Fig. 16.7). These phases are original, before, at, and after the attenuation peak, then approaching failure. The original metal showed low dislocation density. Just before the attenuation peak, the dislocations become very dense, tangling with one another, piling up at obstacles, and making clusters at many places (Fig. 16.7a). At the peak, the dislocations form broad walls and occur rarely between walls. Very long dislocation lines are observed, which appear to be highly mobile (Fig. 16.7b). After the peak, the walls are thinner, making cells. Twins become clearly visible with the dislocations piled up against them (Fig. 16.7c). In the final stage, well-developed cell structures occur, inside which dislocations are rare (Fig. 16.7d).

Possible factors contributing to the change in α and f_n are dislocations, grain scattering, diffraction, cracking, and necking of specimens for a MHz frequency range at room temperature (Truell et al. 1969). The last two factors fail to explain the peak of α and constant α in the subsequent period, because their influences would monotonically expand with the progression of fatigue. More and more cracks emerge as the fatigue advances and these are certainly the direct causes of failure. However, they are principally oriented parallel with the shear wave propagation, thus little influencing the measurements except at the very end of the lifetime. Diffraction loss depends on the thickness, frequency, and transducer size. The specimen thickness is reduced by about 10 % in the final stage and the resonance frequency increased by

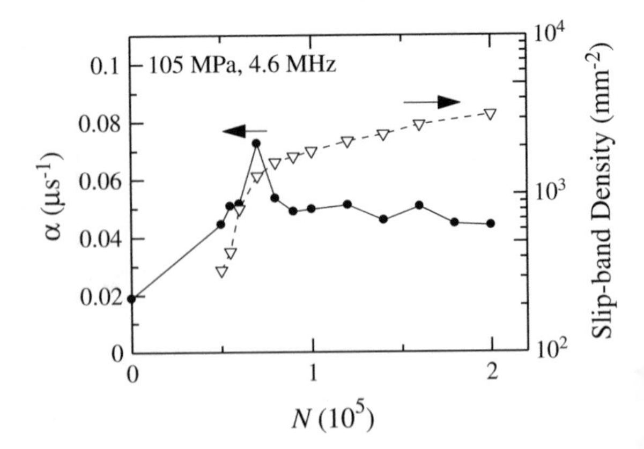

Fig. 16.6 Relationship between the attenuation coefficient (the 28th resonance) and slip band density. Stress amplitude was 105 MPa. Reprinted from Hirao et al. (2000), Copyright (2000), with permission from Elsevier

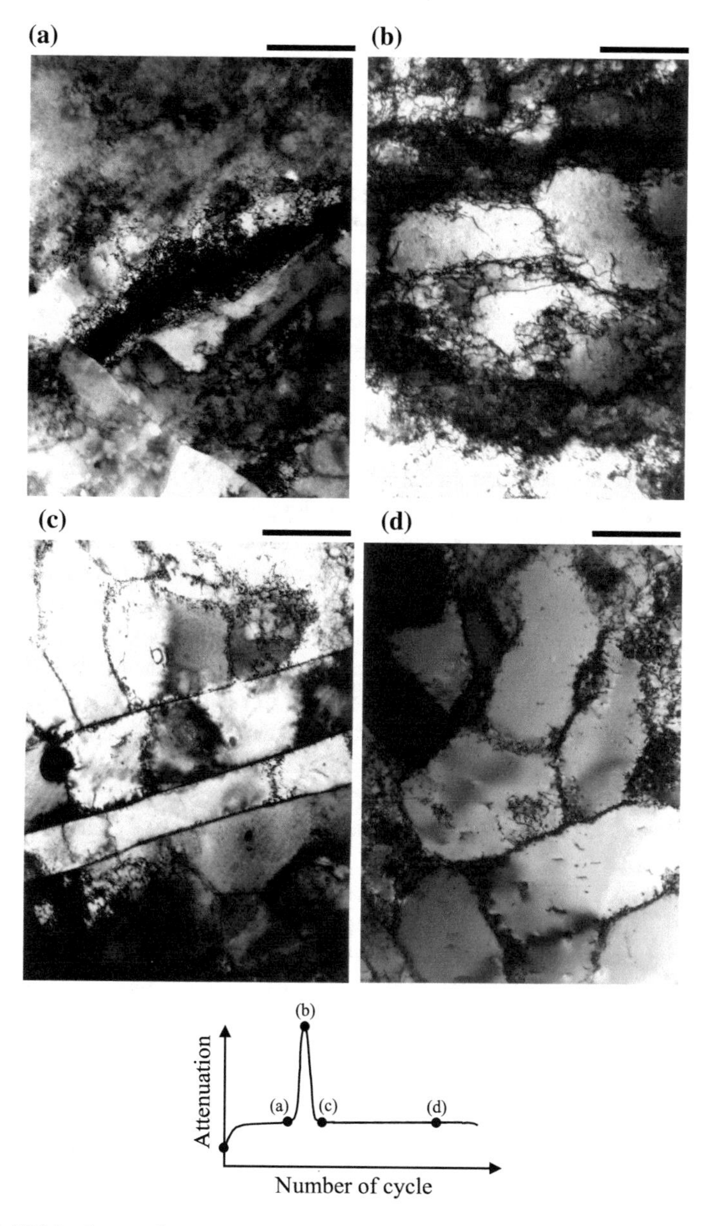

Fig. 16.7 TEM micrographs showing the dislocation structures at four phases in the fatigue lifetime. Stress amplitude was 105 MPa. The bars indicate 1 μm. Reprinted from Hirao et al. (2000), Copyright (2000), with permission from Elsevier

10 %, which gives an estimated diffraction loss change of 1.0×10^{-3} μs^{-1} (see Sect. 6.4.1). When the wavelength is much larger than the grains, grain scattering depends on the fourth power of frequency (see Chap. 15), which explains

the frequency dependence of the background of α values in Fig. 16.5. From the known grain size and the frequency shift, this effect may cause attenuation increases by nearly the same amount as diffraction. These effects then cause only negligible changes in α, and dislocation damping can solely explain the observed acoustic response.

According to the dislocation damping model in Eq. (7.1), α is proportional to the density Λ multiplied by the fourth power of the segment length L of the *effective* dislocations, which are mobile and can vibrate with the ultrasonic stress. Not all dislocations interact with ultrasonic waves. The frequency dependence of the attenuation peak height in Fig. 16.5, which is accompanied by the velocity drop, supports this theoretical model. Because α changes at higher orders of magnitude than f_n, tracing α will be a more practical way to monitor fatigue. Acoustic measurements along with the replica and TEM observations indicate four fatigue stages, I–IV, as indicated in Fig. 16.3.

Stage I: α remains constant (or slightly increases) and f_n slowly increases. There is a balance between the ongoing dislocation multiplication (increase in L) and tangling among them (decrease in L), keeping the dislocation damping nearly constant.

Stage II: α increases steeply after experiencing a small drop. Simultaneously, f_n increases and the slip bands grow at the highest rates, indicating intense dislocation slipping in this short period (10 % of the specimen lifetime). Appearance of slip bands demonstrates dislocation multiplication and movement beneath the surface. The dislocation network is completely changed in such a way that dislocations are released from heavy tangling and piling up at obstacles and then move to form stable cell structures. This is supported by the TEM image in Fig. 16.7b, showing a cell structure in the whole area, though not yet fully developed, which was absent in the previous stage. This transient process produces a larger value of L and increases α by several times.

Stage III: α decreases to a higher level than in Stage I, and the increasing rates of f_n and the slip bands diminish. Deceleration of their increase indicates the beginning of hardening, i.e., suppression of dislocation mobility. Dislocations become stationary even to small amplitudes of ultrasound (of the order of 0.1 nm or less). The shift to a cell structure is essentially completed within this stage. Dislocations pile up at the obstacles such as grain boundaries, precipitates, and twins, resulting in less mobility. TEM images show piled-up dislocations against twin boundaries, which were not observed in Stages I and II. An important observation here is that fatigue hardening, which starts during this stage, tends to diminish the dislocation damping, not gradually, but in a short interval in the early lifetime of the specimen.

Stage IV: α slowly decreases (or remains unchanged) and f_n shows a steady increase except for the final stage, where the resonance peak deteriorates because of uneven thickness and rough surfaces. The formation of cell walls is finished and free dislocations are few (Fig. 16.7d). Dislocation damping is small despite their high density.

The present ultrasonic method detects the through-thickness average of the elastic response. The fatigue process primarily proceeds in the surface layer. However, at low to moderate cycles to fracture of polycrystalline metals, the whole thickness is subjected to plastic deformation. In such cases, tracing the attenuation evolution with EMAR can point out slip band occurrence and the start of hardening at approximately the same fraction of cycles to fracture. This measurement will potentially indicate the remaining lifetime of the component. The EMAR observation is insensitive to crack growth, but it detects crack nucleation through the attenuation peak (Stages II and III). It is an open question as to which occurs first and then triggers the other: crack nucleation or dislocation restructuring.

16.3 Rotating-Bending Fatigue of Low-Carbon Steels

16.3.1 Specimen Preparation and Measuring Method

Specimens were commercial steel rods containing 0.22, 0.35, and 0.45 mass% C. They were heated at 880 °C for 1 h and cooled in air before the fatigue test. Table 16.1 gives their chemical compositions and the yield strengths σ_y. The specimen diameter gradually decreased to the center (14 mm) so as to cause failure at the minimum diameter, where ultrasonic measurement was made. Electropolishing was applied to the specimen surface. Figure 16.8 shows the measurement setup. The axial-shear-wave EMAT was used to excite and detect surface shear waves traveling in the circumferential direction along the specimen surface. The EMAT consists of a solenoid coil to apply the bias magnetic field in the specimen's axial direction and a

Table 16.1 Chemical compositions (mass%) and the yield strengths (MPa) of the carbon steel rods. Reprinted with permission from Ogi et al. (2002), Copyright (2002), AIP Publishing LLC

C	Si	Mn	P	S	Cu	Cr	Ni	Yield strength (σ_y)
0.22	0.19	0.41	0.18	0.15	0.02	0.06	0.01	333
0.35	0.18	0.64	0.2	0.11	0.1	0.15	0.06	449
0.45	0.18	0.64	0.21	0.18	0.13	0.11	0.06	471

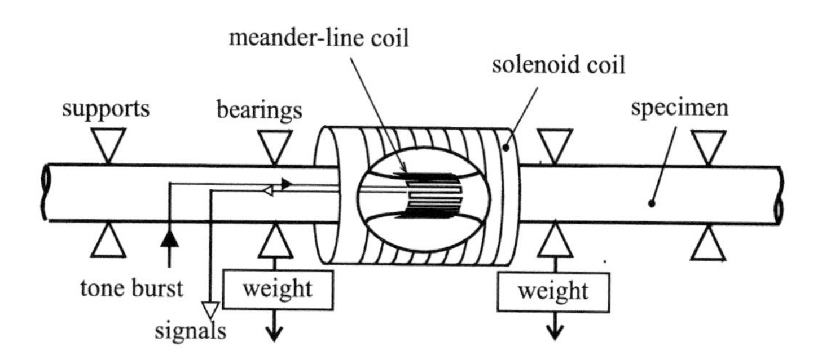

Fig. 16.8 Measurement setup for the velocity, attenuation, and nonlinearity of the axial shear wave during rotating-bending fatigue test of a steel rod. Reprinted with permission from Ogi et al. (2002), Copyright (2002), AIP Publishing LLC

meander-line coil to induce a dynamic field in the circumferential direction (see Fig. 3.17). The physical principle of transduction with this EMAT is given in Sects. 2.4.1 and 3.7 in detail. The minimum diameter of the specimen was wrapped with the meander-line coil, whose axial length was 20 mm. The axially polarized shear wave was radiated from the whole region under the coil. The motor rotated the specimen rod at 240 rpm (4 Hz). The bias field was fixed to 1.4×10^4 A/m. (The presence of the static field hardly affected the fatigue life.) Four-point-bending configuration introduced the maximum bending stresses at the measuring portion, which were varied between 140 and 490 MPa, or 0.42 and 1.2, of the yield strengths σ_y.

Equation (3.3) provides resonance frequencies of the axial shear wave for a cylinder of radius R with the integer n, which corresponds to the number of meandering along the circumference. Two meander-line coils of $\delta = 0.9$ mm (coil A) and 0.44 mm (coil B) were used (see Eq. (3.4)). The fundamental resonance frequency f appears at 3.89 and 7.94 MHz with coils A and B, respectively. The present EMAR study adopted the fundamental modes, because they show the maximum amplitudes at the surface and decay with steep gradients with the radius (see Sects. 3.7 and 13.2). This property is best suited for concentrating the measurement at the thin surface layer and detecting the microstructure evolution there caused by rotating-bending fatigue. Figure 16.9 shows the calculated surface wave energy distribution for the fundamental modes generated by the two coils. The penetration depths are estimated to be 0.5 and 0.2 mm, respectively. Thus, the measured ultrasonic characteristics reflect the microstructure in these surface regions.

Interrupted and continuous measurements were performed. In the interrupted measurements, the fatigue test was stopped at some intervals and replicas of the specimen surface over the whole measuring area were obtained by applying 70-MPa bending stress to open the surface cracks. The load was then released to measure the resonance frequency (Chap. 5), the attenuation coefficient (Chap. 6), and acoustic nonlinearity (Chap. 10). In the continuous measurements, the resonance frequency and attenuation coefficients were measured throughout the lifetime without interrupting the fatigue test.

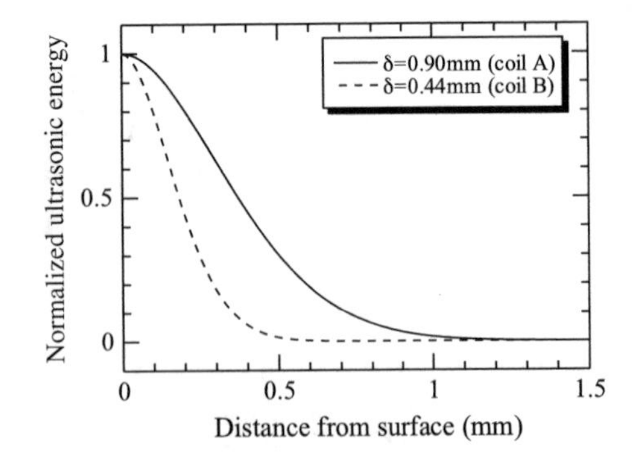

Fig. 16.9 Distribution of the axial-shear-wave energy at the fundamental modes with two wavelengths. Reprinted with permission from Ogi et al. (2002), Copyright (2002), AIP Publishing LLC

16.3.2 Attenuation Peak

Figure 16.10 shows the resonance spectrum, including up to the sixth modes, and the amplitude ring-down curve at the fundamental mode measured using coil A. Figure 16.11a shows typical evolutions of the attenuation coefficient α and the phase velocity v during the fatigue test. Because the specimen diameter remains unchanged throughout, the resonance frequency change equals the velocity change, $(v - v_0)/v_0$, where v_0 denotes the velocity before fatiguing. The horizontal axis in Fig. 16.11 is the cycle number normalized by the failure cycle number, N_F. More than forty specimens were tested with various loading conditions. Common observations are as follows:

(i) Over the first half of the life, α is almost stable, but v monotonically decreases. Approaching 70 % of the lifetime, α starts to decrease below the initial value. This is accompanied by acceleration in the decrease of v.
(ii) After this, α increases to a maximum and immediately drops. At the maximum, v pauses or slightly increases.
(iii) Peak(s) of α is very sharp. The peak width is only a few percent of the life. The peak sometimes appears twice (Fig. 16.20).
(iv) The ratio of the cycle number N_P at the attenuation peak to the failure cycle number, N_P/N_F, appears to be independent of carbon content and the bending stress, as seen in Fig. 16.12, indicating a potential capability of evaluating

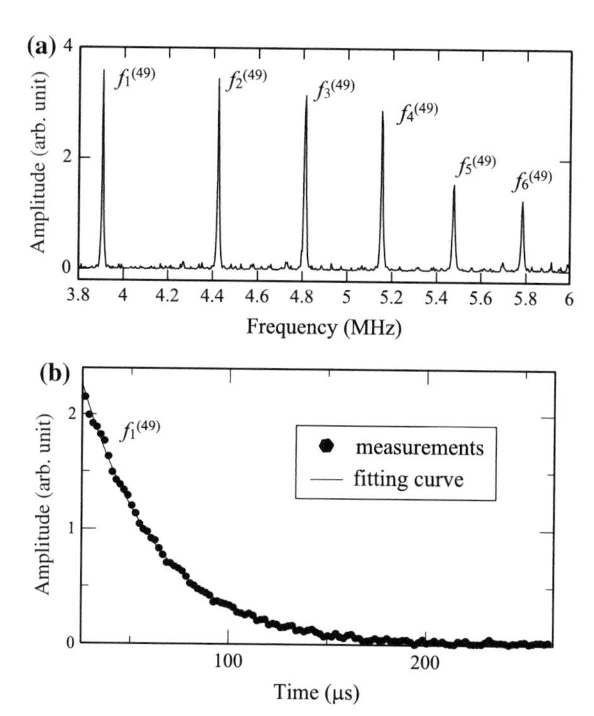

Fig. 16.10 **a** Resonance spectrum of the axial shear wave measured with coil A for a carbon steel rod of 14-mm diameter. **b** Amplitude ring-down curve of the fundamental mode and fitted exponential function. After Ogi et al. (2000)

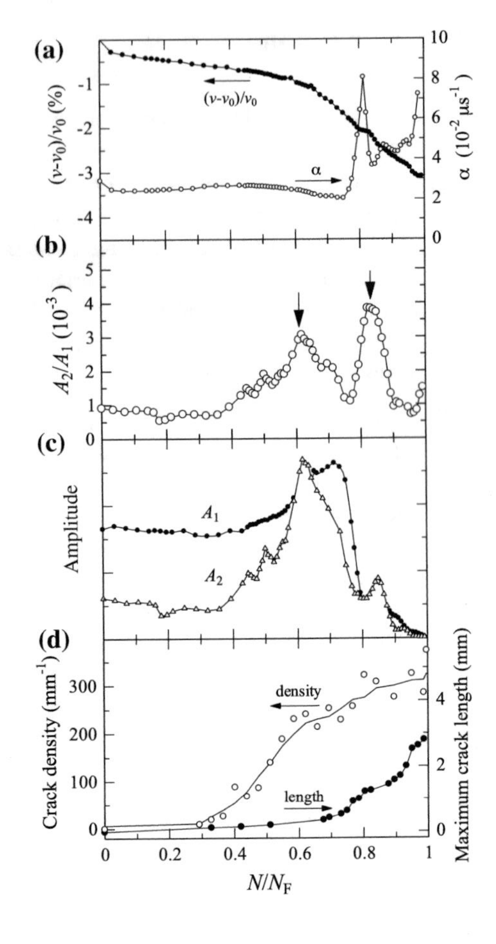

Fig. 16.11 Typical evolutions of **a** velocity and attenuation, **b** the nonlinearity, **c** the fundamental and second-harmonic amplitudes, and **d** surface cracks for 0.25 mass% C steel (N_F = 56,000, $\sigma = 0.84\sigma_y$). *Arrows* indicate the first and the second nonlinearity peaks. Coil A was used. Reprinted with permission from Ogi et al. (2001), Copyright (2001), AIP Publishing LLC

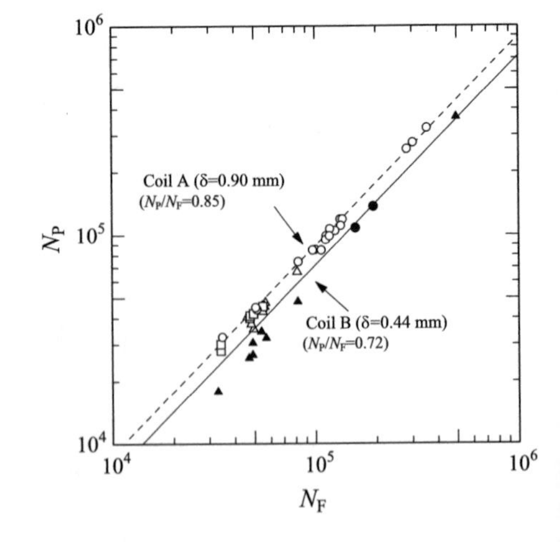

Fig. 16.12 Failure cycle number N_F versus the cycle number at the attenuation peak, N_P. Open marks denote measurements by coil A and closed marks denote those by coil B. *Circles*, *squares*, and *triangles* denote the measurements for 0.22, 0.35, and 0.45 mass% C steels, respectively. Reprinted with permission from Ogi et al. (2002), Copyright (2002), AIP Publishing LLC

the remaining life. The attenuation peak appeared earlier with coil B ($N_P/N_F \approx 0.72$) than with coil A ($N_P/N_F \approx 0.85$). When the bending stress was so small that N_F was beyond one million, α remained unchanged throughout the life as shown in Fig. 16.13 with solid triangles.

(v) When the fatigue loading was aborted at the attenuation peak and the specimen was exposed to a low-temperature heat treatment at 300 °C for 1 h, the attenuation and velocity returned to the values before the peak (Fig. 16.14).

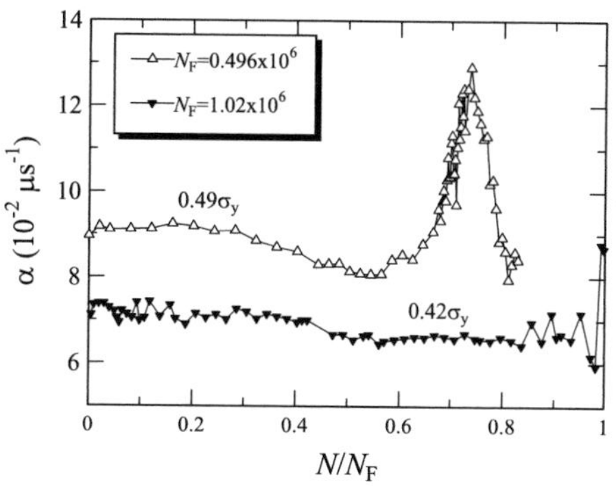

Fig. 16.13 Evolutions of the attenuation coefficient during rotating-bending fatigue and pull–push fatigue for the 0.22 mass% C steel. The number of cycles N is normalized by the failure cycle number N_F. Coil B was used. Reprinted with permission from Ogi et al. (2002), Copyright (2002), AIP Publishing LLC

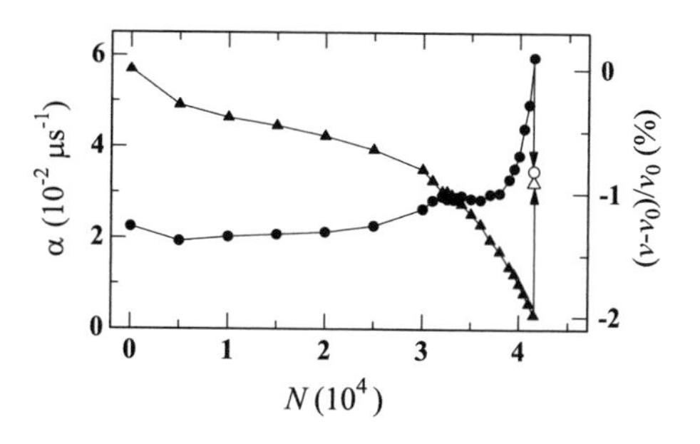

Fig. 16.14 Recovery of attenuation and velocity by a low-temperature heat treatment of 300 °C for 1 h for 0.22 mass% C steel ($\sigma = 0.84\sigma_y$). Open marks denote the recovered values. Coil A was used

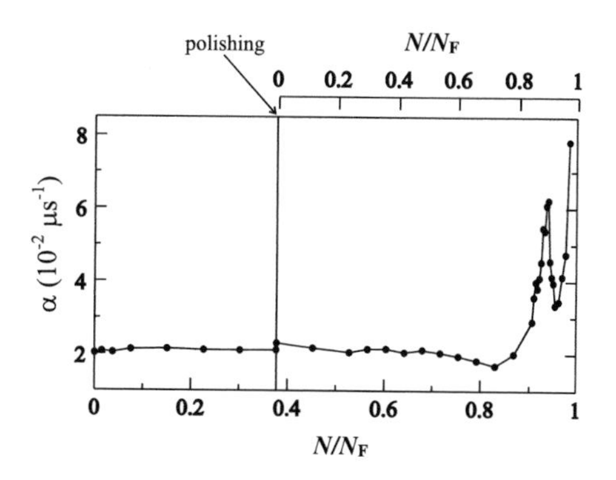

Fig. 16.15 Attenuation evolution after removing the damaged surface layer for 0.45 mass% C steel. Coil A was used. Reprinted with permission from Ogi et al. (2002), Copyright (2002), AIP Publishing LLC

(vi) Removal of deteriorated surface layer delayed the peak occurrence as shown in Fig. 16.15. The fatigue test was interrupted before the attenuation peak, the specimen surface was electrically polished to remove the damaged layer about 0.3 mm thick, and then, the cyclic loading was restarted. The attenuation peak appearance was delayed ($N_P/N_F \approx 0.93$), but the ratio of N_P/N_F was unchanged ($\approx$0.87) if the cycle number was counted from the restarting point.

16.3.3 Nonlinearity Peaks

Figure 16.11b shows the normalized nonlinearity A_2/A_1 for the fundamental axial-shear-wave resonance. (Details of measuring the nonlinearity are given in Chap. 10.) For these measurements, only coil A was used. Individual measurements of A_1 and A_2 are plotted in Fig. 16.11c. The nonlinearity shows two peaks during the fatigue life; the first one appears around 60 % and the second one around 85 % of N_F. Thus, the second nonlinearity peak simultaneously occurs with the attenuation peak. Figure 16.16 plots the cycle numbers at the nonlinearity peaks N_P versus the failure cycle number N_F. The ratio of N_P/N_F is independent of the carbon content and the bending stress. Not shown here, the second nonlinearity peak disappeared when the specimen was annealed at 300 °C for 1 h, as seen with the attenuation peak.

16.3.4 Microstructure Observations

Figure 16.11d shows the replica observations for evolving surface cracks. The crack density was calculated from the total crack length divided by the viewing

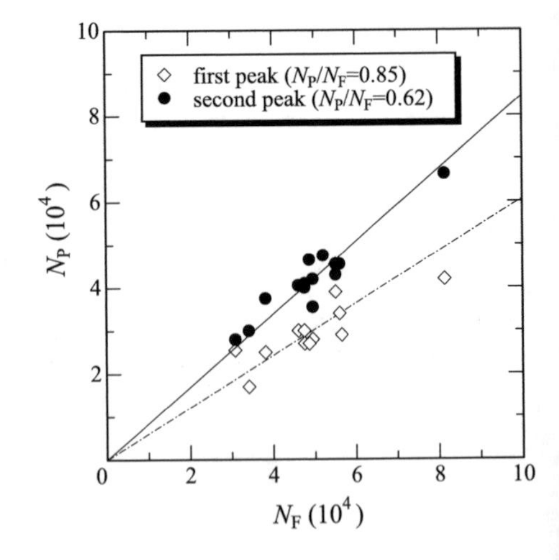

Fig. 16.16 Relationship between the failure cycle number (N_F) and the cycle numbers at the first and the second nonlinearity peaks (N_P) for 0.25 mass% C and 0.35 mass% C steels. Coil A was used. Reprinted with permission from Ogi et al. (2001), Copyright (2001), AIP Publishing LLC

area. They were observable with an optical microscope as early as about 25 % of lifetime; the average length was about 50 μm at this stage. The crack density monotonically increased after that, while the maximum length remained small until about 80 % and rapidly increased to failure. This observation indicates that the cyclic loading between $N/N_F \approx 0.25$ and 0.8 was spent mainly by crack nucleation, not crack growth, producing many small cracks of nearly a uniform size. Just before the attenuation peak, these small cracks coalesce with each other to form longer cracks, resulting in growth inward in the final stage and then fracture.

Referring to the attenuation evolution, the fatigue test was truncated to observe the dislocation structure with TEM. Thin specimens were carefully cut from the fatigued specimens to about 0.4-mm-thick slices involving the specimen surface using a low-speed electrical discharge wire. Polishing on emery paper made them about 100 μm thick. A jet electropolishing apparatus introduced a small hole at the center of the foil with an electrolyte of 8 % perchloric acid, 10 % 2-butoxyethanol, 70 % ethanol, and water. TEM images were obtained around the hole.

Figure 16.17a–c shows the typical dislocation structures at 0.15 mm deep from the specimen surface before, at, and after the attenuation peak. Corresponding attenuation evolutions monitored by coil B are also provided. Since such a TEM micrograph provides highly localized information, many TEM images were observed from the fatigued specimens with various bending stresses and the representative dislocation structure was identified at each stage. There were few dislocations in the annealed specimen before fatigue. Before the attenuation peak (Fig. 16.17a), many dislocation walls, about 0.2 μm thick, occurred, which were aligned parallel with nearly constant spacing about 1 μm. There were few dislocations between the walls. At the attenuation peak (Fig. 16.17b), many single-line dislocations were bridging the walls. After the attenuation peak, the dislocation wall structures were divided into smaller cells with few dislocations inside (Fig. 16.17c). Similar dislocation structure evolution was observed around the attenuation peak measured with coil A at a 0.3-mm-deep region from the surface.

16.3.5 Mechanism of First Nonlinearity Peak

The first nonlinearity peak is attributed to crack nucleation and growth. Richardson (1979) analyzed the second-harmonic amplitude generated by the opening and closing effect of an interface subjected to a bias pressure with a passing acoustic wave and suggested that an optimum pressure existed for second-harmonic generation. Buck et al. (1978) and Morris et al. (1979) demonstrated that Richardson's model is basically applicable to fatigue cracks in metals; the second-harmonic amplitudes showed maxima with a low external compressive stress. Thus, small fatigue cracks can act as harmonics generators. They are partly closed by compressive residual stress in the early stages. Large cracks are fully open and produce no higher harmonics. Thus, the second-harmonic amplitude exhibits a maximum during crack growth.

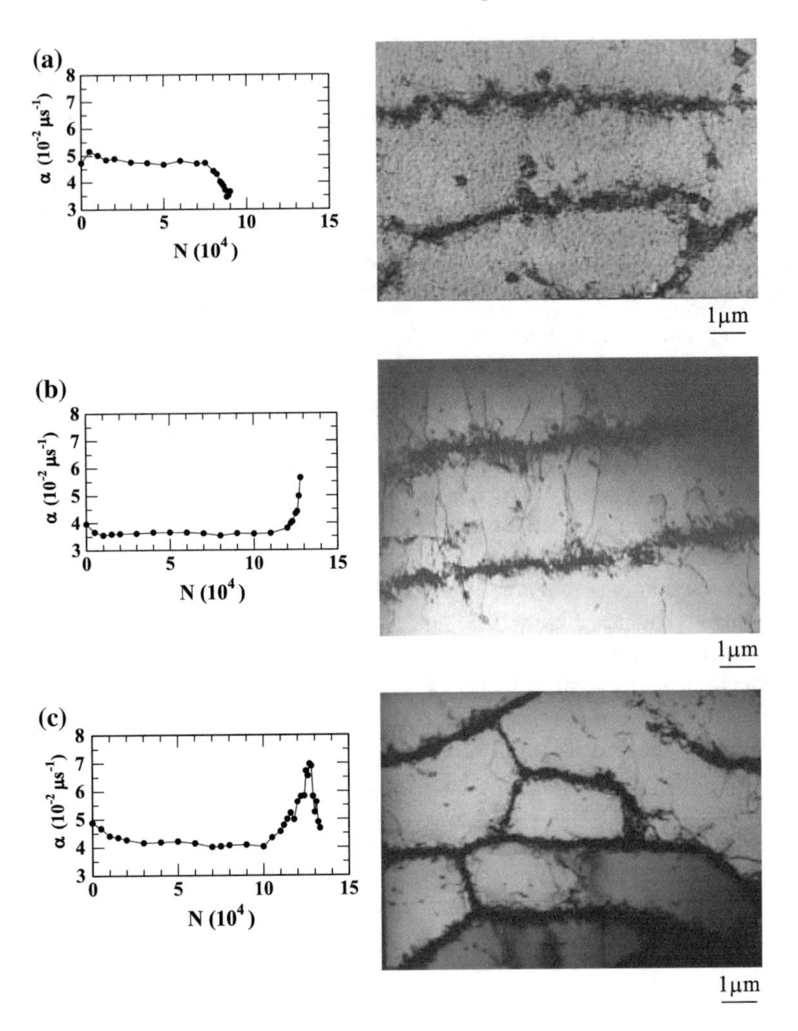

Fig. 16.17 TEM micrographs: **a** before, **b** at, and **c** after the attenuation peak for 0.45 mass% C steel. Attenuation evolutions until the interruption of the fatigue test are also shown ($\sigma = 0.72\sigma_y$). Coil B was used. Reprinted with permission from Ogi et al. (2002), Copyright (2002), AIP Publishing LLC

Indeed, small fatigue cracks are partly closed along the tips because of the compressive residual stress, which arises from local plastic deformation (plasticity-induced closure) (Newman and Elber 1988). The crack closure stress can be measured by applying tensile stress to the specimen surface and monitoring the reflected Rayleigh wave signal (London et al. 1989; Hirao et al. 1993). The residual stress is large near the tip, which tightly closes the crack faces. It diminishes the approaching of the crack mouth, and there is a thin band where the crack faces are in contact at very small pressure. These thin bands open and close when the acoustic wave impinges on them, distort the waveform, and make a strong non-linearity source. In the present case, there are many band sites, since the crack shape is shallow, semi-ellipsoidal, of relatively uniform size. The nonlinearity should then increase as the number of small cracks increases. In Fig. 16.11, the crack density correlates with the nonlinearity before the first peak; in this period, the crack length remains unchanged and the number of cracks monotonically increases. In the course of fatigue, the small cracks coalesce with each other and start to grow inward. This process reduces the area of weakly touching crack faces because of the asperity contact across the shallower faces, resulting in nonlinearity decrease. The crack length starts to increase after this nonlinearity peak in Fig. 16.11d, being compatible with the above interpretation. Thus, the first nonlinearity peak indicates the beginning of crack coalescence.

16.3.6 Mechanism of Attenuation Peak and Second Nonlinearity Peak

First of all, the recovery of the attenuation and the nonlinearity at the second peak with the heat treatment excludes the possibility of cracks' direct contribution to these two peaks. Figure 16.15 indicates that the attenuation change was caused by the microstructure change within the thin surface layer. Thus, these peaks are related to the dislocation structure change in the surface region. On the other hand, these peaks indicate the remaining life (Figs. 16.12 and 16.16), which is undoubtedly controlled by cracking. Seeking an interrelation among attenuation, dislocations, and cracks, the discussion should involve the following factors

- (i) dislocation damping,
- (ii) dislocation mobility,
- (iii) crack tip zone and dislocation evolution, and
- (iv) volume fraction of the crack tip zone to the surface wave probing region.

Of all the crystal defects contributing to attenuation and nonlinearity, only dislocations cause the nonmonotonic evolutions during fatigue. Anelastic behavior of the free dislocations is the common source of attenuation and nonlinearity. Granato-Lücke's theory (1956) relates the attenuation α to the density Λ and the length between pinning points L of the *effective* dislocations that can vibrate responding to the acoustic wave. Because of the phonon viscosity, the vibration is anelastic, causing loss. At frequencies much lower than the dislocation-segment's resonance frequency ($\sim$0.1 GHz), the relationship reduces to $\alpha \propto \Lambda L^4 f^2$ as shown in Eq. (7.1), indicating that α increases with the effective dislocations. The theory also predicts that $v \propto \Lambda L^2$ for the velocity v (Eq. (7.2)). Concerning the relationship between the nonlinearity and dislocations, Hikata et al. (1965) derived a form of $A_2/A_1^2 \propto \Lambda L^4$, which is similar to Eq. (7.1).

Thus, these theories can explain the attenuation and second nonlinearity peaks, provided that the dislocations bridging the walls (Fig. 16.17b) are effective and the densely gathering dislocations in the walls are ineffective (immobile).

In fatigued fcc crystals, very similar dislocation structures to Fig. 16. 17a, b have been reported (Mughrabi et al. 1979). Those are known as persistent slip bands (PSBs), consisting of dense array of dislocation (walls) and the sparse dislocation zone between the walls. The dislocation walls consist mainly of edge dislocation dipoles (pairs of opposite-sign dislocations), which strongly attract each other and will be immobile with the acoustic wave. Such a PSB structure was also observed for low-carbon steels by Pohl et al. (1980); the dislocation structures observed before and at the attenuation peak appear to be the PSBs. The dislocations bridging the walls in Fig. 16.17b appear to be mobile, and they are the very cause of raising attenuation. Indeed, they occur and glide when local plastic deformation occurs (Suresh 1998). These dislocations will gather by further cyclic loading to divide the between-wall areas, resulting in the cells with few dislocations inside. In Sect. 16.2, we saw that the dislocation cells absorb little acoustic energy. The attenuation peak and the second nonlinearity peak can thus be explained by the microstructure rearrangement from the walls to the cells, during which many mobile dislocations temporarily appear to absorb the acoustic energy.

Pohl et al. (1980) indicated that such a rearrangement requires sufficiently large stress amplitude. This can be made possible by crack growth even for lower cyclic stress amplitude. There is high-stress concentration zone ahead of a crack tip, which causes local plastic deformation and then dislocation rearrangement. As shown in Fig. 16.18, many small dislocation cells form around a crack tip, and the microstructure is quite different from those far away from the tip.

Interpretations of those measurements and observations result in understanding the mechanism for the attenuation and second nonlinearity peaks appearing at fixed fractions to N_F. Figure 16.19 illustrates this view. Dislocations will multiply from the beginning of fatigue throughout the lifetime, but most of them pile up to

Fig. 16.18 Dislocation structure around the crack tip (0.45 mass% C steel). Reprinted with permission from Ogi et al. (2002), Copyright (2002), AIP Publishing LLC

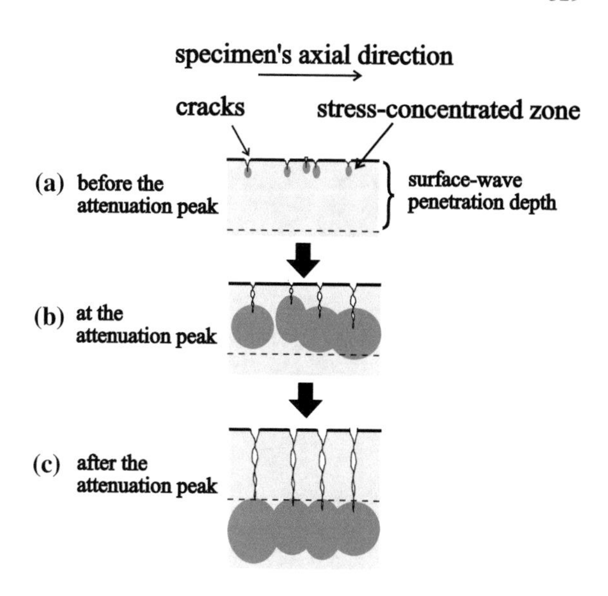

Fig. 16.19 Schematic explanation of the interaction between the dislocations and crack growth to cause the attenuation peak. Reprinted with permission from Ogi et al. (2002), Copyright (2002), AIP Publishing LLC

obstacles and tangle each other to shorten their lengths. The dislocation density increase and the dislocation length decrease balance each other and keep the attenuation unchanged for a long time. Meanwhile, the small cracks nucleate and the density increases. But, the average length changes little (Fig. 16.11d), indicating that the cyclic loading is spent by crack nucleation, not crack growth, because of the steep bending stress gradient. The crack length was less than about 0.1 mm until the attenuation peak. The crack shape is shallow semi-ellipsoidal, and the average crack is much shallower than the surface wave penetration; the resultant tip zones are small. (The stress concentration factor at a crack edge increases with the crack depth). Also, because of fewer cracks, the volume fraction of the total crack tip zone to the whole surface wave probing region should be fairly small until the attenuation peak. The surface shear wave is then insensitive to the microstructure change in such a small volume because ultrasonics provides the averaged microstructure information over the propagation region (Fig. 16.19a). At the later stages, the dense surface cracks coalesce to grow not only along the circumference but also to the interior, extending the tip zones toward the interior (Fig. 16.19b). Then, the crack tip zone's volume fraction becomes sufficiently large to give rise to a temporal increase in attenuation and nonlinearity, responding to the dislocation structure evolution from the walls to the cells. With further cycling, the cracks extend further inside and the tip zones leave the surface wave's probing region (Fig. 16.19c). Therefore, the shallower the surface wave penetration is, the earlier the attenuation peak occurs. The attenuation peak occasionally appeared twice as seen in Fig. 16.20. This may reflect the dislocation rearrangement occurring at different positions within the probing region at different times.

As shown in Fig. 16.13, no attenuation peak appeared in the case of very high-cycle rotating-bending fatigue tests, where the failure cycle number exceeded 10^6. In the

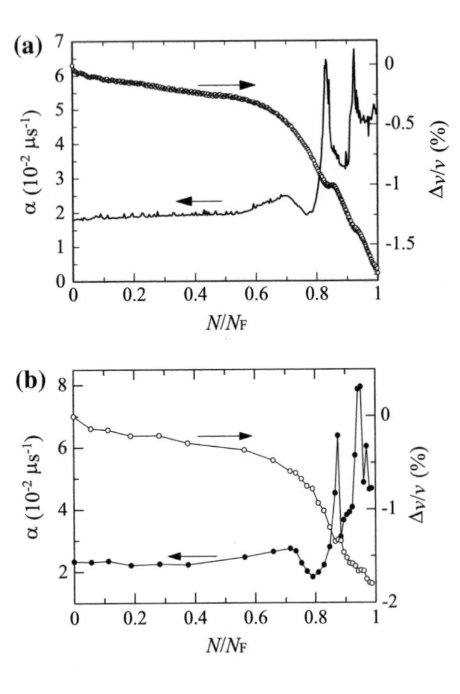

Fig. 16.20 Attenuation and velocity evolutions during fatigue of 0.45 mass% C steel obtained from **a** continuous and **b** interrupted measurements. $N_F = 61{,}600$ and 53,000, respectively. $\sigma = 0.93\sigma_y$. Coil A was used. After Ogi et al. (2000)

high-cycle fatigue tests, very few surface cracks were visible with replication and one of them grew inside to cause failure. In this case, the crack tip zone occupies very little volume in the surface wave propagation region, and the surface wave fails to detect such a highly localized microstructure change. To detect the attenuation peak in this situation, we need to decrease the axial length of the meander-line coil to narrow the surface wave propagation region.

16.3.7 Influence of Interruption of Fatiguing on the Measurements

Figure 16.20 compares the attenuation and velocity evolutions obtained by continuos and interrupted measurements. In the interrupted measurement, the ultrasonic characteristics were obtained after nearly 10 min passed after stopping the fatiguing. There was no noticeable difference between the two evolutions so far as the steel specimens were concerned. However, the aluminum alloy behaved differently as shown in Fig. 16.21 (Ogi et al. 2000). The Lorentz force axial-shear-wave EMAT (Fig. 3.18) was used since the material is nonmagnetic. In the interrupted measurement, both α and v remained unchanged until fracture. In contrast, the continuous measurements showed a sharp attenuation peak at a later stage of fatigue, at which the velocity temporarily increases. Thus, similar evolutions of α and v occur in the continuous measurements for the aluminum alloy and carbon steels.

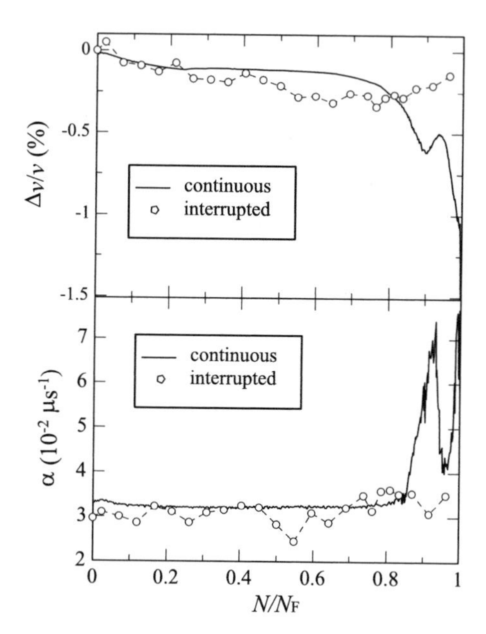

Fig. 16.21 Continuous and interrupted measurements of attenuation and velocity in an aluminum alloy (Al 5052) rod. $\sigma = 1.07\sigma_y$ (σ_y = 140 MPa). N_F = 73,300 and 84,040, respectively. After Ogi et al. (2000)

Point defects such as vacancies and interstitials are more mobile in aluminum than in steels at room temperature. They diffuse toward dislocations to pin them and make dislocation damping ineffective. This phenomena has been studied in details in Sects. 7.3 and 7.4. Considering that the number of vacancies increases with the progress of fatigue, vacancy diffusion to dislocations explains stable attenuation and velocity throughout the fatigue test in the interrupted measurement. Vacancies migrate to and precipitate on the dislocations to suppress their anelastic vibration with ultrasonic wave.

16.4 Tension-Compression Fatigue of Low-Carbon Steels

The EMAR with the axial shear resonance was also applied to tension-compression fatigue of low-carbon steels by Ohtani et al. (2000). Figure 16.22a shows the measurement setup. The data acquisition for the ultrasonic measurements was synchronized with the cyclic stress so as to continuously monitor the attenuation and velocity at the maximum stress (tension) and at the minimum stress (compression) as illustrated in Fig. 16.22b. One set of measurements for attenuation and velocity needed two hundred cycles, which is much smaller than N_F. The meaner-line coil of δ = 0.90 mm (coil A) was used.

Figures 16.23 and 16.24 show typical evolutions of attenuation and velocity. First of all, magnitudes of their changes are much smaller than those observed in rotating-bending fatigue tests, where a peak value of attenuation is 2–5 times higher

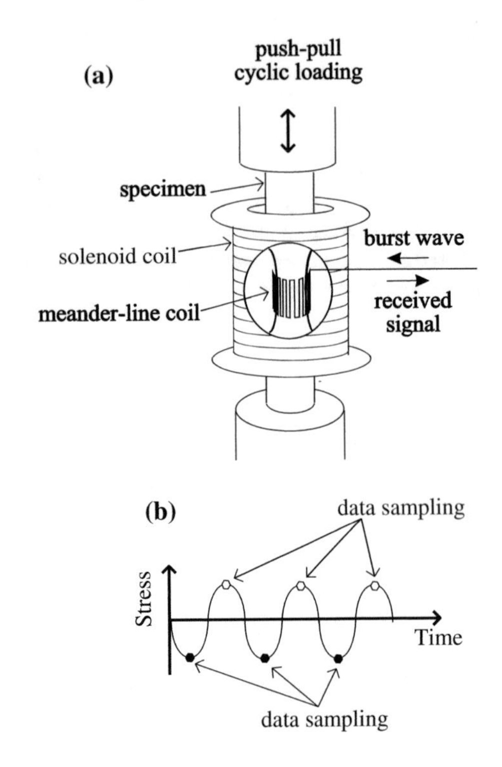

Fig. 16.22 **a** Measurement setup to monitor the attenuation coefficients and phase velocity of axial shear wave during a tension-compression fatigue test. **b** Illustration of data acquisition for the ultrasonic measurements at the maximum (tension) and minimum (compression) stresses. The cyclic stress and the ultrasonic measurements are synchronized to each other. Reprinted from Ohtani et al. (2000), Copyright (2000), with permission from Elsevier

than the initial value and the velocity decrease reaches more than 1 % at the final stage (see Fig. 16.20, for example). Occasionally, we could not detect their indicative changes until failure. The insensitivity in a tension-compression fatigue test is attributed to fewer numbers of cracks. In tension-compression fatigue, only one or two surface cracks were visible, which grew to cause failure as shown in

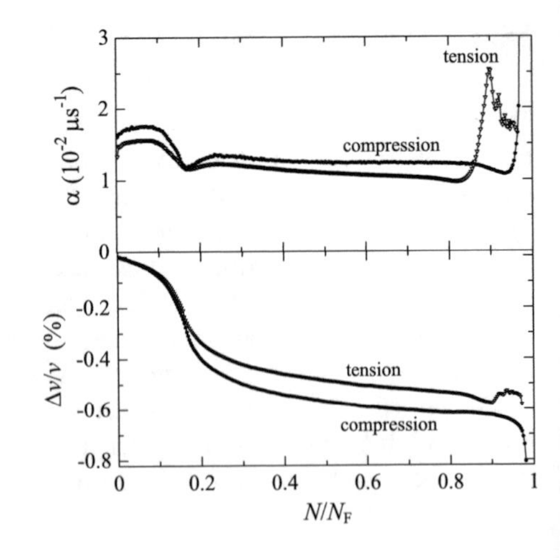

Fig. 16.23 Attenuation (*upper*) and velocity (*lower*) evolutions during a tension-compression fatigue test of a 0.15 mass% C steel. $\sigma_{max} = 0.66\sigma_y$ (σ_y = 294 MPa). N_F = 169,226. Coil A was used

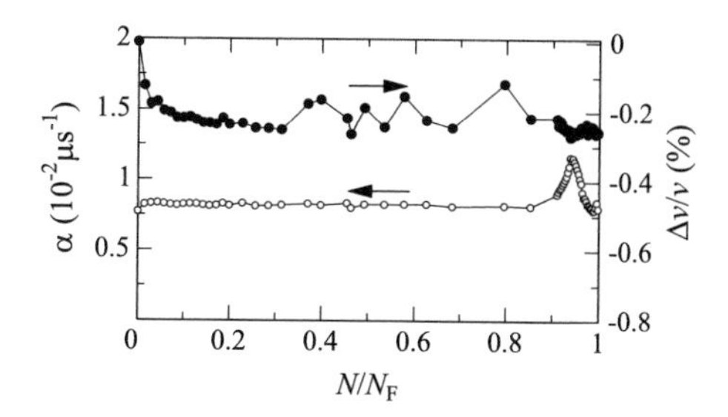

Fig. 16.24 Attenuation and velocity evolutions at the maximum stress (tension) during a tension-compression fatigue test of a 0.45 mass% C steel. $\sigma_{max} = 0.61\sigma_y$ (σ_y = 471 MPa). N_F = 351,780. Coil A was used

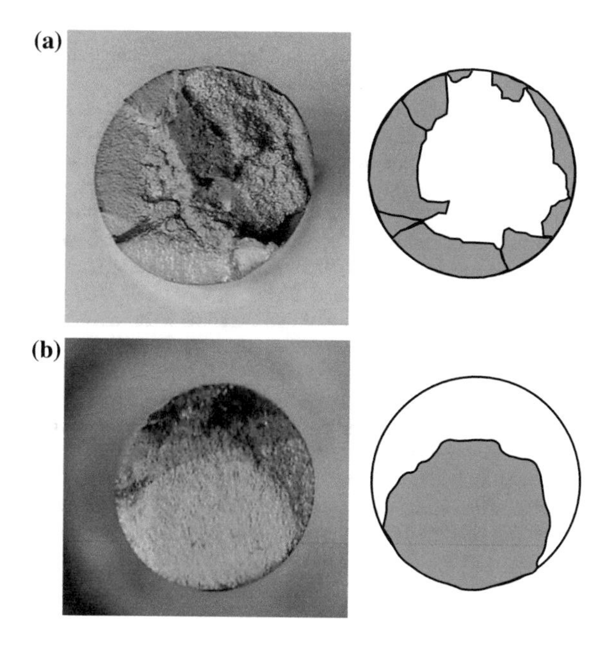

Fig. 16.25 Fracture surfaces of specimens in **a** a rotating-bending fatigue test and **b** a tension-compression fatigue test. Illustrations highlight traces of crack growth

Fig. 16.25b. Because the crack tip zone occupies very little volume in the probing region, any microstructure change in the tip zone will not be detected by the surface wave. This is also the case in the high-cycle rotating-bending fatigue test (Fig. 16.13).

It is interesting to note that in Fig. 16.23, the attenuation peak appears around $N/N_F = 0.9$ in the measurement at the tensile phase, but not at the compression phase. The tensile stress opens the surface cracks and causes the stress concentration in their tip zones. The compressive stress only adds the contact pressure across the crack faces and will not develop high-stress zone ahead of them.

To summarize this chapter, the key for detecting the attenuation peak is the fraction of highly stressed volume to the probing region, which increases with the number of surface cracks.

References

Bratina, W. (1966). Internal friction and basic fatigue mechanisms in body-centered cubic metals, mainly iron and carbon steels. In *Physical Acoustics* (Vol. 3A, pp. 223–291). New York: Academic Press.

Bratina, W. J., & Mills, D. (1962). Study of fatigue in metals using ultrasonic techniques. *Canadian Metallurgical Quarterly: The Canadian Journal of Metallurgy and Materials Science 1*, 83–97.

Buck, O., Morris, W. L., & Richardson, J. M. (1978). Acoustic harmonic generation at unbonded interfaces and fatigue cracks. *Applied Physics Letters, 33*, 371–373.

Fei, G., & Zhu, Z. (1993). The relation between the variation of stress, energy loss, ultrasonic attenuation, and dislocation configuration in aluminum during early stages of fatigue. *Physica Status Solidi (a), 140*, 119–126.

Granato, A., & Lücke, K. (1956). Theory of mechanical damping due to dislocations. *Journal of Applied Physics, 27*, 583–593.

Hikata, A., Chick, B., & Elbaum, C. (1965). Dislocation contribution to the second harmonic generation of ultrasonic waves. *Journal of Applied Physics, 36*, 229–236.

Hirao, M., Ogi, H., Suzuki, N., & Ohtani, T. (2000). Ultrasonic attenuation peak during fatigue of polycrystalline copper. *Acta Materialia, 48*, 517–524.

Hirao, M., Tojo, K., & Fukuoka, H. (1993). Small fatigue crack behavior in 7075-T651 aluminum as monitored with Rayleigh wave reflection. *Metallurgical Transactions A, 24A*, 1773–1783.

Klesnil, M., & Lukáš, P. (1980). *Fatigue of Metallic Materials*. New York: Elsevier.

London, B., Nelson, D. V., & Shyne, J. C. (1989). The influence of tempering temperature on small fatigue crack behavior monitored with surface acoustic waves in quenched and tempered 4140 steel. *Metallurgical Transactions A, 20A*, 1257–1265.

Morris, W. L., Buck, O., & Inman, R. V. (1979). Acoustic harmonic generation due to fatigue damage in high-strength aluminum. *Journal of Applied Physics, 50*, 6737–6741.

Mughrabi, H., Ackermann, F., & Hertz, K. (1979). Persistent slip bands in fatigued face-centered and body-centered cubic metals. In *Fatigue Mechanics* (pp. 69–105). ASTM-STP 675.

Newman, J. C., Jr., & Elber, W. (1988). *Mechanics of Fatigue Crack Closure*. ASTM-STP 982.

Ogi, H., Hirao, M., & Minoura, K. (1997). Noncontact measurement of ultrasonic attenuation during rotating fatigue test of steel. *Journal of Applied Physics, 81*, 3677–3684.

Ogi, H., Hamaguchi, T., & Hirao, M. (2000). Ultrasonic attenuation peak in steel and aluminum alloy during rotating bending fatigue. *Metallurgical and Materials Transactions A, 31*, 1121–1128.

Ogi, H., Hirao, M., & Aoki, S. (2001). Noncontact monitor of surface-wave nonlinearity for predicting the remaining life of fatigued steels. *Journal of Applied Physics, 90*, 438–442.

Ogi, H., Minami, Y., & Hirao, M. (2002). Acoustic study of dislocation rearrangement at later stages of fatigue: Noncontact prediction of remaining life. *Journal of Applied Physics, 91*, 1849–1854.

Ohtani, T., Ogi, H., & Hirao, M. (2000). Ultrasonic attenuation monitoring of fatigue damage in low carbon steels with electromagnetic acoustic resonance (EMAR). *Journal of Alloys and Compounds, 310*, 440–444.

Pawlowski, Z. (1964). Ultrasonic attenuation during cyclic straining. In *Proceedings of Fourth International Conference on Nondestructive Testing* (pp. 192–195). London: Butterworths.

Pohl, K., Mayr, P., & Macherauch, E. (1980). Persistent slip bands in the interior of a fatigued low carbon steel. *Scripta Metallurgica, 14*, 1167–1169.

Richardson, J. M. (1979). Harmonic generation at an unbonded interface—I. Planar interface between semi-infinite elastic media. *International Journal of Engineering Science, 17*, 73–85.

Schenck, H., Schmidtmann, E., & Kettler, H. (1960). Einfluβ einer Verformungsalterung auf die Vorgänge bei der Wechelbeanspruchung von Stahl. *Arch. für Einsenhüttenwesen, 31*, 659–669 (in Germany).

Suresh, S. (1998). *Fatigue of Materials* (2nd ed.). Cambridge: Cambridge University Press.

Truell, R., Elbaum, C., & Chick, B. B. (1969). *Ultrasonic Methods in Solid State Physics*. New York: Academic Press.

Truell, R., & Hikata, A. (1957). In *Fatigue and Ultrasonic Attenuation* (pp. 63–70). ASTM-STP 213.

Zhu, Z., & Fei, G. (1994). Variation in internal friction and ultrasonic attenuation in aluminum during the early stage of fatigue loading. *Journal of Alloys and Compounds,* 211/212, 93–95.

Chapter 17
Creep Damage Detection

Abstract Creep is an important factor in designing metal structures used at elevated temperatures. Dislocations again play a key role in deformation at elevated temperatures; they multiply, slip, form voids and subgrain boundaries, and annihilate. This chapter shows that the shear wave attenuation is highly sensitive to the dislocation activity in crept metals and the measurement of attenuation behavior with EMAR can predict the remaining life, especially the time when tertiary creep begins. This is true for many high-temperature materials. Acoustic nonlinearity is also demonstrated to be sensitive to the creep process, which is the subject of Chap. 10.

Keywords Attenuation peak · Austenitic stainless steel · Cell structure · Cr-Mo-V steel · Dislocation multiplication · Shear-wave EMAT · Ni-based superalloy · SEM/TEM observations

17.1 Aging of Metals

Structural metals are subject to aging from fatigue, creep, corrosion, irradiation, and their combination. Exposure to high temperatures promotes creeping and stress-corrosion cracking. Aged metals lose toughness or ability to absorb energy for stresses above the yield point. They cannot withstand the occasional high loads without fracturing. In-service degradation with creep is one of major concerns in power plants, chemical plants, and oil refineries, being an important issue of worldwide significance. For instance, many of fossil power plants were constructed during 1960s and 1970s and their working time has exceeded more than 100,000 h. They are still operating, while they have undergone progressive damage, mainly from creep as the time proceeds. By shifting the base load of power from fossil power plants to nuclear power plants, they are facing even more severe operating conditions such as daily or weekly startup and shutdown in order to meet the rapid change of electricity demand. Furthermore, the steam pressure and temperature in boiler components are increasing to improve the thermal efficiency for energy saving. As the consequence, the material's degradation is being accelerated.

M. Hirao and H. Ogi, *Electromagnetic Acoustic Transducers*,
Springer Series in Measurement Science and Technology,
DOI 10.1007/978-4-431-56036-4_17

Nondestructive evaluation method for practically measuring metal's creep damage has not been established yet. Ultrasonics is one of the candidates. It can basically provide health monitoring in the bulk of metals through the attenuation measurements. Dislocations dominate the creep mechanism. They also dissipate acoustic energy causing ultrasonic attenuation (or internal friction). The EMAR method is capable of sensing the dislocation behavior in quite a high sensitivity (see Chaps. 7 and 16). This Chapter presents the EMAR measurements of attenuation during creep tests on austenitic stainless steel and steel alloys (Ohtani 2005; Ohtani et al. 2005a, b, c; 2006a, b).

17.2 Creep and Dislocation Damping in Cr-Mo-V Steel

Creep is a thermally activated process under applied stresses. The deformation occurs involving dislocation slipping, subgrain formation, grain boundary sliding, and diffusion of vacancies and interstitials (Dieter 1961; Čadek 1988; Stouffer and Dame 1996). In crept metals, dislocation multiplication and annihilation occur simultaneously under the reduced flow stress (or increased dislocation mobility) at elevated temperatures. Balance between dislocation formation and annihilation determines the creep rate (strain increment per hour), which evolves until fracture, depending on temperature and stress. Attenuation measurement thus detects the creep process and potentially indicates the remaining life of crept metals. Unlike the high cycle fatigue process whose damage is localized at the surface area, the creep damage extends in the bulk of metals, although not necessarily uniform. Ultrasonic assessment of deterioration should be easier than detecting fatigue damage in principle, because the material's information is provided along the propagation path of ultrasound.

The material studied was taken from a commercial plate of ASTM A193-B16; Cr-Mo-V ferritic steel containing 0.42 mass%C, 1.09 mass%Cr, 0.51 mass%Mo, and 0.28 mass%V. It was heated at 1283 K for 2 h, air-cooled, heated at 1223 K for 2 h, oil-quenched, heated at 963 K for 6 h, and then air-cooled. Specimens as shown in Fig. 17.1 were machined for the creep tests. Creep tests were carried out at 923 K in air. Applied tensile stresses were 55, 45, 35, and 25 MPa. Two different creep tests were conducted: (i) continuous test and (ii) interrupted test. In the continuous test, the creep loading was stopped and the sample was furnace-cooled. After the ultrasonic measurements for attenuation and velocity, the creep loading was restarted. This sequence was repeated every 20, 30, or 50 h until rupture. For the interrupted test, 24 specimens were prepared, and they were exposed to the creep test at stresses of 35 and 25 MPa. The creep loading was individually carried out until the creep strain reached a target value. After this, the ultrasonic properties were measured. The interrupted test provided us with a series of samples with different strains and damage state. Thus, the continuous test was conducted to monitor thoroughly the evolution of ultrasonic properties as creep progressed, and

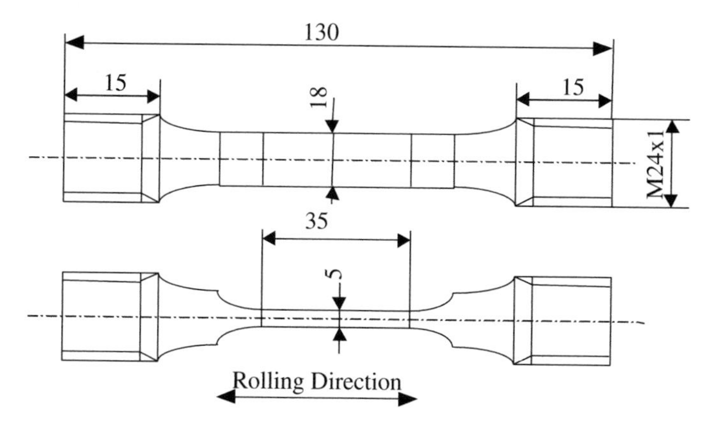

Fig. 17.1 Geometry and dimensions of creep specimens. Reprinted from Ohtani et al. (2005b), Copyright (2005), with permission from Elsevier

the interrupted test allowed microstructure observation, including scanning electron microscopy (SEM) and transmission electron microscopy (TEM) observations.

A shear-wave EMAT (Sect. 3.1) of 10×10 mm^2 active area was oriented so as to measure the velocity and attenuation of the shear wave polarized in the stress direction. The resonant frequencies in the 1–8 MHz range and their attenuation coefficients were measured.

Figure 17.2 shows the typical relationship among the attenuation coefficient α, velocity change $\Delta V/V_0$, creep strain, and creep strain rate as the creep test progresses. t/t_r (creep time/rupture life) indicates the life fraction. The attenuation coefficient increases, showing a peak at $t/t_r = 0.3$, then decreases, showing a minimum near $t/t_r = 0.5$, and finally increases until failure. The velocity gradually decreases and shows a local minimum at the attenuation peak, then remains nearly unchanged until the attenuation minimum, and finally decreases to rupture. The creep strain, on the other hand, monotonically increases until rupture. The creep strain rate slightly decreases from the beginning and increases monotonically until rupture. These trends were commonly observed for other resonant modes. The most important observation was that attenuation always shows a peak at around $t/t_r = 0.3$ and a minimum near $t/t_r = 0.5$, being independent of the creep stress as shown in Fig. 17.3. The identical attenuation and velocity behaviors were observed for the interrupted tests as shown in Fig. 17.4.

For investigating the relationship between the acoustic properties and microstructure evolution, detailed microstructure observations were conducted on interrupted samples, and changes of the block-boundary size, precipitate size, dislocation cell size, subgrain size, dislocation loop length, and dislocation density were measured as the creep damage progresses. It was then revealed that the evolutions of loop length and density of *free* dislocations govern the acoustic parameter changes. Figure 17.5a–d are TEM images of specimens exposed to stress of 25 MPa at $t/t_r = 0$ (original), 0.21, 0.54, and 0.84, respectively. Figure 17.5a

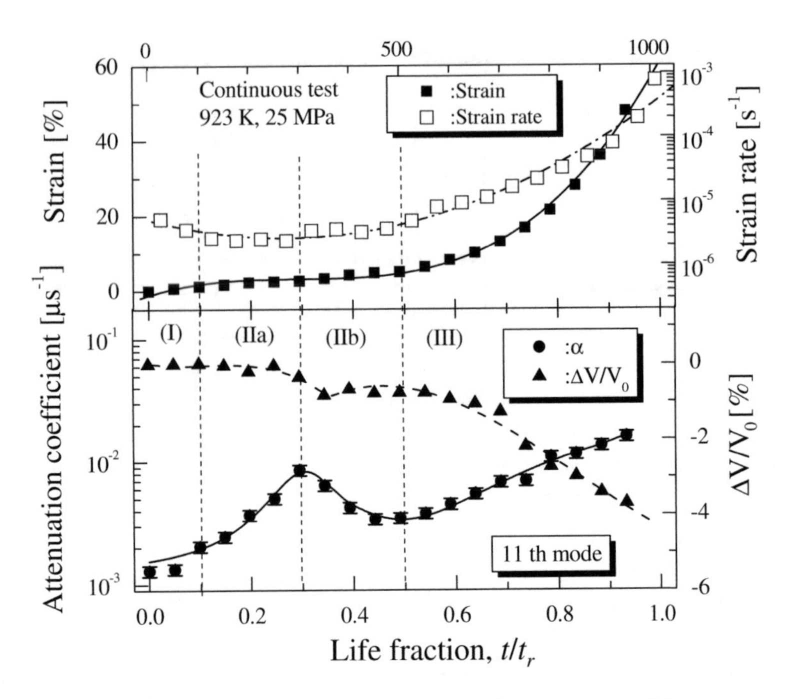

Fig. 17.2 Evolution of α, $\Delta V/V_0$, creep strain, and strain rate at the 11th resonant mode (~3.5 MHz) from the continuous test. The creep test was performed with 25 MPa at 923 K. The rupture life t_r was 1018.1 h. Reprinted from Ohtani et al. (2006a), Copyright (2006), with permission from Elsevier

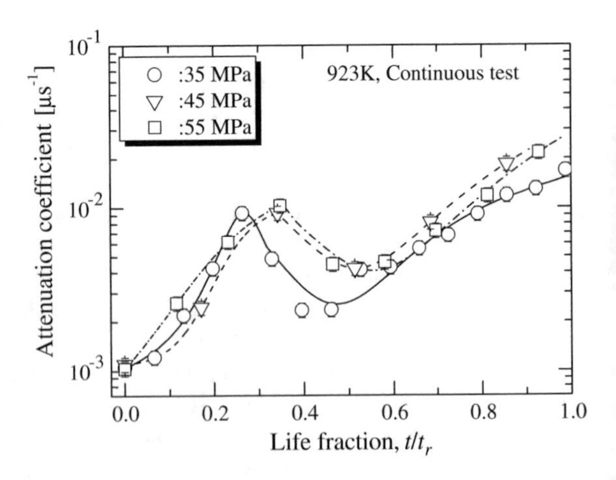

Fig. 17.3 Attenuation evolution at the 11th resonant mode with three different creep stresses of 35, 45, and 55 MPa. The creep temperature was 923 K. The rupture lives were t_r = 759.1, 292.0, and 258.5 h for 35, 45, and 55 MPa, respectively. Reprinted with permission from Ohtani et al. 2006a, Copyright (2006), with permission from Elsevier

shows many elongated cell structures. The dislocations within the cells were tangled, and their density was high. Figure 17.5b shows the microstructure near the attenuation peak. Many equiaxed cell structures were observed, and the number of

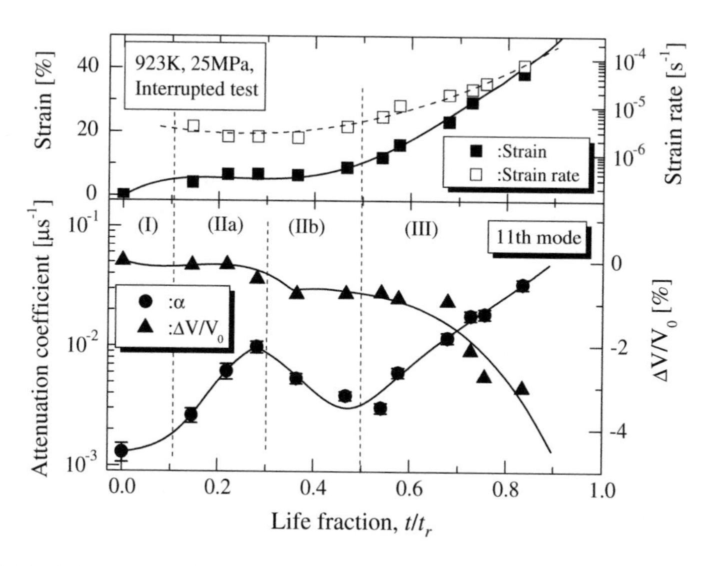

Fig. 17.4 Evolution of α, $\Delta V/V_0$, creep strain, and strain rate at the 11th resonant mode under 25 MPa from the interrupted test. Reprinted from Ohtani et al. (2006a), Copyright (2006), with permission from Elsevier

elongated cells decreased. The dislocation density was still high. Figure 17.5c shows the microstructure when attenuation shows a minimum. A few elongated cell structures and subgrains were observed. The cell size was larger, and the boundaries of the cells and subgrains were clearer than in Fig. 17.5b. The dislocation density within the cells and subgrains was low. Thus, free dislocations in the dislocation cell were spent for forming subgrains. Figure 17.5d shows the microstructure at $t/t_r = 0.84$, when attenuation increased again. The entire area was covered by the subgrain structure, and few cell structures existed. The boundaries of the subgrains were clearer than those in Fig. 17.5c.

From such TEM images, the density Λ_1 and segment length L_1 of effective free dislocations were digitally measured as shown in Fig. 17.6, where dislocations tangling and piled up against boundaries of grains, cells, and subgrains were neglected, because they will not absorb the ultrasonic energy. The attenuation coefficient was then calculated by substituting Λ_1 and L_1 for Λ and L of Eq. (7.1), adjusting the proportional coefficient so as to yield the initial attenuation. The result is compared with the measured attenuation coefficients in Fig. 17.7, where the same trend with the measurement is reproduced. Thus, evolution of free dislocation structure dominates the acoustic properties.

Being based on these observations and analysis with a large number of tests, the relationship between the microstructural change and the acoustic parameters can be discussed by dividing the creep life into four stages as shown in Fig. 17.2: Stage I corresponds to the primary creep, Stages IIa and IIb to secondary creep, and Stage III to the tertiary creep.

Fig. 17.5 TEM images of interrupted creep specimens at t/t_r = 0, 0.21 (near the attenuation peak), 0.54 (near the attenuation minimum), and 0.84 (25 MPa, 923 K). Reprinted from Ohtani et al. (2006a), Copyright (2006), with permission from Elsevier

Stage I ($0 < t/t_r < 0.1$). Precipitation and softening occur just after the creep begins. Meanwhile, the decrease in the strain rate indicates that no significant dislocation movement (multiplication or slip) occurs in this stage, leading to stable α and $\Delta V/V_0$.

Stage IIa ($0.1 < t/t_r < 0.3$). The increase of the creep strain suggests the initiation of dislocation multiplication (increase in Λ). The increase of cell size and

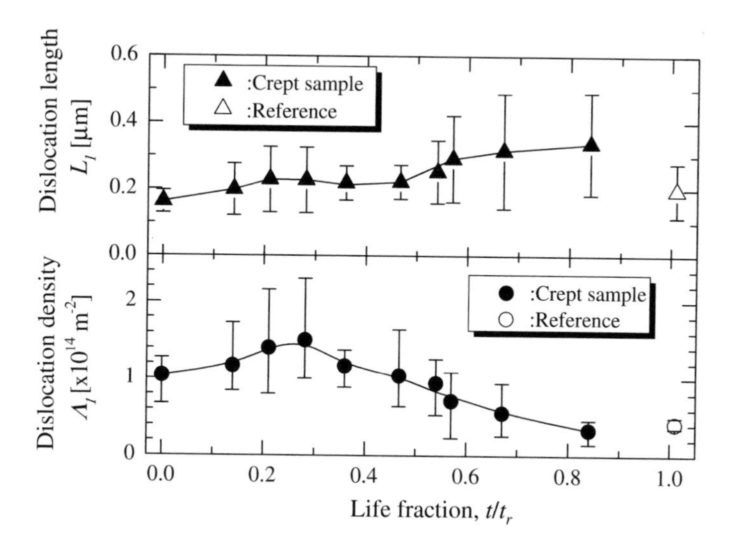

Fig. 17.6 Change of the effective dislocation density and length as creep progresses (25 MPa, 923 K). Open symbols represent data for the heated sample without applying stress (reference). Reprinted from Ohtani et al. (2006a), Copyright (2006), with permission from Elsevier

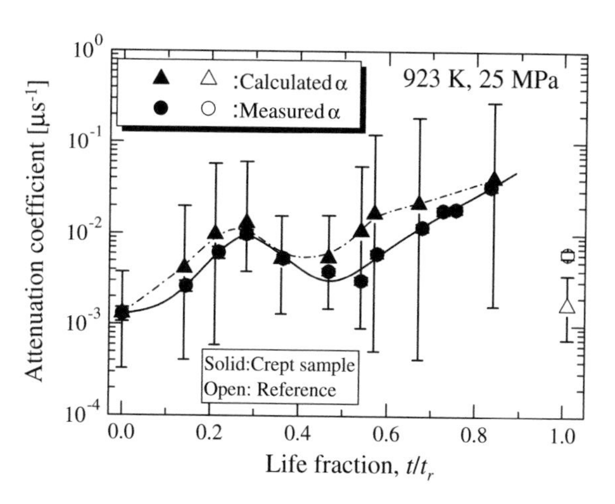

Fig. 17.7 Comparison between calculated and measured attenuation coefficients in the 11th resonant mode (25 MPa, 923 K). Open symbols represent data for the heated sample without applying stress (reference). Reprinted from Ohtani et al. (2006a), Copyright (2006), with permission from Elsevier

transformation from the elongated cell to the equiaxed cell proceed rapidly. The coarsening of precipitates occurs, which decreases the number of pinning points for the dislocation lines. As a result, the distances between the pinning points become larger and L_l increases, causing the increase in α and decrease in V.

Stage IIb ($0.3 < t/t_r < 0.5$). The total dislocation density further increases. The multiplied dislocations are continuously consumed by formation of the cell walls and subgrains, and the dislocation density on the boundary of cell or subgrain increases, while the density in their interior decreases. Thus, the density of the

Table 17.1 Life fraction at attenuation peak during creep lives of various materials

Material	Cr-Mo-V steel	2.25Cr-1 Mo steel	304 stainless steel	316 L stainless steel	Waspaloy
t/t_r at attenuation peak	0.25–0.3	0.5–0.6	0.3–0.4	0.6–0.7	0.35–0.4
Temperature (K)	923	923	973	973	1073
Stress (MPa)	25, 35, 45, 55	45, 65	100, 120	100, 110, 160	140, 150, 160

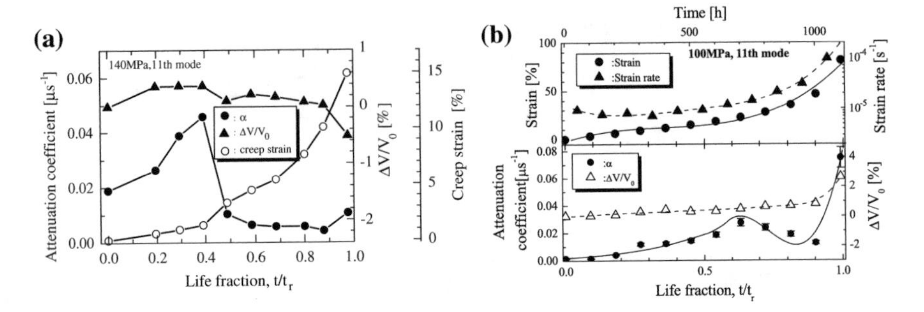

Fig. 17.8 Attenuation and velocity behaviors during the creep tests for **a** nickel-based superalloy (WASPALOY) with 140 MPa at 1073 K (Reprinted from Ohtani et al. 2005b, Copyright (2005), with permission from Elsevier) and **b** 316L austenitic stainless steel with 100 MPa at 973 K. (after Ohtani 2005c)

effective dislocation decreases, being accompanied by the decrease in α. The attenuation peak, therefore, indicates formation of subgrain structure, during which long free dislocations temporarily appear inside the cells and they absorb the acoustic energy.

Stage III ($t/t_r > 0.5$). The strain rate and the total dislocation density increase. The cell structures then occur, and the formation of subgrains is accelerated. They become nuclei for triggering recrystallization. In addition, the coarsened precipitates facilitate the formation of subgrains and recrystallization. Then, Λ and L increase, leading to increase in α and decrease in V. The attenuation minimum therefore corresponds to the beginning point where the size of cell or subgrain increases rapidly and the coarsening and condensation of precipitates suddenly advance.

17.3 Commonly Observed Attenuation Peak

The attenuation peak during creep test (Fig. 17.3) was commonly observed in various materials, including 2.25Cr-1Mo steels (Ohtani et al. 2005a), Ni-based superalloys (Ohtani et al. 2005b), 304-type (Ohtani et al. 2006b) and 316L-type

(Ohtani et al. 2005a) austenitic stainless steels as shown in Table 17.1. For example, Fig. 17.8a and b present the relationship between the acoustic parameters and creep strain for the Ni-based superalloy and 316L-type stainless steel, respectively. They are almost identical to Figs. 17.2, 17.3 and 17.4, although the fractional life ratio at the attenuation peak is dependent on the material, not on the applied stress. These attenuation peaks are consistently explained by the dislocation structure change. The universality of the attenuation behavior during creep progression is practically important, and monitoring the attenuation behavior with EMAR allows us to predict the remaining life, especially the time when the tertiary creep begins.

References

Čadek, J. (1988). *Creep in Metallic Materials*. Amsterdam: Elsevier.

Dieter, G. E, Jr. (1961). *Mechanical Metallurgy*. New York: McGraw-Hill.

Ohtani, T., Ogi, H., & Hirao, M. (2005a). Dislocation damping and microstructural evolutions during creep of 2.25Cr-1Mo steels. *Metallurgical and Materials Transactions A, 36A*, 411–420.

Ohtani, T., Ogi, H., & Hirao, M. (2005b). Acoustic damping characterization and microstructure evolution in nickel-based superalloy during creep. *International Journal of Solids and Structures, 42*, 2911–2928.

Ohtani, T. (2005c). Acoustic damping characterization and microstructure evolution during high-temperature creep of an austenitic stainless steel. *Metallurgical and Materials Transactions A, 36*, 2967–2977.

Ohtani, T., Ogi, H., & Hirao, M. (2006a). Evolution of microstructure and acoustic damping during creep of a Cr-Mo-V ferritic steel. *Acta Materialia, 54*, 2705–2713.

Ohtani, T., Ogi, H., & Hirao, M. (2006b). Creep-induced microstructural change in 304-type austenitic stainless steel. *Journal of Engineering Materials and Technology, 128*, 234–242.

Stouffer, D. C., & Dame, L. T. (1996). *Inelastic Deformation of Metals*. New York: Wiley.

Chapter 18
Field Applications of EMATs

Abstract Many EMAT techniques have already been transferred from laboratory studies to field applications, in particular, for infrastructure (railroad, pipeline, etc.), steel industry, and petrochemical industry. The transfer began in 1970s at the latest, when the theoretical modeling was still under way. This chapter shows some targets of nondestructive inspection that the EMATs have made possible to realize. They are cracks, solid-liquid interfaces at elevated temperatures, wall thickness and thinning of pipes (and tubes), and residual stresses; but much more unpublished cases probably exist. As briefly reviewed, incorporation of phased-array technique and also the laser excitation of elastic waves will expand the horizon of noncontact ultrasound measurements.

Keywords Acoustoelastic birefringence · Axial-shear-wave resonance · Curie temperature · Dual-mode EMAT · Electromagnet · Gas pipeline · Guided waves · In-process inspection · Laser ultrasonics · Meander-line coil · Phased-array technique · PPM EMAT · Rayleigh wave · SH wave · Shear-wave EMAT · Solid-liquid interface · Thermomechanical processing

18.1 Monumental Work of EMAT Field Applications

Ultrasonic testing volume in *Nondestructive Testing Handbook* (2nd ed., 1991), Alers (2008), and Passarelli et al. (2012) provide a good sketch of successful EMAT field applications in early days. Three representative cases are picked up in the following with reverence to the pioneers, including R.B. Thompson and G.A. Alers.

In Chap. 14, a method of using guided SH waves that propagate in the circumferential direction of the pipe was proposed for on-site inspection of pipelines. Much earlier than this, an EMAT-loaded device was tested in a long pipeline, which used guided SH wave that travels in the spiral direction. Figure 18.1 shows the system called *pig* developed for detecting axially oriented stress-corrosion cracks

M. Hirao and H. Ogi, *Electromagnetic Acoustic Transducers*, Springer Series in Measurement Science and Technology, DOI 10.1007/978-4-431-56036-4_18

Fig. 18.1 Commercially operating pipeline inspection pig. Reprinted with permission from Passarelli et al. (2012). Copyright (2012), AIP Publishing LLC

(SCCs). The pig is a self-contained and mobile inspection station that can be placed inside a gas pipeline. It moves along the axial direction of the pipe, transmits SH waves within the pipe wall, receives reflected and transmitted signals, and records their amplitudes. Single pig system carries 10–20 sets of EMAT units around the circumference. Each unit has two magnetic poles (dark colored) on the both ends for applying the magnetic field along the pipe axis, and four meander-line coils (light colored) (two for generation and two for reception), which are arranged in the square positions parallel to the axial direction of the pipe. A generation coil launches the SH wave in the diagonal direction with the magnetostrictive mechanism (see Figs. 3.12 and 3.13), and it is reflected from the SCC and detected by the receiver coil located on the axial direction with the transmitter coil. A part of the SH

Fig. 18.2 An EMAT inspection system designed to inspect installed railroad tracks for several types of defects at speeds of up to 16 km/h. Reprinted with permission from Passarelli et al. (2012). Copyright (2012), AIP Publishing LLC

wave transmits the SCC and is also detected by the second receiver located on the diagonal line; because the SCC scatters the wave energy, the amplitude of the SH wave detected by the diagonally located receiver is lowered. Thus, the flaw is evaluated both by the reflected and transmitted wave amplitudes. The suitable frequency depends on the pipe wall thickness and is between 200 and 400 kHz. Two pigs are used by rotating the angle positions so as to inspect the whole region.

For inspection of installed railroad rails, piezoelectric transducers are usually adopted. A rubber wheel or a rubber-bottomed boat is used in this case for making acoustical coupling with the rail, where water is still needed to make the coupling reliable. Use of water presents problems when inspecting over long distances and in freezing temperatures. Figure 18.2 shows a trailer that was pulled along the railroad track at about 16 km/h while performing an ultrasonic inspection of the rail head, the flange, and the joints between rails by a set of EMATs on each rail. This system uses several EMAT coils placed under a single pulsed electromagnet, which send SH waves along the rail head, to perform an angle beam inspection of the bolt holes at rail joints and to detect cracks in the web of the rail. A portable welding generator was mounted on the trailer to supply the pulsed currents for the electromagnets at the speeds of inspection.

Much of the piping used in the petroleum and chemical processing industries is fabricated from strips of sheet steel bent lengthwise into a circular shape and then welded along a seam that runs the full length of the strip. It is mandatory that this weld line be inspected for defects before the welded tube is cut into the desired lengths. Conventionally, this inspection is performed by piezoelectric transducers mounted on each side of the weld line and focusing angle beam shear waves into the weld. Using guided waves supplied by EMATs, the weld can be inspected when the tube is still hot from the welding process and the critical alignment of the angle beam

Fig. 18.3 A pair of pulsed electromagnet EMATs on each side of longitudinal weld seam in electric resistance welded (ERW) pipe. Reprinted with permission from Passarelli et al. (2012). Copyright (2012), AIP Publishing LLC

transducers can be avoided (see Fig. 18.3). Furthermore, the guided wave traverses the entire circumference, making it possible to detect flaws anywhere in the pipe wall.

18.2 Crack Inspection of Railroad Wheels

The EMAT technique has been developed for nondestructive examination of railroad wheels for ensuring transportation integrity (Salzburger and Repplinger 1983; Schramm et al. 1989). The flaws may result from the contact stress with rails, fatiguing, and tensile residual stress caused by excess braking (see Sect. 18.3). They potentially induce wheel failure and derailment, causing catastrophic damage. Early detection of flaws is required for replacement, which is realized with the automatic inspection of every wheel on a train.

The EMAT's contactless aspect enabled the crack inspection for wheels in an automatic, roll-by mode by generating and receiving a low-frequency Rayleigh wave propagated around the wheel tread. Figure 18.4 shows one of such in-rail EMATs designed by Schramm et al. (1989). They machined a square pocket in the rail head to locate an Nd-Fe-B permanent magnet having a vertical magnetization direction. The opening was covered with two sheet coils separately used for transmission and reception. They are meander-line coils of eight turns and 6-mm periodicity for a Rayleigh wave at 0.5 MHz (see Sect. 3.9). A layer of polymer foam of a few millimeters thick was inserted between the magnet and the coils. This layer compressed under the wheel's weight to conform the coils to the curvature of the wheel tread, thereby minimizing the liftoff for the best efficiency. A membrane switch was embedded in this compliant foam. Responding to the wheel pressure, it triggered the electronic system to capture the transient ultrasound.

Rf burst signals of 5–10 cycles were used to excite the EMAT to generate the Rayleigh wave pulses traveling in both directions along the tread. A 4-m section of track simulated the field condition. Actual wheel set (two wheels mounted on an axle) rolled over the EMAT. Figure 18.5 shows the received signals from a wheel having a 2-mm-deep cut on the tread. The Rayleigh wave pulses traveled several times along the circumference. Impinging on the flaw, they produced reflection echoes, which traveled back to the EMAT. Signals A and C are such echoes that traveled 22 and 78 % of the round path along the wheel circumference and back to the transducer. Signals B and D are the first and the second round-trip signals, being

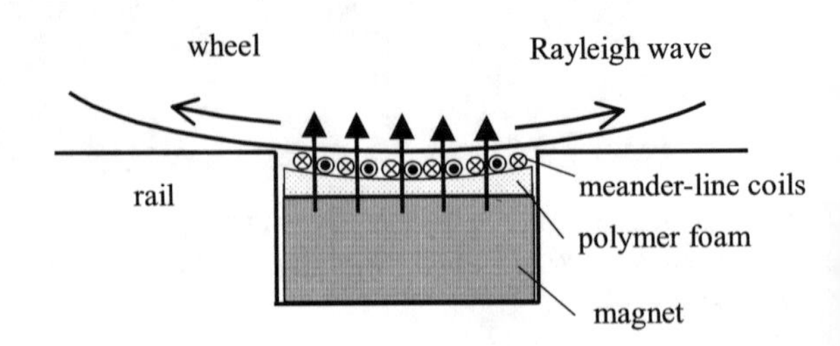

Fig. 18.4 In-rail EMAT structure for roll-by inspection of wheel cracks

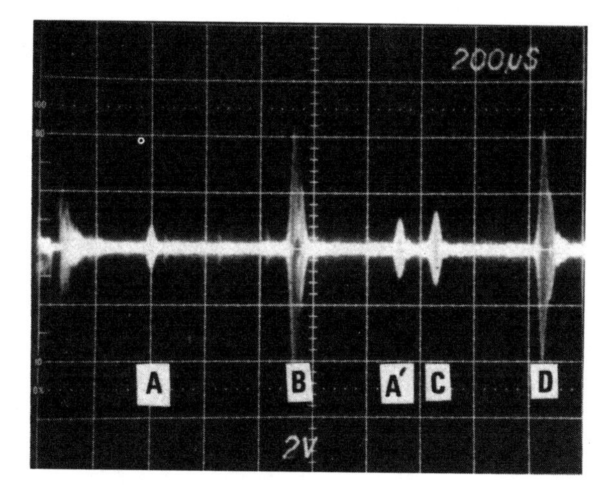

Fig. 18.5 Received trace from a test wheel with 2-mm-deep cut. Signals A and C are the flaw echoes propagated with short and long round paths. B and D are the first and second round-trip signals. A′ is the short-path echo after a complete round-trip (A′ = A + B). After Schramm et al. (1989)

composed of counter-rotating signals. Signal A′ is the short-path echo after traveling the entire circumference of 2.6 m long.

There are two dead zones in this inspection. One occurs when the cracking is right on the EMAT and the flaw signals return to it during the recovery of the electronic devices after high-power excitation. The other occurs when the crack is exactly opposite the transducer. Signals A, B, and C are then overlapped with each other because of the same path length. Schramm et al. (1989) remarked that each dead zone is 24 cm long and approximately 80 % of the tread can be inspected with a single pass. Another pair of EMATs located 0.5–1 m down the track would assure 100 % inspection.

18.3 Residual Stress in Railroad Wheels

The shear-wave EMATs (Sect. 3.1) have been used to measure the acoustoelastic birefringence for characterizing the residual stress in the rims of railroad wheels (Fujisawa et al. 1992; Herzer et al. 1994). The thermal treatment is conducted to introduce the compressive residual stress in the circumferential direction of the rims. It is desirable to prevent crack initiation and shield the flaws from the applied tensile stresses. However, too much braking in service relieves the compressive residual stress and can furthermore induce harmful tensile stress there from the constrained shrinkage after being extended with heating. Safe service then requires the residual stress assessment on the wheel rims set in a dolly. The acoustoelastic birefringence measured with shear-wave EMATs is a promising candidate for the purpose (Chap. 12). One benefit comes from forged steel of wheels, where the grains are randomly oriented to allow assuming $B_0 = 0$ (Fukuoka et al. 1985).

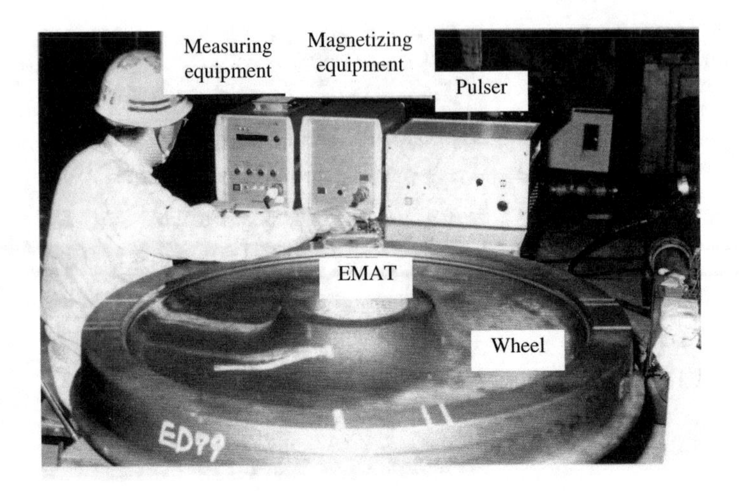

Fig. 18.6 Measurement of residual stress in wheel rim (With courtesy of K. Fujisawa)

Fujisawa et al. (1992) developed a shear-wave EMAT with an electromagnet. The coil has two layers of racetrack shape. They were stacked one over the other to measure the through-thickness transit times for the orthogonal polarizations, the radius, and circumferential directions in the wheel rims, without rotating the EMAT. The zero-crossing points were detected in the received signals of 2-MHz center frequency. They investigated the feasibility first on the as-manufactured wheels (Fig. 18.6), after being shrink-fitted in the wheel sets, then at seven occasions during 140,500-km service over a year; four wheels on both sides of an electric locomotive. The negligible texture-induced anisotropy resulted from the comparison of the birefringence measured before and after sectioning an equivalent wheel to relieve the residual stress. Measurements along the radial lines on the rims (80 mm high) showed that the new wheels possess a compressive stress of 70 MPa at the outer edge and 40 MPa at the inner edge. These circumferential residual stresses were reduced by approximately 20 MPa when shrink-fitted to the shafts (see Sect. 12.4.3). In-service measurements confirmed that the compressive residual stress persisted during the service, although the heat-sensitive paint indicated that the rims had experienced heating up to 280 °C by braking.

18.4 Measurement on High-Temperature Steels

18.4.1 Velocity Variation at High Temperatures

The noncontact EMAT technique has met many demands in the steel industry from the early time of development, because the steel products (billets, plates, pipes, etc.) are heated over 1000 °C for thermomechanical processing while being moved at high speeds. Tight control of dimensions and qualities requires the online and noncontact/nondestructive assessment of interior conditions of intermediate products. Sensing should not perturb the process. Previous researches have demonstrated the usefulness of EMAT techniques for reconstructing the solid-liquid interface in partially solidified steel strands during the continuous casting as well as the internal temperature (Kawashima et al. 1979; Boyd 1990), and for measuring the wall thickness of hot seamless steel tubes (Yamaguchi et al. 1986), to name but a few. These two approaches have proven promising through field testing in commercial production lines and will be reviewed in the following sections. They used EMATs to launch and detect pulsed longitudinal wave in the low-MHz range. The time of flight becomes an interior temperature gauge, enabling noncontact and real-time monitoring.

These applications need the temperature dependence of velocities to deduce the temperature or the wall thickness from the time-of-flight measurements. Nominal temperature dependence of elastic constants is available in the literature (e.g., Simmons and Wang 1971; Dever 1972). But, the velocity-temperature relationship depends on the alloying elements and the heating/cooling rate, because they cause hysteretic phenomena associated with austenite (γ) to ferrite (α) phase transformation. Magnetic transition further complicates the temperature dependence. Calibration tests were therefore often performed to find the temperature dependence of velocities and attenuation on individual steel grades. Indeed, many studies occurred (Kurz and Lux 1969; Papadakis et al. 1972; Parkinson and Wilson 1977; Whittington 1978; Wadley et al. 1986; Boyd 1990; Idris et al. 1994; Spicer 1997; Dubois et al. 1998, 2001). These previous measurements were made using either pulse echo method with piezoelectric transducers and buffer rods, heat-resisted EMATs, remote laser ultrasound, or hybrid laser/EMAT combination. Common observations are the velocity decrease with temperature including a steep fall at the Curie temperature and a nearly linear decrease above this temperature as well as the shear wave's high attenuation and the longitudinal wave's attenuation drop in the same temperature range. (For this reason,the longitudinal wave is chosen for many high-temperature applications.) Seeing that austenitic stainless steels show a linear velocity decrease throughout, the phase transformation is the cause of these characteristics. Although the mechanism is not well understood to date, the shear wave attenuation could be explained with the high sensitivity to grain coarsening and probably interphase boundaries, which will enhance the wave scattering. In the Rayleigh scattering regime, the scattering intensity is governed by $(ka)^4$ with wavenumber k and grain diameter a (Truell et al. 1969). The shear wave then undergoes more grain scattering than the longitudinal wave of the same frequency (see Chap. 15).

Of the existing work, Dubois et al. (2001) provided conclusive measurements on this long-running problem, which will contribute to facilitate the ultrasonic sensing at elevated temperatures. They used laser ultrasound technique to determine the longitudinal wave velocity in the 500–1000 °C temperature range, 0.0–0.72 mass% carbon content range, and 0.1–20 °C/s heating/cooling rate range. Plate samples of 1 mm thickness were measured to minimize the temperature gradient. Figure 18.7

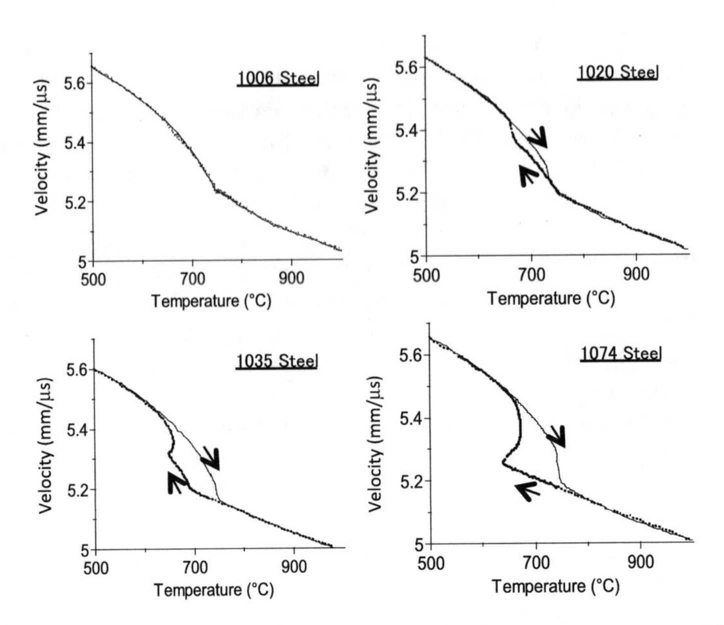

Fig. 18.7 Longitudinal wave velocity in the 500–1000 °C temperature range measured in 1006, 1020, 1035, and 1074 steel samples heated and cooled at 5 °C/s. Reprinted with permission from Dubois et al. (2001). Copyright (2001), AIP Publishing LLC

reproduces a part of their observations to compare the responses for different carbon contents at a common heating/cooling rate. The velocity shows a discontinuous change at the Curie temperature ($\sim$750 °C) for all the tested cases. It then shows a linear dependence on temperature up to 1000 °C. Hysteretic behavior appears just below the Curie temperature, the magnitude of which increases with the carbon content. Discussions are provided as to the effects of phase transformations and exothermal process during cooling.

18.4.2 *Solidification-Shell Thickness of Continuous Casting Slabs*

Solidification-shell thickness of steel slabs and interior temperature are key factors for quality control and effective operation in the process of continuous casting and direct rolling. Real-time and noncontact measurement is required. Such a continuous process has certain benefits of simplifying the process and saving production time and cost. In the operation, the tundish supplies molten metal to the mold, which provides the cooling and shaping to form continuous cast slabs. The pinch rollers draw the slabs out of the mold and feed them down to the rolling stage after further cooling through water spray and radiation.

Ultrasound time-of-flight measurements with EMATs allowed the calculation of shell thickness. Kawashima et al. (1979) and Yoshida et al. (1984) developed an EMAT system to measure the though-thickness time of flight of the longitudinal wave at high temperatures over 1000 °C. Figure 18.8 gives a simplified sketch of the system and operation. To transmit the longitudinal wave, they discharged the capacitor to supply a pulsed current as large as 2000 A to the pancake coil of 40 mm diameter placed over the slab surface. Induced current in the slab then produces a radial dynamic field parallel to the surface beneath the coil element. The Lorentz force arises acting in the thickness direction from the interaction between the induced eddy current and the dynamic magnetic field. Because both intensities are proportional to the driving current, the resulting Lorentz force is proportional to the square of the current (Eq. 2.38). Thus, the generation efficiency remarkably increases with the driving current. (Note that this excitation mechanism without the static field generates the elastic wave of double frequency of the excitation current and is inapplicable for receiving the ultrasonic vibration.) The launched longitudinal wave traverses the slab thickness of ~200 mm and reaches the opposite side of the slab, where the receiver EMAT of the type shown in Fig. 3.3 was placed. The transmitter coil and receiver EMAT were protected from the heat by thermal shielding at the front faces and watercooling. Both were retractable as being inserted close to the hot slab with 1 mm liftoff for about 2 s and then pulled out of the machine. The system used an optical pyrometer to measure the surface temperature and a thickness gauge for the total slab thickness installed at several meters down the mold.

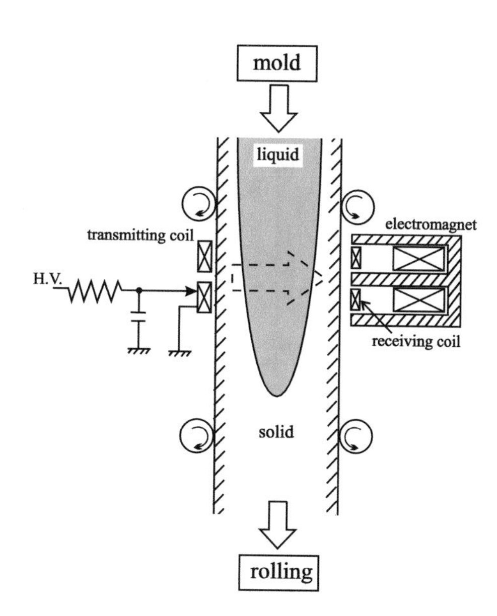

Fig. 18.8 Through-thickness pulsed-EMAT system for continuously cast steel slabs

Measured time-of-flight t was used to calculate the shell thickness d from the total slab thickness D and the wave speeds C_{liq} and C_{sol} in liquid and solid phases;

$$d = \left(t - \frac{D}{C_{liq}}\right) / 2\left(\frac{1}{C_{sol}} - \frac{1}{C_{liq}}\right). \tag{18.1}$$

For the liquid phase (1470 °C), C_{liq} = 3900 m/s was used; for the solid phase, calibration measurements on several grades of carbon steels provided $C_{sol} = 5580 - 0.67T_{ave}$ (m/s) in the 800–1200 °C range. They assumed a parabolic temperature variation across the shell to substitute for the average temperature T_{ave}. Figure 18.9 shows the result of d with the readings of t, D, V, and the surface temperature T_s. The measurement was repeated at every 1 min for more than 2 h, while the cast speed V varied between 0.2 and 0.9 m/min until the slab was stopped for complete solidification. When the cast speed remained constant at 0.9 m/min, the shell thickness was leveled around 50 mm. On a sudden slowdown to 0.5 m/min, the thickness responded to increase but with a notable delay. After the cast run was

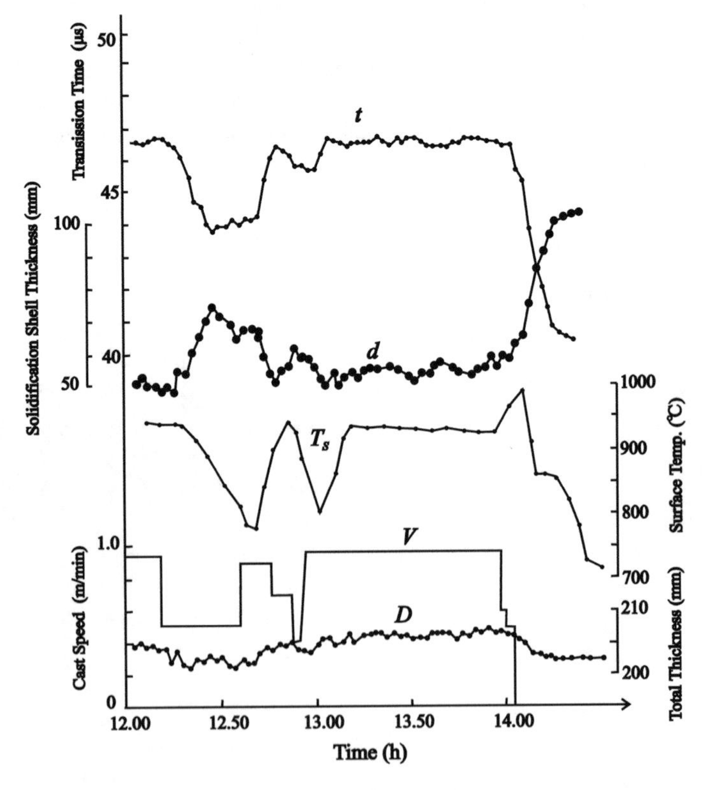

Fig. 18.9 Continuous measurement of shell thickness with other readings. After Kawashima et al. (1979)

stopped, the shell kept thickening to 100 mm (complete solidification) over 15 min. The destructive measurement was carried out to verify the EMAT's nondestructive results by driving aluminum-coated rivets into a slab. They claimed the difference between the two methods to be 2–3 mm.

The above technique can locate the crater end, or solidification completed point, in a hot slab. One of other options for this purpose is to propagate the shear wave in the thickness direction. Liquid will not support the shear deformation and the shear wave is totally reflected at the solid-liquid interface. Takada et al. (1989) observed the amplitude of through-thickness shear wave at an extremely low frequency of 50 kHz to cope with the high attenuation at elevated temperatures. The transmitted signal was absent at partially solidified positions, when the face-to-face paired EMATs scanned over the hot slabs of 230–260 mm thick. It is then possible to detect the two-dimensional profile of the crater end in a hot steel slab.

Recently, Iizuka and Awajiya (2014) further developed this measurement principle by incorporating the chirp pulse compression technique (see Sect. 3.6). A pair of shear-wave EMATs (Sect. 3.1), being operated at 300 kHz and allowing 5-mm liftoff or more, can detect the crater end during continuous casting of steel. The signal intensity is enhanced by the high-power excitation at 2000 $V_{p\text{-}p}$, synchronous averaging over 16 signals, and surface cooling down to 700 °C (below Curie temperature) to include the magnetostriction contribution for transduction (see Sect. 2.2.5). Figure 18.10 presents the trace of through-thickness signal amplitude obtained when the casting speed is changed. The crater end is detected at point A, when the molten metal blocks the shear wave propagation at the measuring site. By decreasing the speed, the signal reappears at point B after a delay time, indicating that the crater end moves to the upstream side.

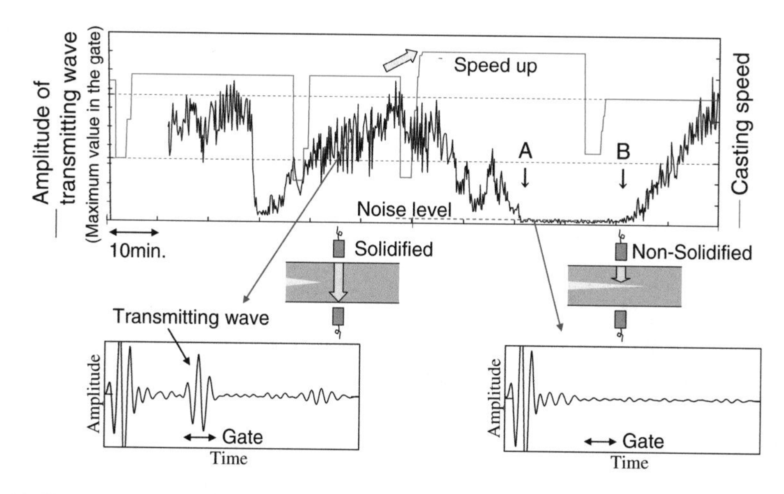

Fig. 18.10 Continuous measurement of shear wave signal amplitude during actual continuous casting. After IIzuka and Awajiya (2014)

18.5 Wall Thickness of Hot Seamless Steel Tubes

This section briefly describes the EMAT method developed for online measurement of wall thickness of hot seamless steel tubes in a production condition (Yamaguchi et al. 1986). These products are widely used as oil well pipes (oil drilling, casing, etc.), boiler tubes, and pipe structures. Recent service environments require the higher performance and dimensional precision, which are best assured through the measurements directly during the rolling process. Requisites for application include automated, continuous, and high-speed measurement of wall thickness and the simultaneous measurements at the multiple points on a circumference of tube to check the local eccentricity. Noncontact nature of EMATs allowed the wall thickness to be measured with the accuracy of ±0.1 mm for the maximum temperature of 1000 °C. The cross-sectional profile was obtained from the results at six sensor locations.

The EMAT system was field-tested in a tube milling plant. The manufacturing sequence of seamless steel tubes is as follows:

(i) Round billets are soaked at about 1250 °C in a rotary furnace and pierced by a plug and two rollers to the outer diameter of 187 mm and wall thickness of 14.9–37.5 mm.
(ii) A mandrel bar is inserted into the pierced billet to roll it with two mutually orthogonal roll assemblies to a specified outer diameter and thickness.
(iii) The tubes are reheated to about 1000 °C and rolled by a series of stretch reducing mills to finish to the specified dimensions. At a stand, three rollers are assembled at 120° apart in the circumferential direction. These assembles are aligned at different angular positions, which differ by 60° alternatively (Fig. 18.11).

The measurement site was at the final rolling stand of the stretch reducer, where the tube travels in a steady motion. Also, the measurement here guarantees the thickness of the final product. The system used a tunnel-type EMAT, which induced a strong magnetic field to measure the through-thickness traveling time in longitudinal wave and can be installed in the mill line. The EMAT's key component is a magnetizing coil that surrounds the tube. The magnetizing coil is in turn surrounded by the yoke so that the flux is concentrated at the poles to induce a strong axial field in the tubes. The separate sensor coils for transmission and reception were used by placing one over the other and positioned at six locations, 60° apart each other, around the circumference. Each of them has an elongated spiral (racetrack) shape, being 41 mm long in the axial direction and 20 mm wide in the circumferential direction of the tube. Integrated combination of poles and the sensor coils are replaceable for accommodating tubes of various diameters. The EMAT designing restricts the liftoff to be within 1.5–3.1 mm range. Watercooling is supplied to the magnetizing coil for its own Joule heating and to the yoke as well as the sensor coil housing to protect for the radiant heat from the hot tube. Tangential magnetic field of 1 T was obtained at 5 mm beneath the poles

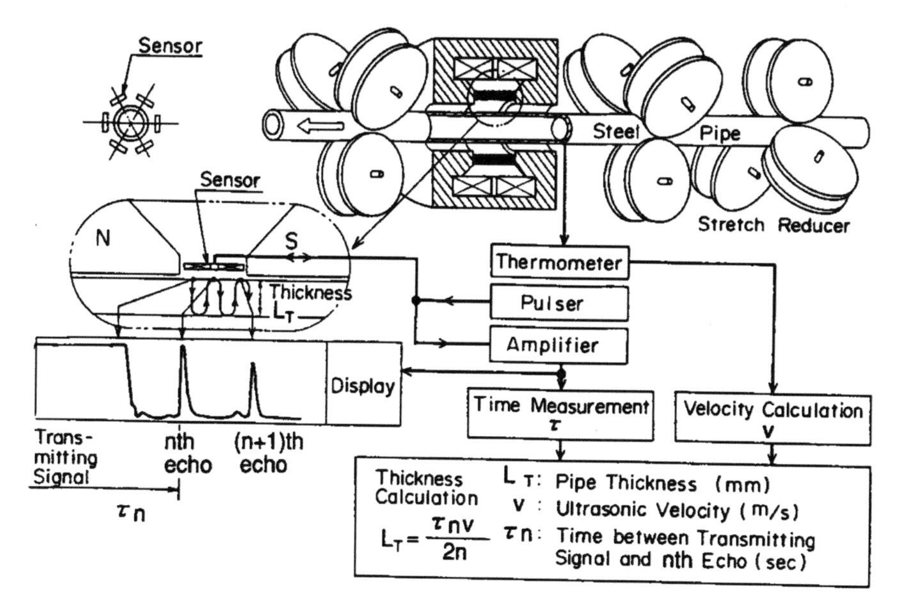

Fig. 18.11 EMAT measuring system installed in a stretch reducing mill line. Reprinted with permission from Yamaguchi et al. (1986)

by feeding a 40-A current to the magnetizing coil. Magnetic saturation transformer pulser was used to generate a short pulse for achieving a good time resolution with 3-MHz longitudinal wave, whose round-trip traveling time through 3-mm-thick wall is approximately 1.2 μm at 1000 °C, for instance. The six transmitter coils were simultaneously excited by the pulser for every 4 ms period (repetition rate of 250 Hz) to generate the longitudinal wave. After a dead time of approximately 6 μs following the excitation, the receiver coils detected a train of reflected echoes. Arrival-time difference between two successive echoes was recorded. Typically, the 6th and 7th echoes were used for 3-mm-thick wall.

Pulse-echo method using conventional transducer and a buffer rod calibrated the temperature dependence of velocity using the raw materials of tubes during the slow cooling process from 1100 °C. The velocity appeared to increase linearly at a rate of 85 m/s per 100 °C, until a discontinuous rise occurred around 720 °C caused by magnetic transition. These measurements provide the basis of temperature compensation along with a suitable algorithm to relate the surface temperature to the through-thickness average temperature, and then to the average velocity.

Figure 18.12 shows the comparison between the online measurement on hot tubes and the usual ultrasonic thickness gauge on cold tubes. Two measurements agree well with each other with an accuracy of ±0.1 mm. For the line speed between 3.3 and 8.2 m/s, a repetition rate of 250 Hz was achieved, which implies measuring the thickness at every 13–33 mm steps along the tubes. Fine features of thickness profile are observed, which includes the thick walls on both ends caused

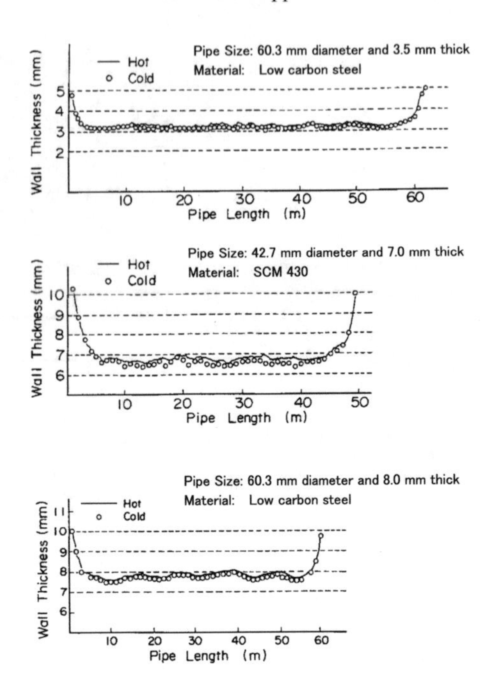

Fig. 18.12 Comparison between the online and off-line wall thickness measurements using one of six sensors on three commercial tubes. Temperature ranged from 700–900 °C. Reprinted with permission from Yamaguchi et al. (1986)

by less tensile stress in the stretch reducer mill and the spiral thickness variation caused by the eccentric rolling in the piercing process.

Before closing this section, the latest EMAT technique for measuring wall thickness of steel pipes at elevated temperatures is addressed. Such a technique was realized by utilizing the laser beam to excite ultrasound and an EMAT for detector. Laser ultrasound is another well-established noncontact technique and is widely applied for nondestructive testing and evaluation of materials (e.g., Scruby and Drain 1990). The laser beam radiated on the solid surface generates ultrasound from the localized thermal expansion or ablation. Laser ultrasound allows much larger liftoff than EMATs, which allow only a few millimeters depending on the frequency and material being tested. It is a merit and demerit at the same time that a laser generates all the possible elastic modes simultaneously into broad angle ranges. Ultrasound is detected also optically, often using interferometer. However, optical detection system is highly affected by unexpected vibrations of the specimen and airflow, making it difficult to be installed for an online measurement. On the other hand, EMATs are useful for detecting acoustic waves contactlessly, although their generation efficiency is quite low. Thus, the combination of a laser ultrasonics for generation and an EMAT for detection establishes a sensitive noncontacting NDT.

Figure 18.13 shows a dual-mode EMAT (see Fig. 3.3) fabricated by Burrows et al. (2014). It contains a cylindrical Nd–Fe–B magnet, 35 mm in diameter, which is perforated at the center for the generation laser beam to pass through onto the sample surface. The coil design is optimized for detection. The EMAT's face is

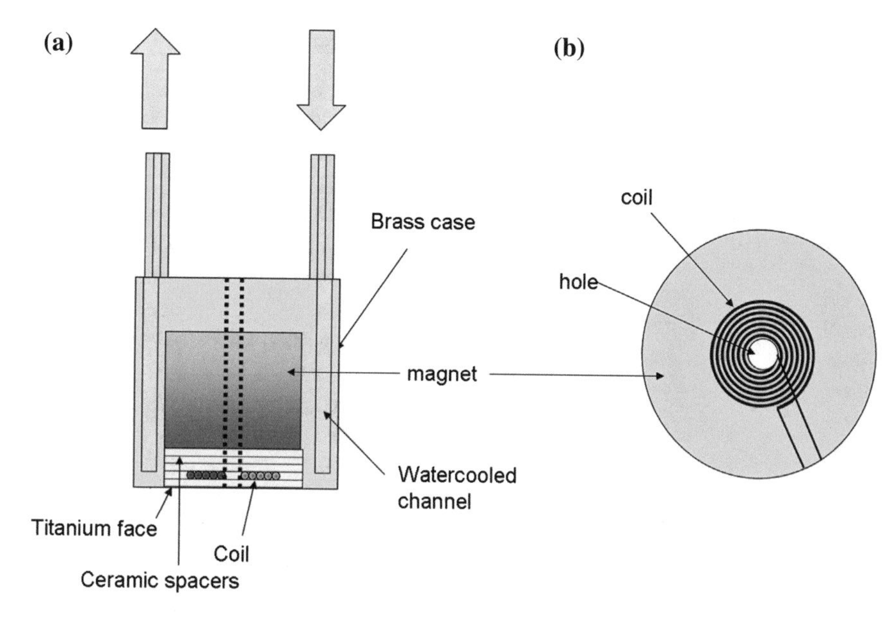

Fig. 18.13 Watercooled and heat-protected dual-mode EMAT. **a** Side view and **b** circular coil. After Burrows et al. (2014)

protected for heat up to 900 °C and the inside of the housing is watercooled. The Nd:YAG laser used has a typical pulse energy of 50 mJ at 1064 nm, and 10 ns duration, coupled from the laser head to the sample via a fiber optic cable, operating in Q-switched mode at a pulse repetition of 20 Hz. The liftoff is less than 0.5 mm. A train of shear wave back-wall echoes are visible with a 316 stainless steel pipe, 70 mm outer diameter and 9 mm thickness, from room temperature up to 900 °C (Fig. 18.14). With the knowledge of temperature dependence of shear wave velocity, the wall thickness can be monitored by this hybrid technique as well.

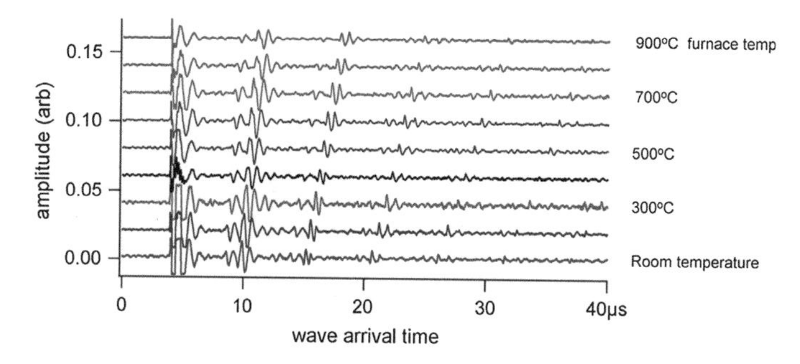

Fig. 18.14 Shear wave echoes from 316 stainless steel pipe at various temperatures. After Burrows et al. (2014)

18.6 Corrosion of Heat Exchanger Tubes

In a petroleum refining plant, distillation unit distills crude oil into many hydrocarbon components, which can be used as fuels and feedstock. Naphtha is the typical light distillate and is produced by condensing while it flows through tubes with air-cooled heat exchanger fins. Since naphtha is a highly flammable fluid having a fire hazard, wall thinning of the tubes is inspected on regular basis to prevent a leakage into the atmosphere. Current integrity management counts on ultrasonic immersion test. After high-pressure water jet removes the scale formed on the inner surface, an ultrasonic sensor makes a spiral movement in the water-filled tubes to detect the wall thinning. This method is time- and cost-consuming.

EMAT technique offers an alternative approach, which can assure the safety in a contactless manner without requiring preprocess. The EMAT built for this purpose is shown in Fig. 18.15 (Kikuchi and Fujimaki 2010). This configuration is a variation of the axial-shear-wave EMAT (see Sect. 3.18) and is composed of a circumferential array of twelve permanent magnets with alternating radial polarity and two separate solenoidal coils wound around the magnet array, each for transmitter and receiver. It is inserted into the tube to send out and receive the axial SH wave that travels in the circumferential directions with the axial polarization. The whole circumference becomes the active aperture and the coil width (10 mm now) equals the spatial resolution along the tube; the wavelength of excited SH wave is fixed to the magnet period.

The target tube is STPG 370 (JIS G 3454), which is a carbon steel tube of 25.4 mm outer diameter and 2.77 mm thickness. (The liftoff is then less than 0.4 mm from the inner surface of uncorroded tubes.) The outer surface has a spiral groove of 2-mm spacing to attach aluminum fins, 0.4 mm thickness, to accelerate cooling. Figure 18.16 show photographs of finned tube, tube after removing the fin, and the reference tube of the same nominal dimensions. Figure 18.17 presents the typical axial-shear-wave resonance spectra with the EMAT inside and outside of the tubes. Only two lowest resonances are measured in case of the finned tube, while

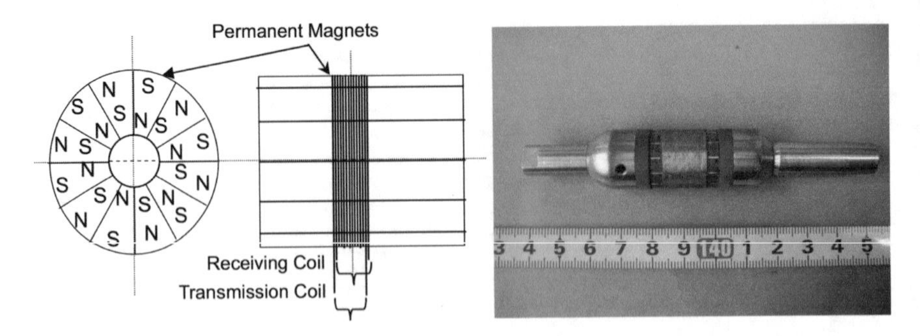

Fig. 18.15 PPM EMAT for axial-shear-wave resonance to inspect tube wall thinning. The magnetic assembly is 18.6 mm in diameter and 20 mm long and the coil is 10 mm wide. Reprinted with permission from Kikuchi and Fujimaki (2010)

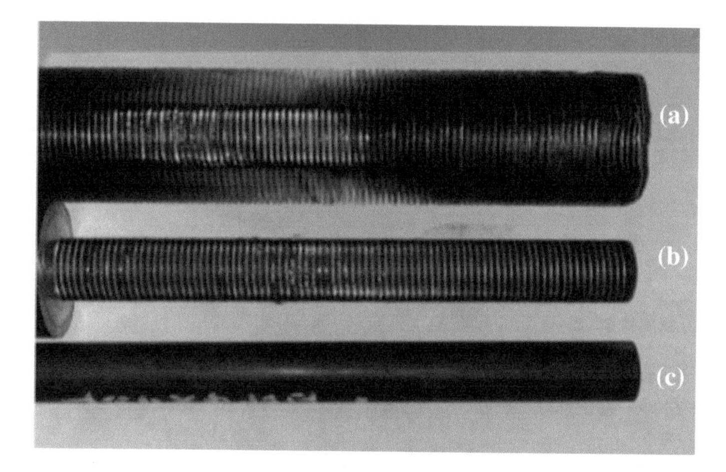

Fig. 18.16 **a** Heat exchanger tube with aluminum cooling fin, **b** tube whose fin is removed, and **c** reference tube of the same nominal diameter and thickness with **a** and **b** (With courtesy of Y. Kohno)

the full spectrum is observed for the reference tube. This occurs because the amplitude at the outer surface is larger than that at the inner surface for the higher modes and the fin absorbs much of the wave energy. The second resonance around 0.6 MHz is adopted for the further feasibility tests, since there is no overlapping with the resonance within the magnets (see Sect. 4.2).

Three sample tubes, labeled A, B, and C, were taken from a refinery unit on the occasion of periodic overhaul. Figure 18.18 shows the spectra around the second axial-shear-wave resonance obtained from them. Corrosion wastage over a long operation causes the wall thinning and shifts the resonance frequency upward from that of virgin tubes. At the same time, the liftoff increases to approximately 1 mm, deteriorating the resonance sharpness. The resonance frequency is nevertheless well measurable and a

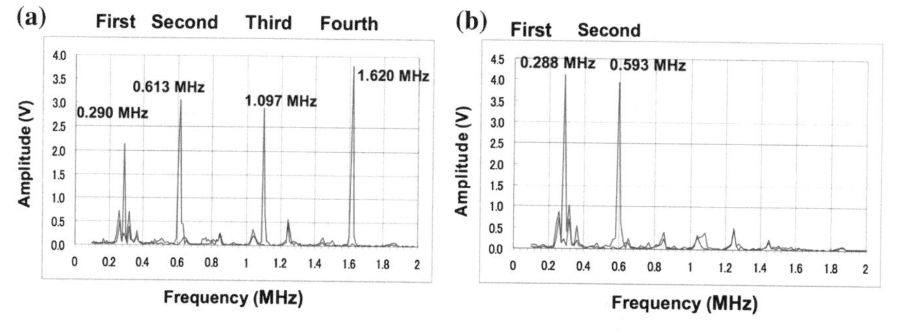

Fig. 18.17 Typical axial-shear-wave resonance spectra **a** from reference tube (Fig. 18.16 **c**) and **b** from the heat exchanger tube with fin. Spectra in *blue* are obtained when the EMAT is outside the tubes. Resonance frequencies shown are calculated with the dispersion relation. Reprinted with permission from Kikuchi and Fujimaki (2010)

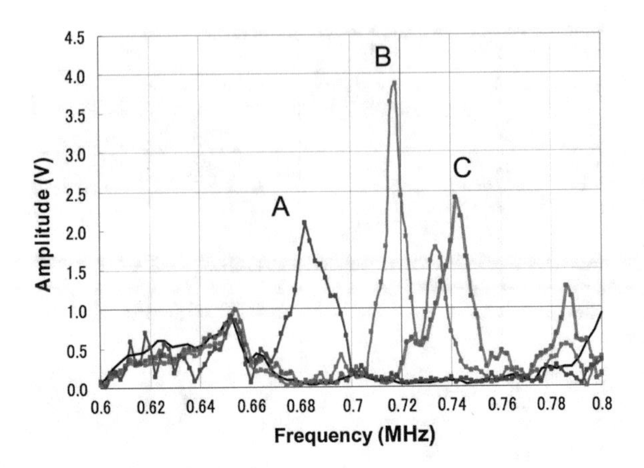

Fig. 18.18 Resonance spectra from three used heat exchanger tubes with fins attached. Reprinted with permission from Kikuchi and Fujimaki (2010)

Table 18.1 Comparison of the tube wall thicknesses between EMAR and destructive measurement (after Kikuchi and Fujimaki 2010)

Sample	Resonant frequency (MHz)	Thickness by EMAT (mm)	Thickness by destructive measurement (mm)
A	0.681	2.5	2.1–2.6
B	0.719	2.4	1.8–2.4
C	0.741	2.2	1.6–2.2

favorable comparison can be made between the thickness reduced from the resonance frequency and that measured after cutting the tubes (Table 18.1). Visual inspection of the inner surfaces indicates overall moderate corrosion and no sharp pitting. Being based on these positive results of feasibility tests, this noncontact inspection technique has been transferred to the on-site operation at the plant (Fig. 18.19). The maximum

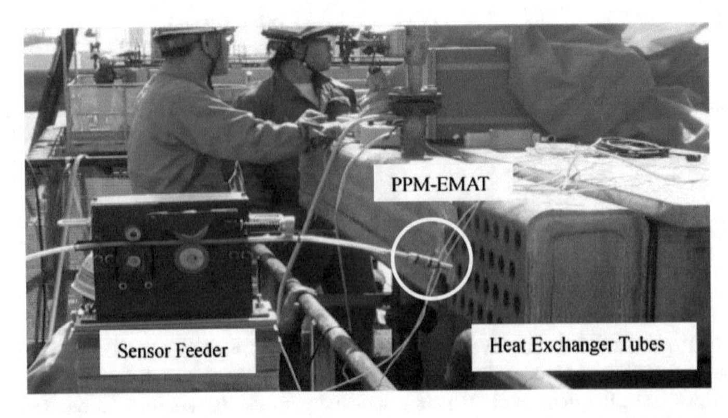

Fig. 18.19 On-site inspection of air-cooled heat exchanger tubes using PPM EMAT (With courtesy of Y. Kohno)

inspection speed at a verification test was found to be 78 mm/s, which is four times faster than the conventional immersion ultrasonic testing.

18.7 In-Process Weld Inspection

Welding to join structural components together is a highly complex process involving thermomechanical, metallurgical, chemical, and operational parameters. It accompanies heating, melting, cooling, and solidifying of metals, which, if not properly operated, may cause various defects (voids, incomplete fusion, etc.) and degraded heat-affected zone with grain coarsening. Residual stress always arises in the weld region from the contraction of molten metal (see Sect. 12.4.4). As a consequence, weld inspection is a primary target of nondestructive testing and evaluation. Weld inspection has been performed relying on radiographic and ultrasonic testing. This section briefly reviews the EMAT techniques applied to weld inspection. Noncontact nature again allows in-process inspection of welding. For further survey, Shao and Yan (2008) gave a review on online and offline ultrasonic monitoring techniques developed over the past years.

18.7.1 Phased-Array SH-Wave EMAT

Low transduction efficiency is only one weakness of EMATs as is widely recognized. There are a number of ideas of compensating for this weakness as thus far discussed, most of which rely on resonance and focusing principles (see Chap. 5 and Sect. 3.10). Electromagnetic acoustic resonance, EMAR, is the central subject of this monograph, which superimposes many weak truncated monochromatic waves emitted by EMATs *in phase* to build up signals of sufficient intensity. Coherency of overlapping waves, or phase matching, is the key point to result in the constructive interference for the focusing EMATs as well.

Coherent overlapping of acoustic waves is achievable with phased-array technique, which is commonly applied in medical imaging and ultrasonic testing (UT) on industrial materials, in particular in nuclear power plants. The phased-array probe is made up of multiple small elements, which are excited independently by the individual pulsers with the computer-generated delay times. The radiation fields from the elements are superposed in the material volume and the constructive interference forms a beam at an intended angle or focusing at an intended position. Beam can be steered electronically and swept through a material at high speed. Mechanical movement is not necessary. The same delay times are used to compose a signal with good time resolution when the received waves at the elements are put together and to provide imaging with aperture synthesis technique.

The phased-array technique has been incorporated with SH-wave EMATs by Sawaragi et al. (2000), Maclauchlan et al. (2008), and Gao and Lopez (2010), where the PPM EMATs (Sect. 3.3) or the meander-line coil EMATs (Sect. 3.5) are used to assemble multi-element probes. Use of SH wave offers notable benefits along with simplification. When the polarization is parallel to the reflecting surface, the reflection and refraction occur without accompanying a longitudinal wave (no mode conversion). When a line source oscillates in the line direction on the surface of an isotropic half-space, the SH wave radiates into the medium with equal amplitude for all directions from –90° to 90°. The SH waves suffer less beam skewing and attenuation due to the microstructural inhomogeneity at welds. Finally, EMATs can generate and receive SH waves with much ease. Fortunko and Schramm (1982) had already demonstrated the advantages of employing SH wave EMATs to detect defects during butt welding. Alers and Maclauchlan (1983) designed a meander-line coil with a pulsed electromagnet to generate focusing SH waves and demonstrated clearly resolved echoes from the corner and drilled hole of a steel plate. Petcher and Dixon (2015) recently compared guided SH_0 mode generated by PPM EMAT with bulk longitudinal wave generated by one-dimensional piezoelectric phased-array probe for detectability of six different types of defects in 22.3-mm-thick 316L stainless steel plate. They found that all the defects were detected by the SH wave, regardless of which side of the plate, while that is not the case with the phased-array probe.

Maclauchlan et al. (2008) developed a phased-array SH-wave EMAT system for in-process weld inspection. Figure 18.20 compares the calculated and measured SH-wave radiation pattern from a 2-loop meander-coil EMAT. The coil was designed to have maximum amplitude at 60° from the normal direction. Using this coil configuration for the individual elements, beam angles from approximately 45°–90° (grazing angle) were available for ultrasonic testing with the phased-array EMAT probe. Figure 18.21 shows the calculated beam focusing pattern at a V-shaped groove in 1-inch-thick steel plate. It is claimed that spatial resolution as good as 2-mm focal spot is obtained, allowing through wall sizing of planer defects

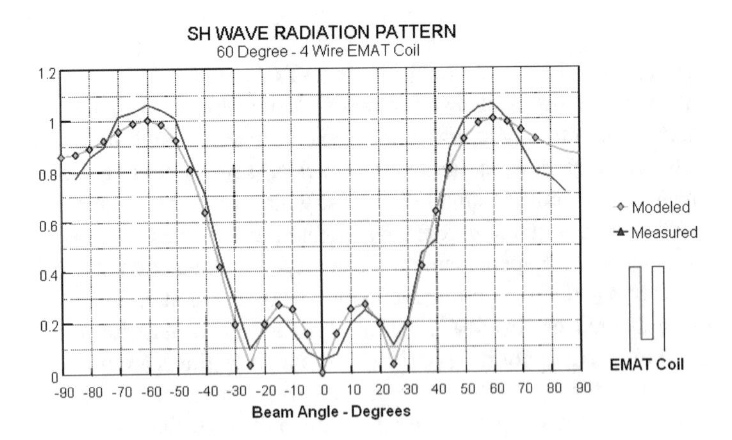

Fig. 18.20 Modeled and measured radiation pattern for a 2-loop (4 wire) meander-line coil EMAT. Reprinted with permission from Maclauchlan et al. (2008), Copyright (2008), AIP Publishing LLC

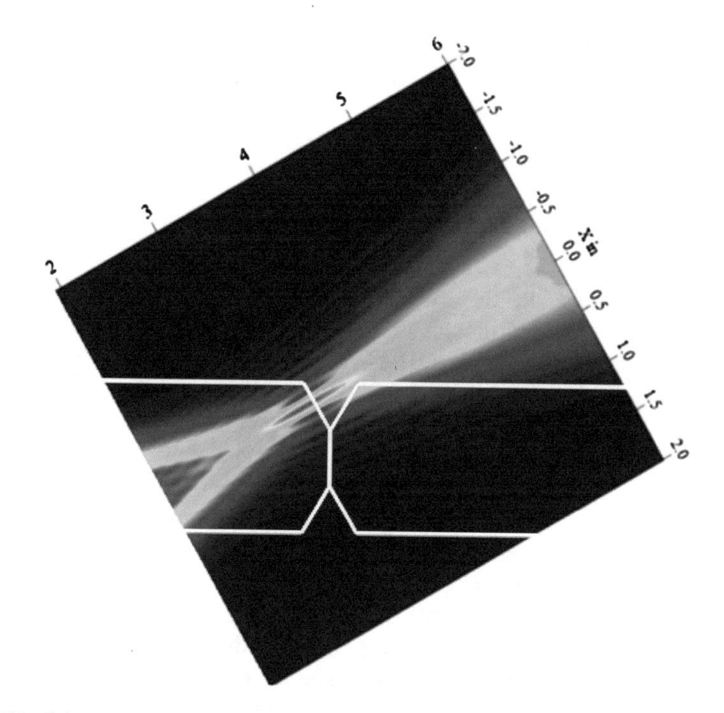

Fig. 18.21 Submerged arc weld prep with modeled beam pattern. Reprinted with permission from Maclauchlan et al. (2008), Copyright (2008), AIP Publishing LLC

such as incomplete sidewall fusion in welds. The system was comprised of 32 active channel UT instrument, EMAT test head, and a magnet pulser. High-power pulser, matching circuitry, and preamplifier for each channel are installed on the test head (Fig. 18.22a). It operated at 2.25 MHz. The phased-array EMAT system was programmed to sweep a focused beam along the grooves and into the volume of weld. The EMAT was attached to the torch head translator and followed behind the

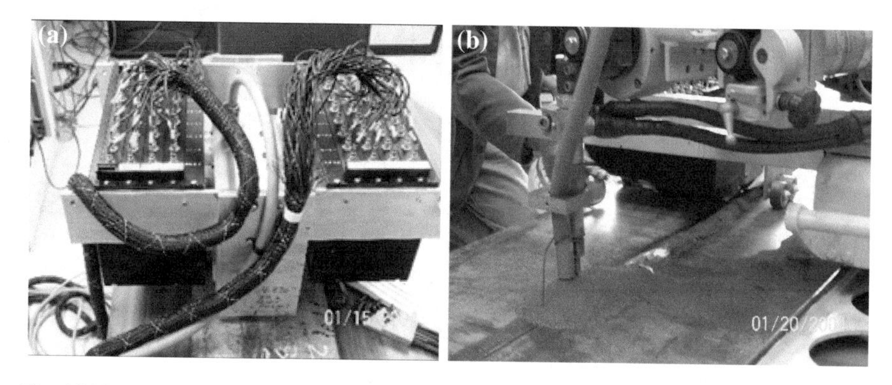

Fig. 18.22 **a** EMAT test head of 32 channels and **b** in-process inspection of submerged arc weld. Reprinted with permission from Maclauchlan et al. (2008), Copyright (2008), AIP Publishing LLC

welding torch (Fig. 18.22b). The weld was inspected shortly after solidification and while it was still hot. Maximum surface temperature at the EMAT coil was in excess of 150 °C.

18.7.2 Laser-EMAT Hybrid Systems

The ultrasonic inspection on welded steel plates was made possible by a laser-EMAT hybrid system, in which a laser beam forms the source of ultrasound on the workpiece's surface and an EMAT receives the reflected and diffracted waves off the welded region (Carlson et al. 1992; Oursler and Wagner 1995; Dixon et al. 1999). A high-energy pulsed laser is used to generate relatively intense elastic waves in the workpiece and an EMAT detects some signals after experiencing a number of internal reflections.

Mi and Ume (2006) have developed a real-time monitoring system based on the laser-EMAT hybrid configuration (Fig. 18.23). The sensing elements are shown in Fig. 18.24, where the laser generation source and the receiving EMAT are located on both sides of the weld seam. The fiber-phased array is composed of three line sources, each of which consists of seven fibers. The fibers have different lengths to give a delay time between the adjacent line sources. The laser beam from the pulsed Nd:YAG laser is coupled into the fiber bundle directly.

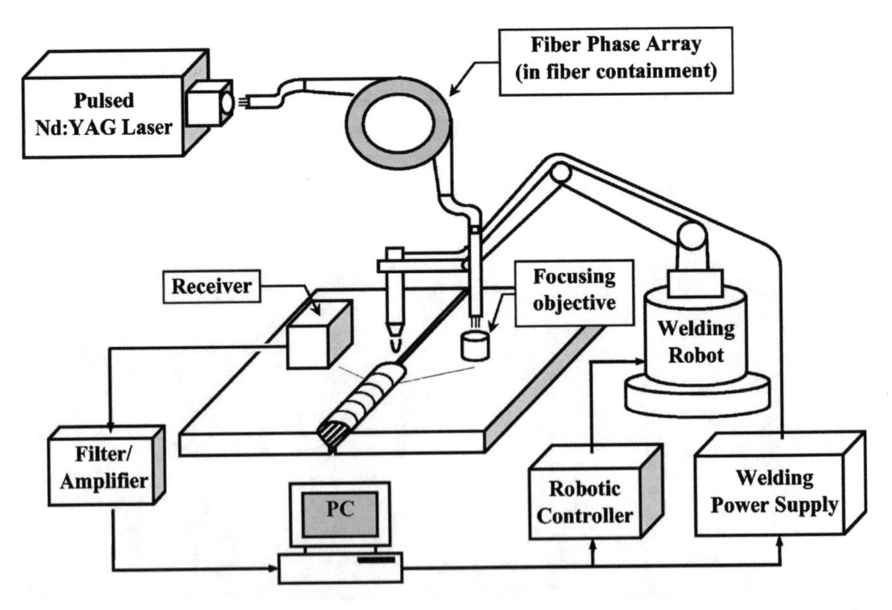

Fig. 18.23 System setup for laser-EMAT hybrid weld process monitoring. Reprinted with permission from Mi and Ume (2006)

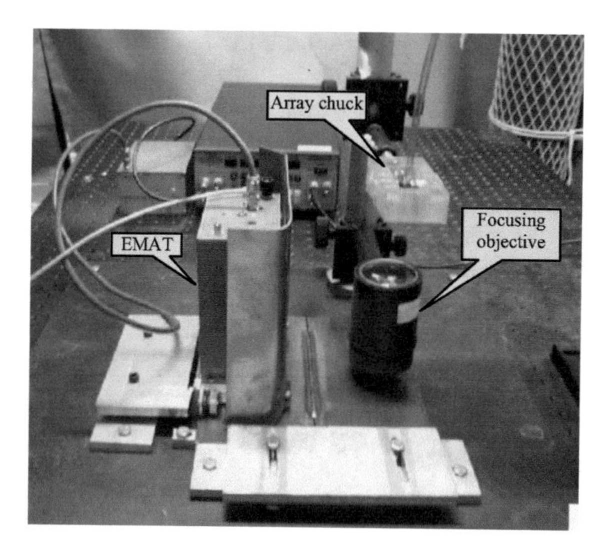

Fig. 18.24 Experimental setup of fiber-phased array laser unit and receiver EMAT. Reprinted with permission from Mi and Ume (2006)

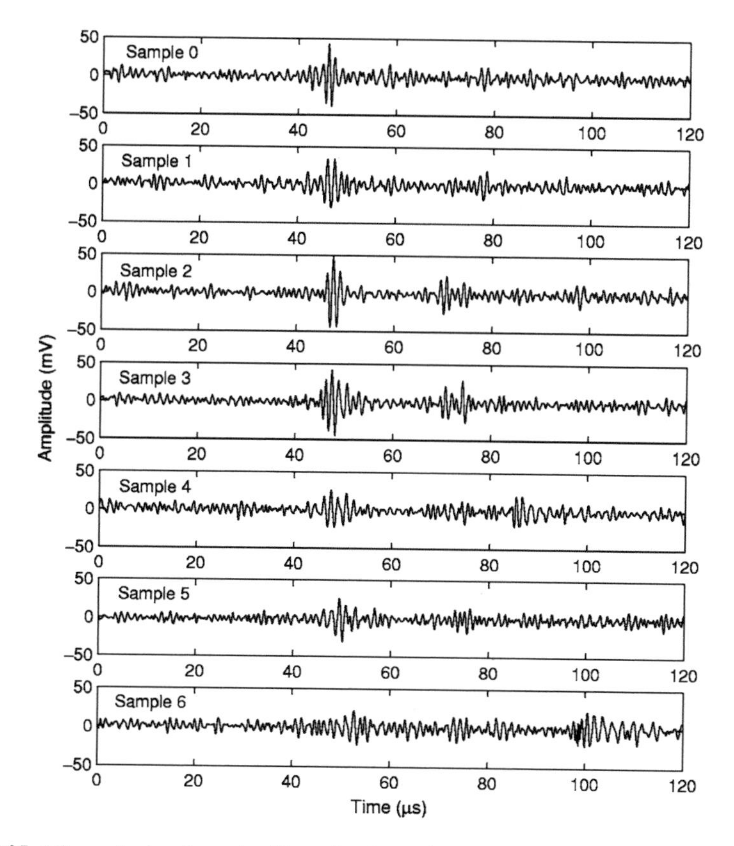

Fig. 18.25 Ultrasonic signals received just after the torch passed the sensing system. Seven welds of different penetration depths are examined. Reprinted with permission from Mi and Ume (2006)

Figure 18.25 presents the typical waveforms received shortly after the torch passed the sensing system. These received signals are processed to deduce the time of flight between the source and the receiver, which is then related to the penetration depth. A ray-tracing algorithm takes account of the curved ray path that occurs with the velocity change due to the temperature distribution around the weld pool. "... At the beginning, the signal was noisy and the amplitude was small. The signal grew gradually as the solid shell formed around the bottom of the weld and built up as it cooled off. ...." It is concluded that the real-time monitoring of weld penetration depth is feasible and this noncontact technique will be useful for in-process welding control.

References

Alers, G. (2008). A history of EMATs. In *Review of Progress in Quantitative Nondestructive Evaluation* (Vol. 27, pp. 801–808).

Alers, G. A., & MacLauchlan, D. T. (1983). High frequency, angle beam EMATs for weld inspection. In *Review of Progress in Quantitative Nondestructive Evaluation* (Vol. 2, pp. 271–281).

Boyd, D. M. (1990). Pulsed EMAT acoustic measurements on a horizontal continuous caster for internal temperature determination. In *Intelligent Processing of Materials* (pp. 253–266). The Minerals, Metals, & Materials Society.

Burrows, S. E., Fan, Y., & Dixon, S. (2014). High temperature thickness measurements of stainless steel and low carbon steel using electromagnetic acoustic transducers. *NDT&E International, 68*, 73–77.

Carlson, N. M., Johnson, J. A., Lott, L. A., & Kunerth, D. C. (1992). Ultrasonic NDT methods for weld sensing. *Materials Evaluation, 50*, 1338–1343.

Dever, D. J. (1972). Temperature dependence of the elastic constants in α-iron single crystals: Relation to spin order and diffusion anomalies. *Journal of Applied Physics, 43*, 3293–3301.

Dixon, S., Edwards, C., & Palmer, S. B. (1999). A laser-EMAT system for ultrasonic weld inspection. *Ultrasonics, 37*, 273–281.

Dubois, M., Moreau, A., & Bussière, J. F. (2001). Ultrasonic velocity measurements during phase transformations in steels using laser ultrasonics. *Journal of Applied Physics, 89*, 6487–6495.

Dubois, M., Moreau, A., Militzer, M., & Bussière, J. F. (1998). Laser-ultrasonic monitoring of phase transformations in steels. *Scripta Materialia, 39*, 735–741.

Fortunko, C. M., & Schramm, R. E. (1982). Ultrasonic non-destructive evaluations of butt welds using electromagnetic-acoustic transducers. *Welding Journal, 61*, 39–46.

Fujisawa, K., Murayama, R., Yoneyama, S., & Sakamoto, H. (1992). Nondestructive measurement of residual stress in railroad wheel by EMAT. In *Residual Stress-III Science and Technology* (pp. 328–333). New York: Elsevier.

Fukuoka, H., Toda, H., Hirakawa, K., Sakamoto, H., & Toya, Y. (1985). Nondestructive assessment of residual stresses in railroad wheel rim by acoustoelasticity. *Journal of Engineering for Industry, 107*, 281–287.

Gao, H., & Lopez, B. (2010). Development of single-channel and phased array electromagnetic acoustic transducers for austenitic weld testing. *Materials Evaluation, 68*, 821–827.

Herzer, R., Frotsher, H., Schillo, K., Bruche, D., & Schneider, E. (1994). Ultrasonic set-up to characterize stress states in rims of railroad wheels. In *Nondestructive Characterization of Materials* (Vol. 6, pp. 699–706).

Idris, A., Edwards, C., & Palmer, S. B. (1994). Acoustic wave measurements at elevated temperature using a pulsed laser generation and an electromagnetic acoustic transducer detector. *Nondestructive Testing and Evaluation, 11*, 195–213.

Iizuka, Y., & Awajiya, Y. (2014). High sensitivity EMAT system using chirp pulse compression and its application to crater end detection in continuous casting. *Journal of Physics: Conference Series, 520*, 012011.
Kawashima, K., Murota, S., Nakamori, Y., Soga, H., & Suzuki, H. (1979). Electromagnetic generation of ultrasonic waves in absence of external magnetic field and its applications to steel productions lines. In *Proceedings of 9th World Conference on Non-destructive Testing* (pp. 1–8).
Kikuchi, T., & Fujimaki, K. (2010). Study on a detection method of corrosion on air-cooled heat exchanger tubes by EMAT. *Journal of Solid Mechanics and Materials Engineering, 4*, 1398–1409.
Kurz, W., & Lux, B. (1969). Die Schallgeschwindigkeit von Eisen und Eisenlegierungen im festen und flüssigen Zustand. *High Temperature-High Pressures, 1*, 387–399 (in Germany).
Maclauchlan, D. T., Clark, S. P., & Hancock, J.W. (2008). Application of large apperture EMATs to weld inspection. In *Review of Progress in Quantitative Nondestructive Evaluation* (Vol. 27, pp. 817–822).
McIntire, P. (Ed.). (1991). *Ultrasonic Testing: Nondestructive Testing Handbook* (2nd ed., Vol. 7). ASNT.
Mi, B., & Ume, C. (2006). Real-time weld penetration depth monitoring with laser ultrasonic sensing system. *Journal of Manufacturing Science and Engineering, 128*, 280–286.
Oursler, D. A., & Wagner, J. W. (1995). Narrow-band hybrid pulsed laser/EMAT system for noncontact ultrasonic inspection using angled shear waves. *Materials Evaluation, 53*, 593–597.
Papadakis, E. P., Lynnworth, L. C., Fowler, K. A., & Carnevale, E. H. (1972). Ultrasonic attenuation and velocity in hot specimen by the momentary contact method with pressure coupling, and some results on steel to 1200 °C. *The Journal of the Acoustical Society of America, 52*, 850–857.
Parkinson, J. G., & Wilson, D. M. (1977). Non-contact ultrasonics. *British Journal of Non-Destructive Testing, 19*, 178–184.
Passarelli, F., Alers, G., & Alers, R. (2012). Electromagnetic transduction of ultrasonic waves. In *Review of Progress in Quantitative Nondestructive Evaluation* (Vol. 31, pp. 28–37).
Petcher, P. A., & Dixon, S. (2015). Weld defect detection using PPM EMAT generated shear horizontal ultrasound. *NDT and E International, 74*, 58–65.
Salzburger, H. J., & Repplinger, W. (1983). Automatic in-motion inspection of the head of railway wheels by E.M.A. excited Rayleigh waves. In *Conference Proceedings of Ultrasonic International* (pp. 497–501). London: Butterworths.
Sawaragi, K., Salzburger, H. J., Hubschen, G., Enami, K., Kirihigashi, A., & Tachibana, N. (2000). Improvement of SH-wave EMAT phased array inspection by new eight segment probes. *Nuclear Engineering and Design, 198*(17), 153–163.
Schramm, R. E., Shull, P. J., Clark, A. V., Jr., & Mitrakonic, D. V. (1989). EMATs for roll-by crack inspection of railroad wheels. In *Review of Progress in Quantitative Nondestructive Evaluation* (Vol. 8, pp. 1083–1089).
Scruby, C. B., & Drain, L. E. (1990). *Laser Ultrasonics: Techniques and Applications*. London: Taylor & Francis Group.
Shao, J., & Yan, Y. (2008). Ultrasonic sensors for welding. In Y. M. Zhang (Ed.), *Real-Time Weld Process Monitoring*. Cambridge: Woodhead Publishing.
Simmons, G., & Wang, H. (1971). *Single Crystal Elastic Constants and Calculated Aggregate Properties: A Handbook* (2nd ed.). Cambridge, Massachusetts: The M.I.T Press.
Spicer, J. B. (1997). In situ laser-ultrasonic monitoring of stainless steel microstructural evolution during heat treatment. *High Temperature and Materials Science, 37*, 23–41.
Takada, H., Ichikawa, F., Kitaoka, H., Momoo, A., & Nishikawa, H. (1989). Detection of liquid crater end in strand cast slab by shear wave EMATs. In *Proceedings of CAMP ISIJ* , (Vol. 2, p.1214) (in Japanese).
Truell, R., Elbaum, C., & Chick, B. B. (1969). *Ultrasonic Methods in Solid State Physics*. New York: Academic Press.

Wadley, H. N. G., Norton, S. J., Mauer, F., & Droney, B. (1986). Ultrasonic measurement of internal temperature distribution. *Philosophical Transactions of the Royal Society of London A, 320*, 341–361.

Whittington, K. R. (1978). Ultrasonic inspection of hot steel. *British Journal of Non-Destructive Testing, 20*, 242–247.

Yamaguchi, H., Fujisawa, K., Murayama, R., Hashimoto, K., Nakanishi, R., Kato, A., Ishikawa, H., Kadokawa, T., & Sato, I. (1986). Development of hot seamless tube wall thickness gauge by electromagnetic acoustic transducer. *Transactions of the Iron and Steel Institute of Japan, 26*, 61–68.

Yoshida, T., Atsumi, T., Ohashi, W., Kagaya, K., Tsubakihara, O., Soga, H., & Kawashima, K. (1984). On-line measurement of solidification shell thickness and estimation of crater-end shape of CC-slabs by electromagnetic ultrasonic method. *Tetsu to Hagane (Iron and Steel), 70*, 1123–1130 (in Japanese).

Index

M. Hirao and H. Ogi, *Electromagnetic Acoustic Transducers*,
Springer Series in Measurement Science and Technology,
DOI 10.1007/978-4-431-56036-4

Printed in the United States
By Bookmasters